Exercises in Graph Theory

Kluwer Texts in the Mathematical Sciences

VOLUME 19

A Graduate-Level Book Series

The titles published in this series are listed at the end of this volume.

Exercises in Graph Theory

by

O. Melnikov
Department of Mathematics,
Belarus State University,
Minsk, Belarus

V. Sarvanov
Institute of Mathematics,
Belarus Academy of Sciences,
Minsk, Belarus

R. Tyshkevich

V. Yemelichev

and

I. Zverovich
Department of Mathematics,
Belarus State University,
Minsk, Belarus

KLUWER ACADEMIC PUBLISHERS

DORDRECHT / BOSTON / LONDON

A C.I.P. Catalogue record for this book is available from the Library of Congress.

ISBN 0-7923-4906-7

Published by Kluwer Academic Publishers,
P.O. Box 17, 3300 AA Dordrecht, The Netherlands.

Sold and distributed in the U.S.A. and Canada
by Kluwer Academic Publishers,
101 Philip Drive, Norwell, MA 02061, U.S.A.

In all other countries, sold and distributed
by Kluwer Academic Publishers,
P.O. Box 322, 3300 AH Dordrecht, The Netherlands.

Printed on acid-free paper

Printed in the Netherlands.

Contents

Introduction

This book supplements the textbook of the authors "Lectures on Graph Theory" [6] by more than thousand exercises of varying complexity. The books match each other in their contents, notations, and terminology. The authors hope that both students and lecturers will find this book helpful for mastering and verifying the understanding of the peculiarities of graphs.

The exercises are grouped into eleven chapters and numerous sections according to the topics of graph theory: paths, cycles, components, subgraphs, reconstructibility, operations on graphs, graphs and matrices, trees, independence, matchings, coverings, connectivity, matroids, planarity, Eulerian and Hamiltonian graphs, degree sequences, colorings, digraphs, hypergraphs.

Each section starts with main definitions and brief theoretical discussions. They constitute a minimal background, just a reminder, for solving the exercises. Proofs of the presented facts and a more extended exposition may be found in the mentioned textbook of the authors, as well as in many other books in graph theory.

Most exercises are supplied with answers and hints. In many cases complete solutions are given.

At the end of the book you may find the index of terms and the glossary of notations.

The "Bibliography" list refers only to the books used by the authors during the preparation of the exercisebook. Clearly, it mentions only a fraction of available books in graph theory. The invention of the authors was also driven by numerous journal articles, which are impossible to list here.

The authors are greatly indebted to Professors Eberhard Girlich, Michael Hazewinkel, Uno Kaljulaid, and Alexander Mikhalev, who have stimulated and promoted publishing this book in English.

The authors wish to give their credit to Dr. Nikolai Korneenko for improving the text during the translation from Russian.

Chapter 1

ABC of Graph Theory

1.1 Graphs: Basic Notions

Let $V^{(2)}$ be the set of all two-elements subsets of a nonempty set V. An ordered pair (V, E), where $E \subseteq V^{(2)}$, is called a *graph*. The elements of V are called the *vertices* and the elements of E are called the *edges*. The sets of vertices and edges of a graph G will be denoted by VG and EG respectively.

A graph is called *finite* if the set of its vertices is finite. This book deals with finite graphs only; therefore in the sequel we shall omit the qualifier "finite".

The number $|VG|$ of vertices of a graph G is called its *order* and is denoted by $|G|$. A graph of order n is called an *n-vertex graph*.

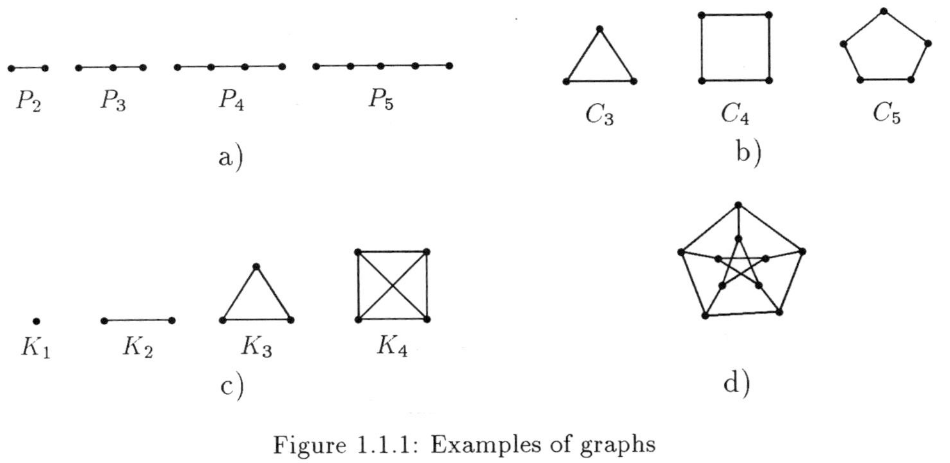

Figure 1.1.1: Examples of graphs

Graphs are conveniently depicted by drawings that consist of points and continuous lines connecting some pairs of the points. The points correspond to the vertices of a graph and the connecting lines correspond to the edges. Fig. 1.1.1 shows the following graphs: a) the simple paths P_n, $n = 2, \ldots, 5$, b) the simple cycles C_n, $n = 3, \ldots, 5$, c) the complete graphs K_n, $n = 1, \ldots, 4$, d) the *Pe-*

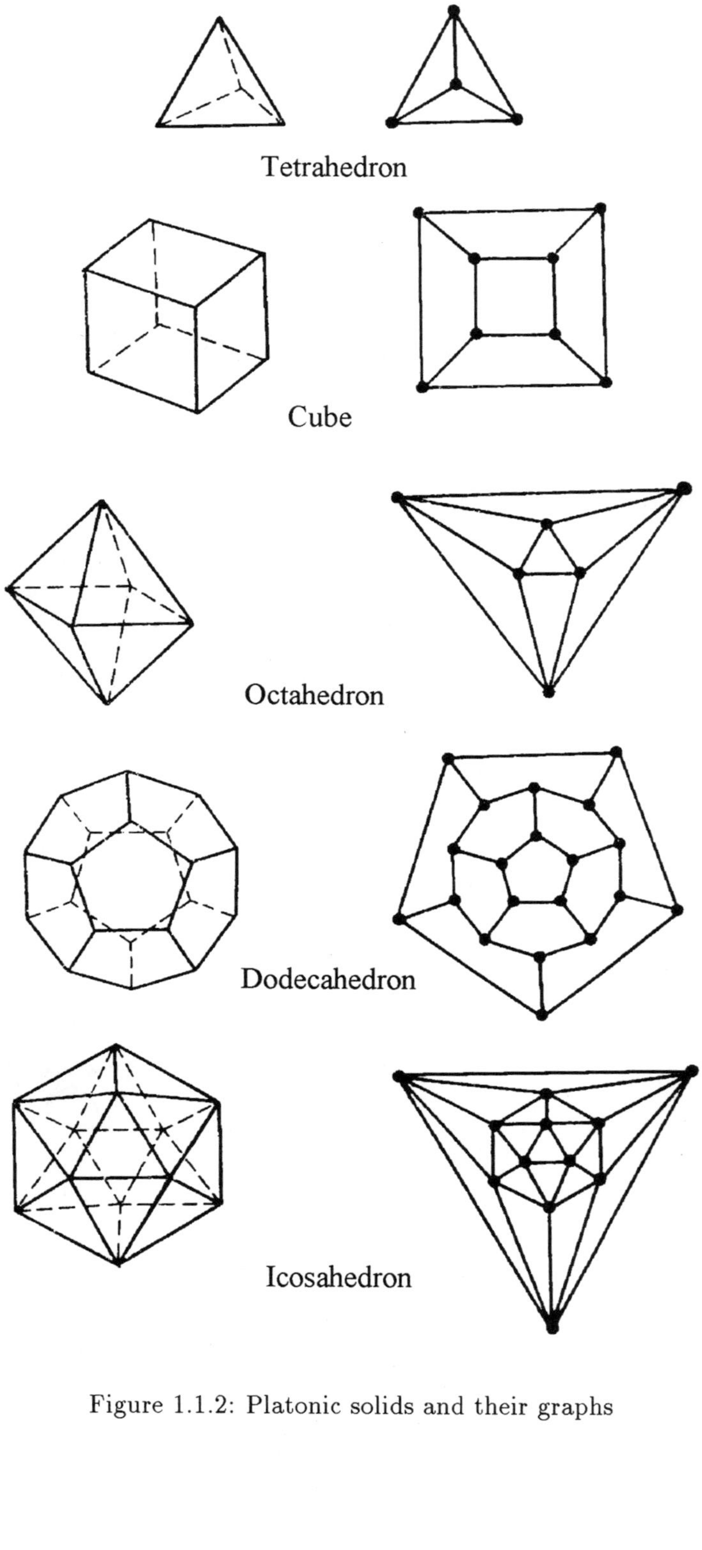

Figure 1.1.2: Platonic solids and their graphs

tersen graph. Fig. 1.1.2 shows five *Platonic solids* (tetrahedron, cube, octahedron, dodecahedron, icosahedron) and the graphs of these solids, or *Platonic graphs.*

Two vertices u, v of a graph are said to be *adjacent* if the set $\{u, v\}$ is an edge of the graph. An edge $\{u, v\}$ is often denoted by uv (or, equivalently, vu.) The vertices u and v are called the *endpoints,* or *ends* of the edge uv. It is also said that the edge uv *connects* the vertices u and v.

Two edges are called *adjacent* if they share an end.

A vertex v and an edge e are called *incident* if v is an end of e.

The set of vertices of a graph incident to a vertex v is called the *neighborhood* of v and denoted by $N(v)$. Similarly, for a subset $X \subseteq VG$, the *neighborhood* of X is defined as follows:

$$N(X) = (\bigcup_{x \in X} N(x)) \backslash X.$$

Let G and H be graphs and let $\varphi : VG \rightarrow VH$ be a bijection. If the images $\varphi(u)$, $\varphi(v)$ of any two vertices u, v of G are adjacent in H if and only if u, v are adjacent in G, then φ is called *isomorphism* of the graph G onto the graph H. If such bijection exists, we say that G and H are *isomorphic* and write $G \cong H$ (and hence, $H \cong G$).

A graph is called *complete* if every two its vertices are adjacent. A graph with no edges is called *empty.* The complete and the empty graphs of order n are denoted by K_n (Fig. 1.1.1c) and O_n respectively.

A single-vertex graph is called *trivial.*

The number of edges incident to a vertex v is called the *degree of* v and is denoted by $\deg v$. The minimal and maximal degrees of the vertices of a graph G are denoted by $\delta(G)$ and $\Delta(G)$ respectively. The list of vertex degrees of a graph is called its *degree sequence.* The order of elements in this sequence is indifferent.

Proposition 1.1.1 (Handshake Lemma) *The sum of degrees of the vertices of a graph is twice the number of the edges.*

A vertex of degree 0 is called *isolated;* a vertex of degree 1 is called *pendant.* The edge incident to a pendant vertex is also called *pendant.* A vertex incident to all other vertices is called *dominating.*

Figure 1.1.3: Bipartite graphs

A graph is called *bipartite* if there is a partition if its vertex set into two subsets sets (called *parts*) such that the ends of each edge are in different parts, see Fig. 1.1.3. Such edge partition is called *bipartition.* A bipartite graph will be denoted by $G = (A, B; E)$, where A, B are the parts of G and E is the edge set. (The trivial graph is by definition bipartite, one of its parts being the empty set.) If a bipartite graph $G = (A, B; E)$ with $|A| = p \geq 1$, $|B| = q \geq 1$ is such that every vertex from the part A is adjacent to all vertices of the part B then G is called the *complete bipartite graph* and is denoted by $K_{p,q}$. In addition, the graph $K_{1,q}$ is called the *star graph,* or *star.*

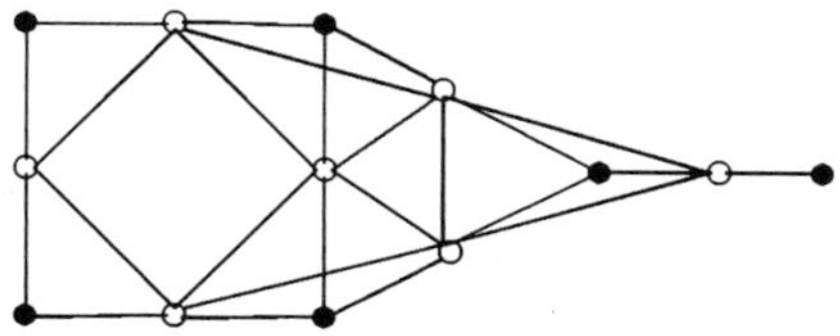

Figure 1.1.4: Graph and its line graph

A graph is called *regular* if all its vertices are of the same degree. For example, the complete graphs K_n, Fig. 1.1.1c, the Petersen graph, Fig. 1.1.1d, the Platonic graphs, Fig. 1.1.2,are regular. The *degree* $\deg G$ *of a regular graph* G is the degree of its vertices. Regular graphs of degree r are called *r-regular.* Graphs of degree three are called *cubic graphs.*

For a graph G, the *complementary graph* $\overline{G}$ is defined as follows:

1. $V\overline{G} = VG$;

2. Two vertices are adjacent in $\overline{G}$ if and only if they are not adjacent in G.

A graph G is called *selfcomplementary* if $\overline{G} \cong G$. For example, the path graph P_4 and the cycle graph C_5 are selfcomplementary.

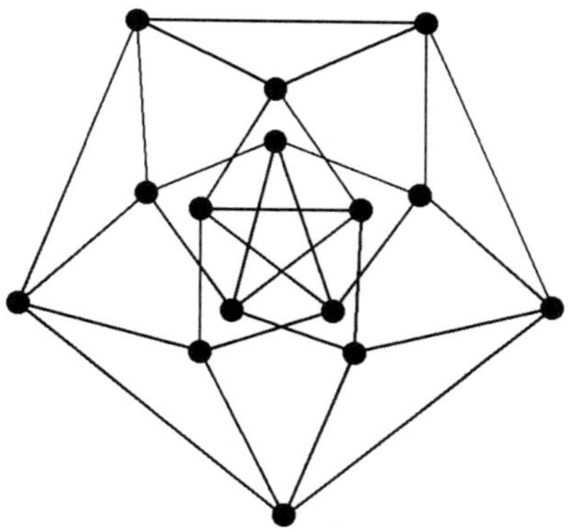

Figure 1.1.5: Line graph of Petersen graph

For a graph G the *line graph* $L(G)$ is defined as follows:

1. The vertices of $L(G)$ bijectively correspond to the edges of G.

2. Two vertices of $L(G)$ are adjacent if and only if the corresponding edges of G are adjacent.

Fig. 1.1.4 shows graphs G and $L(G)$ overimposed: the vertices of G are denoted by dots and the edges of G by thin lines; the vertices of $L(G)$ are denoted by circles and the edges of $L(G)$ by bold lines. Fig. 1.1.5 shows the line graph of the Petersen graph.

A *multigraph* is a pair (V, E), where V is a nonempty set (*vertices*) and E is a collection of subsets of the set $V^{(2)}$ (*edges.*) We use the term "collection" to mean that E may contain several copies of elements of $V^{(2)}$; in such cases equal elements of E are considered to be different edges and are called *multiple edges.*

A further generalization is to allow *loops,* i.e., edges connecting a vertex with itself. A *pseudograph* is a pair (V, E), where E is a collection of unordered pairs of vertices (called *edges*), not necessarily different. Fig. 1.1.6 shows a multigraph and a pseudograph.

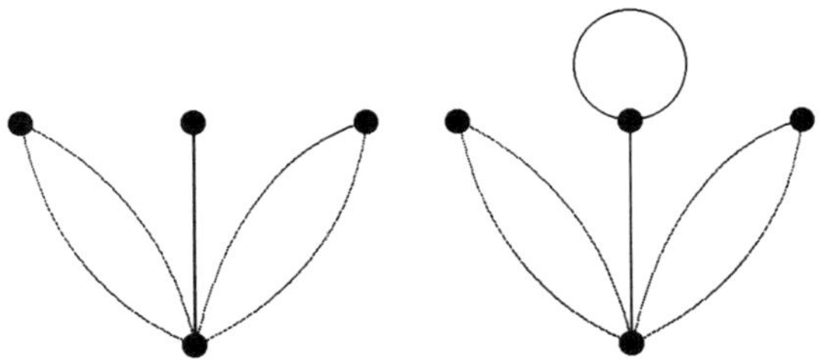

Figure 1.1.6: Multigraph and pseudograph

For graphs of all these types it is natural to define *isomorphism* to be a bijection between the sets of vertices that preserves the adjacency relation and the multiplicities of edges and loops.

The Handshake Lemma holds also for pseudographs, provided that each loop contributes the count two into the degree of the corresponding vertex.

Graphs in the sense of our initial definition will be called *simple,* or *ordinary,* in order to distinguish them from the above generalizations.

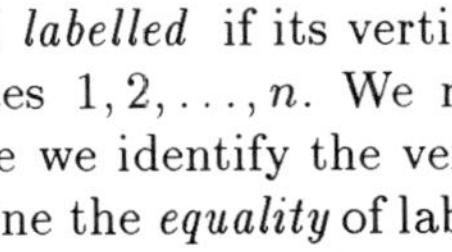

Figure 1.1.7: Labelled graph

A graph of order n is called *labelled* if its vertices are provided with pairwise different labels, e.g., the indices $1, 2, \dots, n$. We may identify each vertex of a graph with its index (and hence we identify the vertex set with the set of indices $\{1, 2, \dots, n\}$). Then we may define the *equality* of labelled graphs of the same order: $G = H$ if and only if $EG = EH$. Fig. 1.1.7 shows a labelled $(5, 6)$-graph G with $EG = \{12, 14, 23, 24, 34, 45\}$.

* * *

1.1.1. Draw:

a) a simple graph;

b) a multigraph that is not a simple graph;

c) a pseudograph that is not a multigraph.

1.1.2. Find the neighborhoods $N(u)$ and $N(v)$ of the vertices u and v for each graph depicted at Fig. 1.1.8.

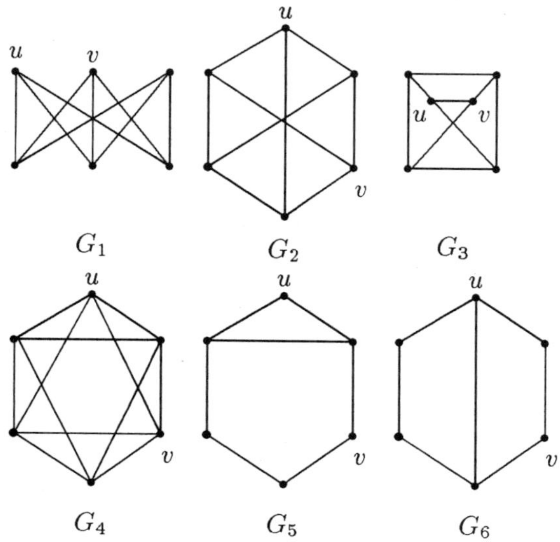

Figure 1.1.8: To Exercise 1.1.2

1.1.3. Describe a graph satisfying one of the following conditions:

a) every its vertex is isolated;

b) every its vertex is dominating;

c) the neighborhoods of any two vertices are equal.

1.1.4. Describe the structure of a graph G that satisfies the following condition: $N(u) \cup N(v) = VG$ for every pair of different vertices u, v.

1.1.5. How many edges has K_n?

1.1.6. a) Does there exist a $(4, 7)$-graph?

b) Prove that if there exists an (n, m)-graph then

$$0 \leq m \leq C_n^2. \tag{1.1.1}$$

c) Prove that for any two positive integers n, m satisfying the inequalities (1.1.1) there exists an (n, m)-graph.

1.1.7. Does there exist a nontrivial graph that has both isolated and dominating vertices?

1.1.8. Find the number of edges in a graph with the degree sequence

a) $(3, 3, 3, 3, 3, 3)$;

b) $(3, 3, 2, 2, 2, 2)$;

c) $(3, 3, 3, 2, 2)$;

Can you draw such graphs?

1.1.9. Prove that every nontrivial graph has two vertices of equal degrees.

1.1.10. Does there exist a bipartite graph G such that $\delta(G) + \Delta(G) > |G|$?

1.1.11. a) Prove that for a nonempty regular bipartite graph the numbers of vertices in both parts are equal.

b) In a group of people, each girl knows exactly k boys ($k > 0$) and each boy knows exactly k girls. Is it true that there is as many boys as girls in the group?

c) Prove that if G is a bipartite r-regular graph then $r \leq |G|/2$.

1.1.12. a) Prove that an n-vertex graph with more than $n^2/4$ edges is not bipartite.

b) Find all n-vertex bipartite graphs with $n^2/4$ edges.

1.1.13. Characterize the bipartite graphs with the following property: for every pair nonadjacent vertices there exists a vertex adjacent to both of them.

1.1.14. Draw the complementary graph for each of the following graphs: O_n, K_n, $K_{n,m}$, $2K_2$ (Fig. 1.1.9), C_4, C_5, P_4. Which of the above graphs are selfcomplementary?

Figure 1.1.9: To Exercise 1.1.14

1.1.15. Prove that if graphs G and H are isomorphic then the complementary graphs $\overline{G}$ and $\overline{H}$ are isomorphic.

1.1.16. Give an example (if possible) of

a) a bipartite r-regular graph of order n;

b) a cubic graph of order 9;

c) an (n, m)-graph with $m = C_{n-1}^2$.

1.1.17. Find the number of pairwise nonisomorphic $(n-2)$-regular graphs of order n.

1.1.18. Find the number of pairwise distinct labelled

a) graphs of order n;

b) (n, m)-graphs.

1.1.19. Prove that the isomorphism relation for graphs is an equivalence relation, i.e.,

a) every graph is isomorphic to itself;

b) if $G \cong H$ then $H \cong G$;

c) if $G \cong H$ and $H \cong I$ then $G \cong I$.

1.1.20. Prove that isomorphic graphs have

a) equal orders;

b) equal numbers of edges;

c) coinciding degree sequences (up to the orders of elements in them).

Are the conditions a)-c) sufficient for graphs to be isomorphic?

1.1.21. Prove that each isomorphism of graphs G and H naturally defines an isomorphism of line graphs $L(G)$ and $L(H)$.

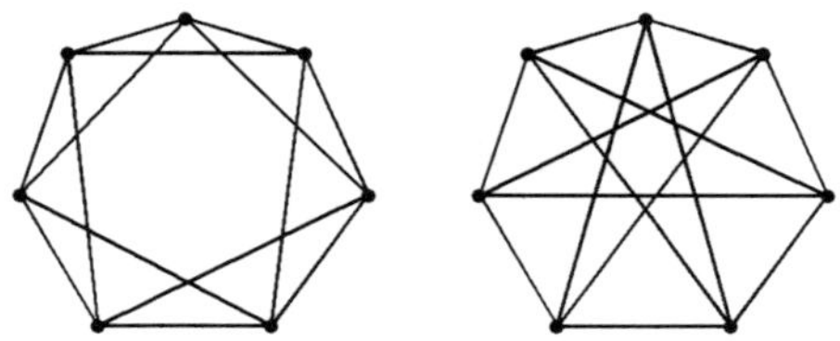

Figure 1.1.10: To Exercise 1.1.22a

1.1.22. a) Prove that the two graphs from Fig. 1.1.10 are isomorphic and those from Fig. 1.1.11 are nonisomorphic;

b) Which pairs of graphs at Fig. 1.1.8 are isomorphic?

1.1.23. Prove that if a graph G is bipartite and $G \cong H$ then H is also bipartite.

1.1.24. Draw all pairwise nonisomorphic graphs of orders $3, 4, 5$.

1.1.25. Draw all pairwise nonisomorphic cubic graphs of orders $4, 6, 8$.

1.1.26. *Circulant graph:* Let $n > 1$ be an integer and let $\mathbf{Z_n}$ be an additive group of modulo n residuals. For a set $U \subset \mathbf{Z_n}$ that satisfies the following two conditions:

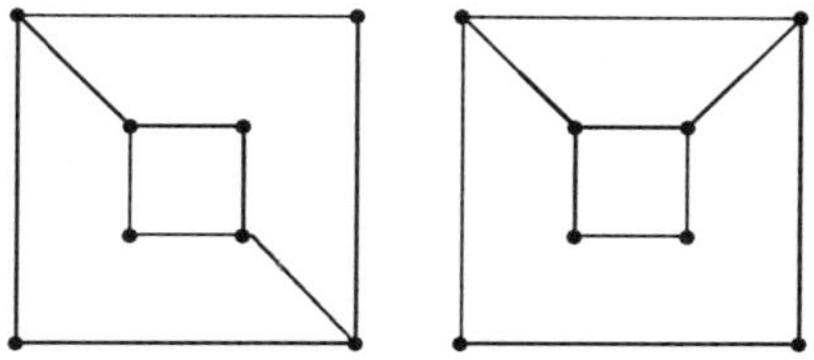

Figure 1.1.11: To Exercise 1.1.22a

(1) $0 \notin U$,

(2) $x \in U$ implies $-x \in U$

we may define the graph $G = G_{n,U}$ by setting $VG = \mathbf{Z_n}$ and defining a pair of vertices x, y to be adjacent if and only if $x - y \in U$. Such graph is called a *circulant graph*.

a) Prove that

1. the above conditions do specify a graph;
2. a circulant graph $G_{n,U}$ is a regular graph of degree $|U|$;
3. a graph complementary to a circulant graph is also circulant;
4. for any integers n, r, $0 \leq r \leq n-1$, such that at least one of them is even there exists an r-regular circulant graph of order n.

b) Draw a cubic circulant graph of order 6. Determine the number of pairwise nonisomorphic cubic circulant graphs of order 6.

c) Do there exist regular graphs that are not circulant?

d) Let G_{n,U_i}, $i = 1, 2$, be two circulant graphs such that

$$U_2 = kU_1 = \{ku : \quad u \in U_1\}$$

for some positive integer k coprime to n. Prove that the mapping

$$\mathbf{Z}_n \to \mathbf{Z}_n, \ x \mapsto kx, \ x \in \mathbf{Z_n}$$

is an isomorphism of graphs G_{n,U_1} and G_{n,U_2}.

1.1.27. *n-Permutation graph:* A graph of order $n!$ whose vertices bijectively correspond to the permutations of the numbers $1, 2, \ldots, n$, and whose two vertices are adjacent if and only if the corresponding permutations are transformed into each other by a single transposition is called the *n-permutation graph*.

a) Draw the 3-permutation graph;

b) What is the degree of the n-permutation graph? How many edges does it have? Is it regular?

c) Prove that the n-permutation graph is bipartite.

1.1.28. *Common divisor graph:* Let V be a nonempty finite subset of the set of positive integers. A graph G with the vertex set V is called a *common divisor graph* if two vertices $u, v \in V$ are adjacent in G if and only if the numbers u an v are not coprime. Prove that every graph is isomorphic to a common divisor graph.

1.1.29. Propose the definitions of isomorphism for multigraphs and pseudographs similar to that for ordinary graphs. Give an example of nonisomorphic multigraphs with the same numbers of vertices and with the same numbers of edges of the same multiplicity.

1.1.30. a) How many edges has a graph complementary to an (n, m)-graph? How many edges has a selfcomplementary graph of order n?

b) Prove that if n is the order of a selfcomplementary graph then

$$\text{either } n \equiv 0 \pmod 4 \text{ or } n \equiv 1 \pmod 4. \qquad (1.1.2)$$

c) Let G be a graph. Let us construct two new graphs H_G and F_G as follows. Take $G_1 \cong \overline{G}$, $G_2 \cong G$, $G_3 \cong G$, $G_4 \cong \overline{G}$ with disjoint vertex sets. H_G is produced from $G_1, \ldots, G_4$ by connecting all vertices from VG_i, with all vertices from VG_{i+1}, $i = 1, 2, 3$, by edges. F_G is produced from H_G by adding a new vertex and connecting it with all vertices of VG_1 and VG_4. Prove that H_G and F_G are selfcomplementary graphs for all G.

d) Prove that for any positive integer n satisfying the condition (1.1.2) there exists a selfcomplementary graph of order n.

e) Prove that the problem of recognition of isomorphism for arbitrary graphs is reducible to that for selfcomplementary graphs, i.e., for any graphs G_1, G_2 one may construct selfcomplementary graphs S_{G_1} and S_{G_2} such that $G_1 \cong G_2$ if and only if $S_{G_1} \cong S_{G_2}$.

1.1.31. Find all selfcomplementary graphs of orders: (a) 4; (b) 5; (c) 6.

1.1.32. Prove that if $(d_1, d_2, \ldots, d_n)$ is the degree sequence of a selfcomplementary graph and $d_1 \geq d_2 \geq \ldots \geq d_n$, then

$$d_i = n - 1 - d_{n+1-i}, \; i = 1, 2, \ldots, \lfloor n/2 \rfloor.$$

1.1.33. Prove that if there exists an r-regular selfcomplementary graph of order n then $n \equiv 1 \pmod 4$ and $r = (n-1)/2$.

1.1.34. Is it possible for a selfcomplementary graph of order 100 to have exactly one vertex of order 50?

1.1.35. Find the numbers of vertices and edges for the line graph $L(G)$ of a graph G with the degree sequence

$$(d_1, d_2, \ldots, d_n).$$

1.1.36. Prove that the line graph $L(G)$ cannot be isomorphic to $K_{1,q}$, $q \geq 3$, for any graph G.

1.1.37. Give examples of

a) a graph with $L(G) \cong G$;

b) a Platonic graph isomorphic to the line graph of another Platonic graph.

1.1.38. Is it true that the line graph

a) of a regular graph is regular;

b) of a bipartite graph is bipartite?

1.1.39. Prove that the graph $L(K_n)$ is a $(2n-4)$-regular graph of order C_n^2. State and prove the similar statement for the graph $K_{p,q}$.

1.1.40. Prove that the graph $\overline{L(K_5)}$ is isomorphic to the Petersen graph.

1.2 Walks, Paths, Components

An alternating sequence

$$(v_1, e_1, v_2, e_2, \ldots, e_k, v_{k+1}) \tag{1.2.1}$$

of vertices and edges of a graph such that $e_i = v_i v_{i+1}$, $i = 1, 2, \ldots, k$, is called a *walk* connecting the vertices v_1 and v_{k+1} (or simply an (v_1, v_{k+1})-*walk*). In (1.2.1) it is possible that $v_{k+1} = v_1$, i.e., a walk may connect a vertex with itself. The vertices v_1 and v_{k+1} are called the *terminal vertices* (or *endvertices*) of the walk. A walk is called a *path* if all its edges are different. A path is called a *simple path* if all its vertices are different, with the possible exception of $v_{k+1} = v_1$. A walk is called *cyclic* if its endvertices coincide and *noncyclic* otherwise. A cyclic path is called a *cycle,* a cyclic simple path is called a *simple cycle.*

The number k of edges in a walk is called its *length.* A simple cycle of length k is called a k-*cycle* and is denoted by C_k. The 3-cycle is often called *triangle.*

A simple noncyclic path with k vertices (i.e., of length $k-1$) is denoted by P_k.

A graph without cycles is called *acyclic.*

In the case of ordinary graphs the walk (1.2.1) may be specified in a simpler way by writing down its vertices (or edges) in the order of their occurrence in (1.2.1):

$$(v_1, v_2, \ldots, v_{k+1}) \quad \text{or} \quad (e_1, e_2, \ldots, e_k).$$

Any vertex of a cycle may be taken as a "starting" vertex. Moreover, the sequences

$$(v_1, v_2, \ldots, v_k, v_1) \quad \text{and} \quad (v_1, v_k, v_{k-1}, \ldots, v_2, v_1)$$

specify the same cycle. For example, the sequences $(1,2,3,4,1)$, $(3,4,1,2,3)$, and $(1,4,3,2,1)$ specify the same cycle. The cycles $(1,2,3,4,1)$ and $(1,2,4,3,1)$ are different: while the sets of their vertices coincide, the orders of their traversal are different.

The length of the shortest cycle in a graph is called the *girth of the graph.* The girth of an acyclic graph is defined to be infinite.

There exists a biparticity criterion for graphs in terms of lengths of cycles:

Theorem 1.2.1 (König's Theorem) *A graph is bipartite if and only if it has no cycles of odd length.*

A graph H is called a *subgraph* of a graph G if $VH \subseteq VG$ and $EH \subseteq EG$.

A graph in which any two vertices are connected by a walk is called *connected.* A maximal with respect to inclusion connected subgraph of a graph G is called a *connected component,* or simply *component,* of G. The vertex set of a connected component is called a *connected domain* of a graph.

The length of the shortest (u, v)-walk in a connected graph G is called the *distance* between the vertices u and v and is denoted by $d_G(u, v)$, or even simply $d(u, v)$.

The value

$$e(u) = \max_{v \in VG} d(u, v)$$

is called the *excentricity* of the vertex u. The maximum among the excentricities of the vertices of a graph G is called the *diameter* of the graph and is denoted by $d(G)$. If $e(v) = d(G)$ then the vertex v is called *peripherical.* A simple path of length $d(G)$ with distance $d(G)$ between its endpoints is called a *diametral path.*

The minimum among the excentricities of the vertices of a graph G is called the *radius* of the graph and is denoted by $r(G)$. If $e(v) = r(G)$ then the vertex v is called *central.* The set Z_G of all central vertices of a graph G is called the *center* of the graph G. The subgraph $G(Z_G)$ with the vertex set Z_G such that two its vertices are adjacent if and only if they are adjacent in G will be called the *central subgraph,* and sometimes simply the *center.*

Let us describe a simple procedure of analysis of a graph called *breadth-first search.* It may be used, e.g., to determine whether a graph is connected or bipartite, to find the distance between two vertices, vertex excentricities, diameter and radius of a graph, to determine the connected domains of a graph.

Breadth-First Search. Given a graph G, we start from an arbitrary vertex v and assign label 0 to it. Each vertex from the neighborhood of v obtains label 1. Each unlabelled vertex from the neighborhoods of the vertices with label 1 obtains label 2. Labels 3 are assigned in a similar way. This process continues until possible. If G is connected then eventually the breadth-first search will label all its vertices, see Fig. 1.2.1.

* * *

1.2.1. Find all simple paths and simple cycles in the graph shown at Fig. 1.2.2.

1.2.2. How many labelled simple cycles are there in the graphs a) K_n and b) $K_{p,q}$?

1.2.3. Find cycles of lengths 5,6,8,9 in the Petersen graph.

1.2.4. Find the girths of a) K_n, b) $K_{p,q}$, c) Petersen graph.

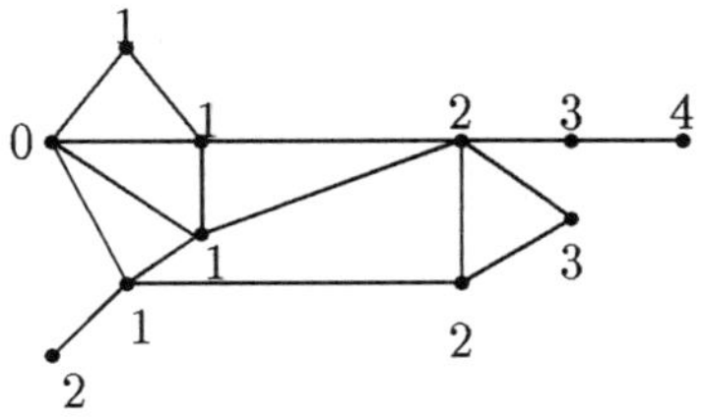

Figure 1.2.1: Breadth-first search labelling of graph

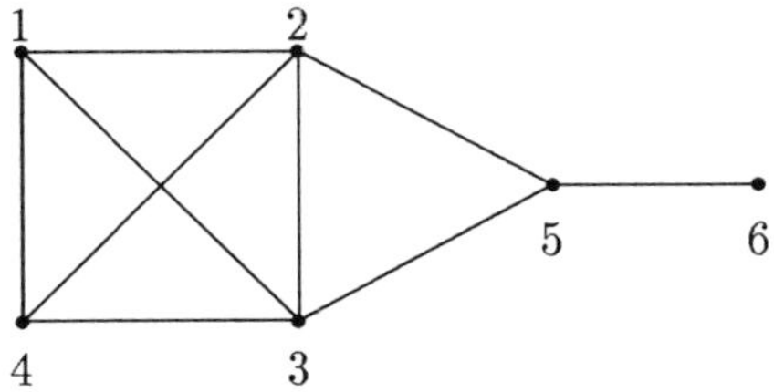

Figure 1.2.2: Example graph

1.2.5. a) Prove that every noncyclic (u,v)-walk contains a simple (u,v)-path. Is this statement true for cyclic walks?

b) Is it true that every (u,v)-walk passing through a vertex w distinct from u and v contains a simple (u,v)-path passing through w?

c) Prove that every cyclic walk of odd length contains a simple cycle. Is this statement true for cyclic walks of even length?

d) Prove that every cycle contains a simple cycle.

1.2.6. Prove that in any graph G the distance $d(u,v)$, $u,v \in VG$, satisfies the following *axioms of metric:*

1) $d(u,v) \geq 0$;
2) $d(u,v) = 0$ if and only if $u = v$;
3) $d(u,v) = d(v,u)$;
4) $d(u,v) + d(v,w) \geq d(u,w)$ (*triangle inequality*).

1.2.7. Prove that a graph G is connected if and only if

a) given a fixed vertex u, there exist (u,v)-paths in G to all other vertices $v \in VG$;

b) for every partition of the vertex set $VG = V_1 \bigcup V_2$, $V_1 \bigcap V_2 = \emptyset$, there exists an edge $ab \in EG$ such that $a \in V_1$, $b \in V_2$.

1.2.8. Prove that each vertex of a graph G lies in exactly one of its components. Prove the similar statement for edges.

1.2.9. a) Prove that in every connected graph any two simple paths of maximal length share a vertex.

b) With the help of the graph from Fig. 1.2.3 verify that the following statement is false: in every connected graph all simple paths of maximal length share a vertex.

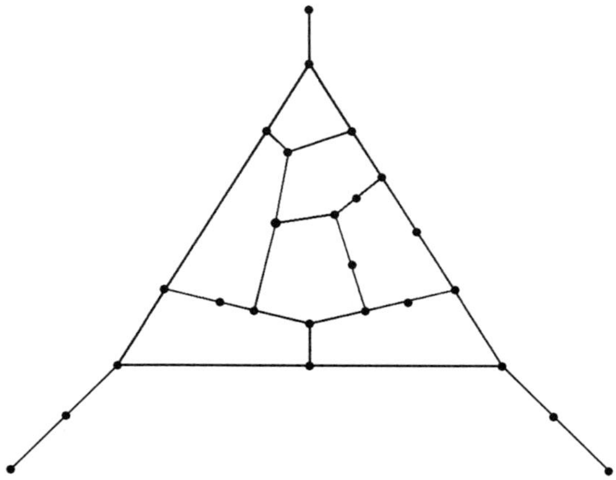

Figure 1.2.3: To Exercise 1.2.9

1.2.10. Prove that if a graph has exactly two vertices of odd degrees then they are connected by a path.

1.2.11. Prove that if the minimal degree $\delta(G)$ of the vertices of a graph G is greater than 1 then G has a cycle.

1.2.12. Define the union $G \cup H$ of a pair of graphs as follows:

$$V(G \cup H) = VG \cup VH, \quad E(G \cup H) = EG \cup EH.$$

Prove that:

a) if P, Q are different simple (u, v)-paths then the graph $P \cup Q$ contains a simple cycle;

b) if C, D are different simple cycles with a common edge e then the graph $(C \cup D) - e$ obtained by the deletion of the edge e from the graph $C \cup D$ contains a simple cycle.

1.2.13. a) Prove that a graph is bipartite if and only if it has no simple cycles of odd length.

b) Prove that the bipartition of a connected bipartite graph is unique. Is this true for disconnected bipartite graphs?

1.2.14. Find all bipartite graphs whose complementary graphs are also bipartite. Indicate the selfcomplementary ones among them.

1.2.15. Find all bipartite graphs of diameter 2.

1.2.16. Prove that for any integer number $r > 1$ there exists an r-regular graph of order $2r$ and diameter 2.

1.2.17. Draw all pairwise nonisomorphic connected graphs with exactly four edges.

1.2.18. Describe the structure of a connected graph G that satisfies one of the following conditions:

a) the length of any maximal with respect to inclusion simple path is 2;

b) for any vertex $v \in VG$ all vertices from its neighborhood are pairwise adjacent;

c) every vertex is pendant;

d) every edge is pendant;

e) all edges are pairwise adjacent;

f) the number of edges is equal to the number of pendant vertices;

g) $\delta(G) = 1$, $\Delta(G) = 2$;

h) $G \cong L(G)$.

1.2.19. Describe the structure of a connected r-regular graph for a) $r = 0$, b) $r = 1$, c) $r = 2$.

1.2.20. Prove that the deletion of any edge that belongs to a cycle of a connected graph produces a connected graph. What will occur if we delete an edge which does not belong to any cycle?

1.2.21. What is the minimal possible number of edges in a connected graph of order n?

1.2.22. Prove that for any two edges e_1, e_2 of a connected graph there exists a simple path such that e_1 is its first edge and e_2 is its last edge.

1.2.23. Prove the following statements.

a) If an (n, m)-graph has k components then

$$n - k \leq m \leq \frac{(n-k)(n-k+1)}{2}. \tag{1.2.2}$$

Describe the cases for which these bounds are attained.

b) If in addition the graph is bipartite then

$$n - k \leq m \leq \begin{cases} \dfrac{(n-k)(n-k+2)}{2}, & \text{if } n-k \text{ is even} \\ \dfrac{(n-k+1)^2}{2}, & \text{if } n-k \text{ is odd.} \end{cases} \tag{1.2.3}$$

c) Prove that for any positive integers satisfying the inequalities (1.2.2) there exists an (n, m)-graph with k components, and if (1.2.3) holds then the required graph exists among the bipartite ones.

1.2.24. a) Prove that a graph G of order n is connected if $\delta(G) \geq (n-1)/2$. Will the statement remain valid if we replace $(n-1)/2$ by $\lfloor (n-1)/2 \rfloor$?

b) Prove that a graph G of order $n > 2$ is connected if the number of edges is greater than $(n-1)/2$. Can we replace here "greater" by "not less"?

c) Prove that an (n, m)-graph is connected if it has no cycles of odd length and $m > ((n-1)/2)^2$. Will this statement remain true if we replace ">" by "$\geq$"?

1.2.25. a) Prove that the n-permutation graph (see Exercise 1.1.27) is connected for any n.

b) Find the diameter of this graph.

c) Find all central and peripherical vertices of this graph.

1.2.26. Prove that

a) the diameter of a graph is at most twice its radius;

b) a diametral path of a graph contains a central vertex.

1.2.27. For any $n \geq 4$ give an example of an n-vertex graph other than the complete graph such that

a) its radius and diameter are equal;

b) each vertex is peripherical;

c) each vertex is central.

1.2.28. Prove that

a) for any graph G either G or $\overline{G}$ is connected;

b) if G is not connected then $d(\overline{G}) \leq 2$;

c) if G and $\overline{G}$ are connected and $d(G) \geq 3$ then $d(\overline{G}) \leq 3$;

d) $1 \leq d(G) \leq 3$ for any nontrivial selfcomplementary graph G.

1.2.29. The *metrical dimension* $\mu(G)$ of a connected graph G is the minimal cardinality of the subsets $A \subseteq VG$ with the following property: for any two vertices $u, v \in VG$ there is a vertex $a \in A$ such that $d(u, a) \neq d(v, a)$. Find the metrical dimensions of the following graphs: a) K_n, b) $K_{p,q}$, c) C_n, d) P_n.

1.2.30. Prove that $\mu(G) \leq |G| - d(G)$ (see Exercise 1.2.29) for any connected graph G.

1.2.31. Give an example of a cubic graph of order $2n$, $n \geq 3$, with girth at least 4.

1.2.32. Prove that $|G| \geq \Delta(G) + \delta(G)$, if the girth of G is 4.

1.2.33. Prove that the order of a k-regular graph with girth 4 is at least $2k$.

1.2.34. Give an example of a graph G whose center is not the set of all vertices and consists of a) one vertex, b) two vertices, c) three vertices.

1.2.35. Prove that for any graph G there exists a graph H such that $G \cong Z(H)$, where $Z(H)$ is the central graph of H.

1.2.36. The *König representation of a graph* G is the bipartite graph $K(G)$ such that the vertices of one part bijectively correspond to the vertices of G, the vertices of the other part bijectively correspond to the edges of G, and two vertices of $K(G)$ are adjacent if and only if the corresponding vertex and edge of G are incident to each other. Prove the following statements.

a) $K(G)$ may be obtained from G by *subdivisions* of all its edges, i.e., for each edge $e = uv \in EG$ we add a new vertex w_e and the edge e is replaced by the two edges uw_e and w_ev.

b) Two graphs are isomorphic if and only if its König representations are isomorphic.

1.2.37. Determine whether the graphs from Fig. 1.2.4 are bipartite, using breadth-first search. Partition the vertex sets of the bipartite graphs into parts.

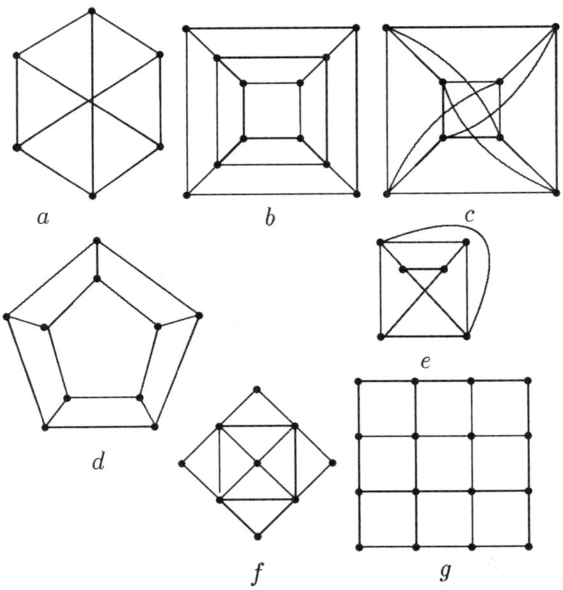

Figure 1.2.4: To Exercises 1.2.37, 1.2.39

1.2.38. Using breadth-first search, find the diameter, radius, central and peripherical vertices of the following graphs: a) P_n; b) C_n; c)K_n; d) $K_{1,n}$, $n \geq 2$; e) $K_{m,n}$, $n \geq 2, m \geq 2$.

1.2.39. Using breadth-first search, find the radius, diameter, central and peripherical vertices of the graphs from Fig. 1.2.4.

1.2.40. Let $f : VG \to VH$ be an isomorphism of graphs G and H. Prove the following statements.

a) If $VG = V_1 \cup V_2 \cup \ldots \cup V_k$ is the partition of VG into the connected domains of G and

$$f(V_i) = \{f(v) : v \in V_i\} = W_i, \quad i = 1, 2, \ldots, k,$$

then $VH = W_1 \cup W_2 \cup \ldots \cup W_k$ is the partition of VH into the connected domains of H.

b) For any two vertices u, v of G,

$$d(u, v) = d(f(u), f(v)); \quad e(u) = e(f(u)).$$

c) $d(G) = d(H)$, $r(G) = r(H)$.

d) If Z_G denotes the center of G then $f(Z_G) = Z_H$.

1.2.41. Prove that the number of components of a circulant graph $G_{n,U}$ (see Exercise 1.1.26) is equal to the greatest common divisor $\mathrm{GCD}(u_1, u_2, \ldots, u_k, n)$, where $\{u_1, u_2, \ldots, u_k\} = U$. In particular, $G_{n,U}$ is connected if and only if

$$\mathrm{GCD}(u_1, u_2, \ldots, u_k, n) = 1.$$

1.3 Subgraphs and Hereditary Properties of Graphs. Reconstructibility

A graph H is called a *subgraph* of a graph G if $VH \subseteq VG$ and $EH \subseteq EG$. For $X \subseteq VG$, we define the *induced subgraph* $G(X)$ to be the subgraph of G such that $VG(X) = X$ and $u, v \in VG(X)$ are adjacent in $G(X)$ if and only if they are adjacent in G. It is said that $G(X)$ is *induced by* X. A subgraph is called *a spanning subgraph* if $VG = VH$. A spanning induced subgraph of a graph G is clearly the graph G itself.

In the cases when no confusion is possible the term "subgraph" will be used to denote the "graph isomorphic to a subgraph" of the graph involved.

A property of graphs will be called *graph-theoretical* if for any pair of isomorphic graphs both of them either possess the property or not.

Formally, a *property* of graphs may be treated as a *class* of graphs. Therefore we shall use the notation $G \in \mathcal{P}$ in order to say "G has a property $\mathcal{P}$".

A graph-theoretical property $\mathcal{P}$ is called *hereditary* if the fact $G \in \mathcal{P}$ implies $H \in \mathcal{P}$ for every induced subgraph H of G. It is said that the set $\mathcal{I}$ of graphs *characterizes* a graph-theoretical property $\mathcal{P}$ *in terms of forbidden induced subgraphs (FIS)* if $G \in \mathcal{P}$ if and only if no induced subgraph of G belongs to $\mathcal{I}$.

Theorem 1.3.1 *The hereditary properties are exactly the ones which may be characterized in terms of forbidden induced subgraphs.*

In an investigation of a hereditary property, a natural intention is to find a minimal under inclusion set of forbidden induced subgraphs (minimal FIS set) characterizing it.

Theorem 1.3.2 *For a given hereditary property $\mathcal{P}$ a minimal FIS set is a set of graphs G satisfying the following conditions:*

1. $G \notin \mathcal{P}$;

2. *every induced subgraph of G other than G itself has the property $\mathcal{P}$.*

Let v be a vertex of G. The graph $G_v = G - v$ is obtained from G by the deletion of v together with the edges incident to it; it is called a *vertex-deleted subgraph*.

The graphs G_v are related with the *Kelly-Ulam reconstruction conjecture*, stated as follows. For a graph G the collection of graphs $P(G) = (G_v : v \in VG)$ is called the *deck* of the graph and the graphs G_v are called the *cards* of the deck. The *equality* of decks is defined up to permutations of cards in decks. It is important to emphasize that there are no correspondences among the vertex sets of cards, i.e., cards are considered to be unlabelled graphs. Alternatively, one may consider the *labelled deck* $P_l(G)$ of a labelled graph G.

For example, if $G = P_3$ then $P(G) = (K_2, K_2, O_2)$. An example of a labelled deck is shown at Fig. 1.3.1.

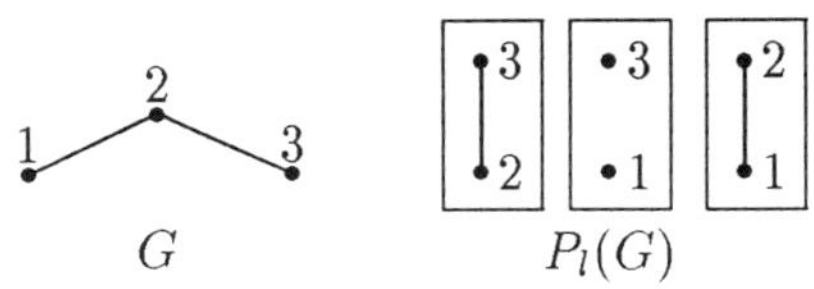

Figure 1.3.1: Labelled deck

A graph G is called a *reconstruction* of a graph H if $P(G) = P(H)$. For example, $P(K_2) = P(O_2) = (O_1, O_1)$, therefore K_2 and O_2 are reconstructions of each other. A graph G is called *reconstructible* if it is isomorphic to every its reconstruction.

Reconstruction Conjecture (P.Kelly, S.Ulam, 1941). *The graphs of order $n > 2$ are reconstructible.*

The Kelly-Ulam conjecture is often called the *vertex reconstruction conjecture.* In addition, the *edge reconstruction conjecture* due to F.Harary is known. It is stated in a similar way in terms of *edge-deleted subgraphs:* the subgraph $G_e = G - e$ is obtained from G by the deletion of the edge e (the ends of e are not deleted, i.e., G_e is a spanning subgraph). Accordingly, the collection $P'(G) = (G_e : e \in EG)$ is called the *edge deck* of the graph G. A graph G is called an *edge reconstruction* of a graph H if $P(G) = P(H)$. A graph G is called *edge reconstructible* if it is isomorphic to every its edge reconstruction.

Edge Reconstruction Conjecture (F.Harary, 1964). *All graphs with more than three edges are edge reconstructible.*

A nonempty subset of vertices of a graph is called a *clique* if the subgraph induced by it is a complete graph. In particular, all single-element vertex subsets are cliques. A clique is called *maximal* if it is not contained in a larger clique of the graph. For example, the graph shown at Fig. 1.3.2 has three maximal cliques.

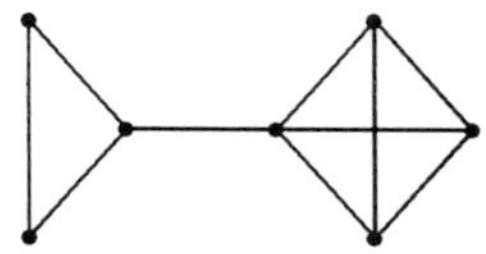

Figure 1.3.2: Example graph

Let G be a graph. The *clique graph* $Q(G)$ is defined as follows. The vertices of $Q(G)$ bijectively correspond to the maximal cliques of G; two vertices of $Q(G)$ are adjacent if and only if the corresponding cliques have a nonempty intersection. For example, the clique graph for the graph from Fig. 1.3.2 is P_3.

Theorem 1.3.3 *For every graph H there exists a graph G such that H is isomorphic to an induced subgraph of the clique graph $Q(G)$.*

* * *

1.3.1. Prove that

a) a subgraph of a subgraph of a graph G is a subgraph of G;

b) a spanning subgraph of a spanning subgraph of a graph G is a spanning subgraph of G;

c) an induced subgraph of an induced subgraph of a graph G is an induced subgraph of G.

1.3.2. Prove that

a) any subgraph of a graph G may be obtained by deletion of some vertices and/or edges of G;

b) any spanning subgraph of a graph G may be obtained by deletion of some edges of G;

c) any induced subgraph of a graph G may be obtained by deletion of some vertices of G.

1.3.3. (a) For a graph from Fig. 1.3.3 determine whether the cycles C_5, C_6, C_7, C_8 are among its

(1) subgraphs;

(2) spanning subgraphs;

(3) induced subgraphs.

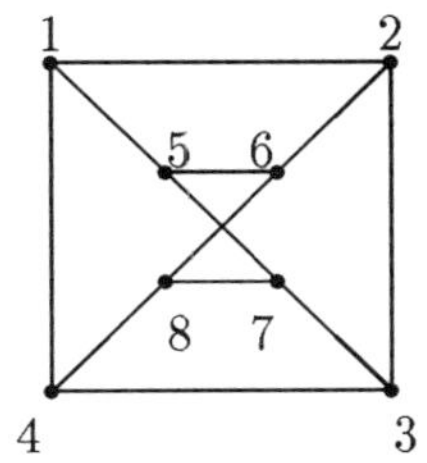

Figure 1.3.3: To Exercise 1.3.3

(b) Is it true that $G(X) = P_5$ for $X = \{1, 2, 4, 6, 8\}$ for a graph G shown at Fig. 1.3.3?

1.3.4. Let G_n be a graph with the vertex set $\{v_1, \ldots, v_n\}$ such that v_i, v_j are adjacent if and only if i, j are coprime. Draw the graphs G_4 and G_8. Prove that if $m < n$ then G_m is an induced subgraph of G_n.

1.3.5. Let F and H be subgraphs of a labelled (n, m)-graph G. We shall say that F is equal to H if $VF = VH$ and $EF = EH$. Determine the numbers of different

(a) induced subgraphs of G;

(b) spanning subgraphs of G;

(c) subgraphs of G for $G = K_n$.

1.3.6. Suppose that $VG = VH$ for graphs G, H and $\deg_G v \geq \deg_H v$ for every vertex v. Is it true that H is a subgraph of G?

1.3.7. Is it true that $\delta(G) \leq \delta(H)$ if H is

(a) a subgraph of G;

(b) a spanning subgraph of G?

1.3.8. Find all graphs of order $n \geq 4$ for which all induced subgraphs of order 3 are isomorphic to (a) O_3; (b) $\overline{P_3}$; (c) P_3; (d) K_3.

1.3.9. What is the structure of a graph in which every simple path is an induced subgraph?

1.3.10. Prove that for every graph G there exists a selfcomplementary graph that has an induced subgraph isomorphic to G.

1.3.11. Is it true that the vertex set of any graph G may be partitioned into parts V_1 and V_2 such that $G(V_1)$ and $G(V_2)$ are regular graphs?

1.3.12. Prove that there are no graphs of order $n \geq 5$ such that every four-vertex induced subgraph is the cycle graph C_4.

1.3.13. Prove that every graph G is an induced subgraph of a $\Delta(G)$-regular graph.

1.3.14. Characterize connected bipartite graphs that have no induced subgraphs P_5 and C_4.

1.3.15. Prove the theorems 1.3.1 and 1.3.2.

1.3.16. Characterize all graphs in which the sets $N(v)$ induce complete graphs for all vertices v.

1.3.17. Find a graph of order 10 in which the neighborhoods of all vertices induce the graph $\overline{P}_3$.

1.3.18. Prove that there are no regular graphs in which the neighborhoods of all vertices induce the star $K_{1,n}$, $n \geq 2$.

1.3.19. Find all connected graphs in which the sets $N(v)$ for all vertices v induce the graph

(a) C_3;

(b) C_4.

1.3.20. Let $\{H_1, H_2, \ldots\}$ be the set of forbidden induced subgraphs for some fixed hereditary property $\mathcal{P}$. Find a condition under which $G \in \mathcal{P}$ implies $\overline{G} \in \mathcal{P}$.

1.3.21. Determine which of the following graph-theoretical properties are hereditary. For every hereditary property find a minimal FIS set.

(a) to be an empty graph;

(b) to be a complete graph;

(c) to be a connected graph;

(d) to be a disconnected graph;

(e) to be a complete bipartite graph;

(f) every component of a graph in question is a complete graph;

(g) every component of a graph in question is a path graph;

(h) every component of a graph in question is either a star or a trivial graph;

(i) to have a vertex of even degree;

(j) $\delta(G) \leq 2$;

(k) $\Delta(G) \leq 2$;

(l) to have an isolated vertex;

(m) to be a selfcomplementary graph;

(n) to be a bipartite graph;

(o) to have no cycles;

(p) to be regular;

(q) to have an induced subgraph isomorphic to a given graph $H \neq K_1$.

1.3.22. Let $\mathcal{P}$ be a hereditary graph-theoretical property with at least one graph $G \in \mathcal{P}$. Are the following properties hereditary?

(a) to have not the property $\mathcal{P}$;

(b) to have no subgraphs with the property $\mathcal{P}$.

1.3.23. Prove that a graph is complete bipartite or empty if and only if it has no induced graphs K_3 and $\overline{P}_3$.

1.3.24. Let $\mathcal{P}$ be a hereditary property. Define $\tilde{\mathcal{P}}$ to be the property such that $G \in \tilde{\mathcal{P}}$ if and only if $\overline{G} \in \mathcal{P}$. Is $\tilde{\mathcal{P}}$ a hereditary property?

1.3.25. Is it true that for hereditary properties $\mathcal{P}_1$, $\mathcal{P}_2$

(a) the *disjunction*

$$\mathcal{P}_1 \vee \mathcal{P}_2 = \{G : G \in \mathcal{P}_1 \text{ or } G \in \mathcal{P}_2\}$$

is a hereditary property?

(b) the *conjunction*

$$\mathcal{P}_1 \wedge \mathcal{P}_2 = \{G : G \in \mathcal{P}_1 \text{ and } G \in \mathcal{P}_2\}$$

is a hereditary property?

1.3.26. (a) Let $\mathcal{K}$ be a class of graphs that contains a graph K_1. Prove that it has a unique maximal under inclusion hereditary subclass $[\mathcal{K}] \subseteq \mathcal{K}$.

(b) Let $\mathcal{K}$ be a class of graphs that contains all disconnected graphs, their complements, and the graph K_1. Characterize the subclass $[\mathcal{K}]$ defined above, in terms of forbidden induced subgraphs.

1.3.27. Prove that a labelled graph of order at least 3 is reconstructible from its labelled deck.

1.3.28. Are labelled graphs of order at least 3 reconstructible from

(a) two,

(b) three

cards of the labelled deck?

1.3.29. Determine whether the following collection of graphs constitute the deck of some graph:

(a) $K_1 \cup P_3$, $K_1 \cup P_3$, $\overline{K_1 \cup P_3}$, $\overline{K_1 \cup P_3}$, P_4;

(b) $K_1 \cup P_3$, C_4, C_4, $K_2 \cup K_2$, $K_2 \cup K_2$.

1.3.30. Prove that all regular graphs of order at least 3 are reconstructible.

1.3.31. Determine the degrees of all vertices of a graph of order at least 3 in terms of its deck.

1.3.32. Prove that given a deck of a graph of order at least 3 without isolated vertices, one may determine the numbers of edges and components of the graph.

1.3.33. Prove the *Kelly lemma:* if the order of a graph F is less than the order of a graph G then the number $S(F, G)$ of subgraphs of G isomorphic to F satisfies the relation:

$$S(F, G) = \sum_{v \in VG} \frac{S(F, G_v)}{|G| - |F|}.$$

1.3.34. Find all graphs whose deck contains only simple cycles.

1.3.35. Is the property "to be reconstructible" hereditary?

1.3.36. Does there exist a graph of order at least 3 whose cards in the deck are pairwise nonisomorphic?

1.3.37. Find a clique graph for the graph shown at Fig. 1.3.4.

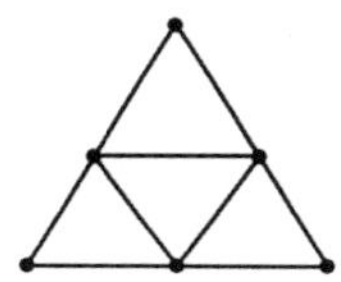

Figure 1.3.4: To Exercise 1.3.37

1.3.38. Find the graphs $Q(G)$ and $L(G)$ for (a) $G = P_n$, (b) $G = C_n$.

1.3.39. Prove that if a graph G has no isolated vertices and triangles then $Q(G) \cong L(G)$.

1.3.40. Describe the structure of $Q(L(K_5))$ (the clique graph of the line graph of K_5).

1.3.41. Prove that

(a) every graph G with minimal degree $\delta(G)$ greater than 2 is isomorphic to an induced subgraph of $Q(L(G))$.

(b) $G \cong Q(L(G))$ if $\delta(G) > 1$ and G is triangle-free.

1.3.42. If a graph G is isomorphic to the line graph of some graph then G is also called a line graph. Prove the following statements.

(a) The property "to be a line graph" is hereditary.

(b) The property "to be a line graph of a triangle-free graph" is hereditary.

(c) A graph G is the line graph of a triangle-free graph if and only if G has no induced subgraphs $K_{1,3}$ and $K_4 - e$.

1.3.43. Let B^r be the set of $(0,1)$-sequences of length r. The *Hamming distance* between the sequences

$$\alpha = (\alpha_1, \ldots, \alpha_r),\ \beta = (\beta_1, \ldots, \beta_r)$$

is the number

$$d_H(\alpha, \beta) = \sum_{i=1}^{r} |\alpha_i - \beta_i|.$$

The *Hamming graph* $H_{r,k}$ is defined as follows. $VH_{r,k} = B^r$, and two vertices α, β are adjacent if and only if $d_H(\alpha, \beta) \leq k$. Prove that every (n,m)-graph is isomorphic to an induced subgraph of the Hamming graph $H_{r,k}$ with $r = n^2 - m$, $k = 2n - 2$.

1.4 Operations on Graphs

We have already defined the operations of vertex deletion and edge deletion in Sect. 1.3. The *deletion* of any set X of elements of a graph G is defined as the successive deletion of all vertices and edges contained in X from G. The resulting graph is denoted by $G - X$.

Let two vertices u, v of G be nonadjacent. Then one may construct the graph $G' = G + e$, where $e = uv$, such that $VG' = VG$ and $EG' = EG \cup \{e\}$. This operation is called the *addition of the edge e*.

Sometimes the notations $G - v$, $G - e$, $G + e$ are used without specifying the elements v, e. In such cases it is assumed that these elements are arbitrary, and either the result is independent of the choice, see Fig. 1.4.1, or the resulting graph is indifferent for us.

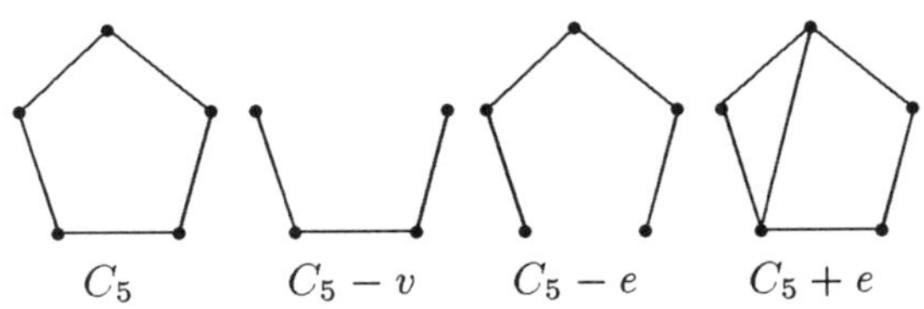

Figure 1.4.1: Operations on graph

The *union of graphs* G, H is the graph $G \cup H$ with the vertex set $VG \cup VH$ and the edge set $EG \cup EH$. If $VG \cap VH = \emptyset$ then the union is called *disjoint.* If VG and/or VH are not specified then the union is assumed to be disjoint. Fig. 1.4.2 shows the graph $K_2 \cup K_2 \cup K_4 \cup K_4 \cup K_4$.

For the sake of brevity, one may use the notation nG for the disjoint union of n graphs isomorphic to G (*n copies* of G). Thus, the graph from Fig. 1.4.2 may be denoted as $2K_2 \cup 3K_4$.

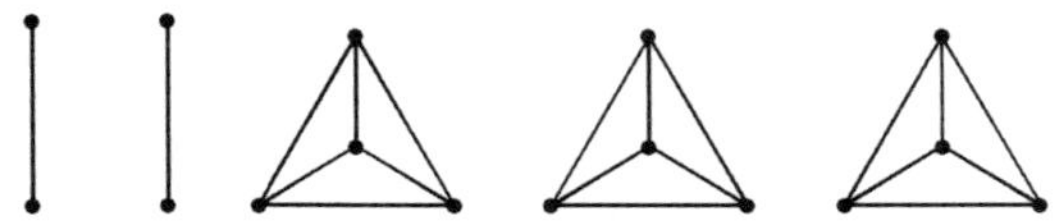

Figure 1.4.2: Disjoint union of graphs

The *intersection* $G \cap H = (VG \cap VH,\ EG \cap EH)$ is defined for any pair of graphs G, H with common vertices.

The *join* $G + H$ is obtained from the disjoint union $G \cup H$ by the addition of the set of all edges of type $uv :\ u \in VG,\ v \in VH$.

The *merging of vertices* u, v of a graph G is the addition of a new vertex w to the graph $G - u - v$, together with all edges connecting w with all vertices from $N_G(u) \cup N_G(v) \backslash \{u, v\}$.

The *contraction of an edge* uv is the merging of the adjacent vertices u, v. For example, the contraction of an edge of C_5 produces C_4.

Consider the unary operation of *raising a graph G to the k-th power.* The k-th power G^k is defined as follows. $VG^k = VG$; two vertices $u, v \in VG^k$ are adjacent if and only if the distance $d(u, v)$ in G is at most k. Clearly, $G^1 = G$.

For the following four binary operations of *(Cartesian) product* $G_1 \times G_2$, *composition* $G_1[G_2]$, *conjunction* $G_1 \wedge G_2$, *and modular product* $G_1 \diamond G_2$ the vertex set of the resulting graph is the Cartesian product $VG_1 \times VG_2$. The adjacencies of vertices $u = (u_1, u_2)$ and $v = (v_1, v_2)$ are defined according to the following rules. For $G_1 \times G_2$:

- either $u_1 = v_1$ and $u_2v_2 \in EG_2$,
- or $u_2 = v_2$ and $u_1v_1 \in EG_1$.

For $G_1[G_2]$:

- either $u_1v_1 \in EG_1$,
- or $u_1 = v_1$ and $u_2v_2 \in EG_2$.

For $G_1 \wedge G_2$:

- $u_1v_1 \in EG_1$ and $u_2v_2 \in EG_2$.

For $G_1 \diamond G_2$:
$u_1 \neq v_1$ and $u_2 \neq v_2$ and

- either $u_1v_1 \in EG_1$ and $u_2v_2 \in EG_2$,
- or $u_1v_1 \notin EG_1$ and $u_2v_2 \notin EG_2$.

Using the product operation, an important class of graphs called *n-dimensional cubes,* or *n-hypercubes* Q_n may be defined in a recurrent way: $Q_1 = K_2$, and $Q_n = K_2 \times Q_{n-1}$, $n > 1$, see Fig. 1.4.3.

* * *

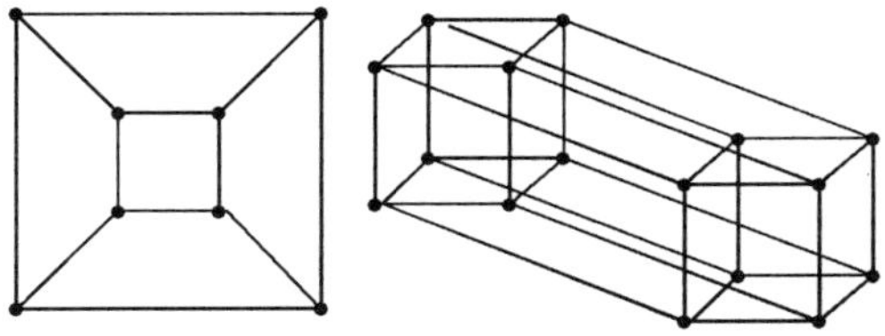

Figure 1.4.3: Cube graphs Q_3 and Q_4

1.4.1. Demonstrate that the operation $G-X$ is defined correctly, i.e., the resulting graph does not depend on the order of deletions of the elements of X.

1.4.2. Prove that the union of a connected graph G and any graph H is connected if and only if every component of H contains a vertex of G.

1.4.3. Express the operation of join of graphs in terms of the operations of union and complement.

1.4.4. Is it true that the intersection $G \cap H$ is an induced subgraph of either G or H?

1.4.5. Let $\mathcal{P}$ be a hereditary property. Is it true that if $G \in \mathcal{P}$ and $H \in \mathcal{P}$ then $G \cap H \in \mathcal{P}$?

1.4.6. Let G' be a graph obtained from a bipartite graph G by the contraction of an edge e. Find necessary and sufficient conditions under which G' is bipartite.

1.4.7. Let u, v be two nonadjacent vertices from different parts of a connected bipartite graph. Prove that the graph obtained by merging u with v is not bipartite.

1.4.8. Find the diameter of P_n^2.

1.4.9. What is the minimal number k such that the k-th power of a connected graph is the complete graph?

1.4.10. Is it true that (a) $(G^m)^n = G^{m+n}$, (b) $(G^m)^n = G^{mn}$?

1.4.11. How many edges has $G + H$?

1.4.12. Prove that if a graph G has $n \geq 2$ vertices and $m \geq \dfrac{(n-1)^2}{2}$ edges then $G = H_1 + H_2$ for some graphs H_1, H_2.

1.4.13. Construct the graphs (a) $C_4 \times P_3$; (b) $C_4[P_2]$; (c) $C_4 \wedge P_3$; (d) $C_4 \diamond P_3$.

1.4.14. Let G_i be (n_i, m_i)-graphs, $i = 1, 2$. Find the numbers of edges of the graphs (a) $G_1 \times G_2$; (b) $G_1[G_2]$; (c) $G_1 \wedge G_2$; (d) $G_1 \diamond G_2$.

1.4.15. Which of the operations listed in the Exercise 1.4.14 are commutative?

1.4.16. Let $d_1, \ldots d_n$ be the degree sequence of a graph G and let $d'_1, \ldots d'_n$ be the degree sequence of a graph G'. What are the degrees of the vertices of the graph $G \times G'$?

1.4.17. Prove that $G \times H$ is connected if and only if both G and H are connected.

1.4.18. Is it true that $G \times H$ is bipartite if and only if both G and H are bipartite.

1.4.19. Prove that

(a) $G_1 \wedge G_2$ is a spanning subgraph of $G_1 \diamond G_2$.

(b) $G_1 \times G_2$ is a spanning subgraph of $G_1[G_2]$.

1.4.20. Prove that

(a) $C_{2m+1} \wedge C_{2m+1}$ is isomorphic to $C_{2m+1} \times C_{2m+1}$;

(b) for connected graphs G_1, G_2, the conjunction $G_1 \wedge G_2$ is connected if and only if one of these graphs has a simple cycle of odd length.

1.4.21. The *major modular product* $G \bar{\diamond} G'$ is the graph specified as follows.

(1) $V(G \bar{\diamond} G') = VG \times VG'$;

(2) two vertices (u, u'), (v, v') of $G \bar{\diamond} G'$ are adjacent if and only if $u \neq v$, $u' \neq v'$ and either $uv \in EG$, $u'v' \in EG'$ or $uv \notin EG$.

a. Construct the major modular product of pairs of graphs $\{P_4, P_3\}$ and $\{2K_2, C_3\}$.
b. Prove that $G \diamond G'$ is a subgraph of $G \bar{\diamond} G'$.
c. What is the degree of a vertex (u, v) of $G \bar{\diamond} G'$?
d. Find graphs G, G' such that $G \bar{\diamond} G'$ and $G' \bar{\diamond} G$ are nonisomorphic.

1.4.22. Let us define the operation of *extension of a graph G by a graph H* denoted by $G \leftarrow H$. For every vertex $v \in VG$ we take a copy H_v of the graph H; all copies are pairwise disjoint. We set $V(G \leftarrow H) = \bigcup_{v \in VG} VH_v$. Two vertices $x \in H_v$, $y \in H_w$ are adjacent in $G \leftarrow H$ if either $v = w$ and x, y are adjacent in H_v or $v \neq w$ and v, w are adjacent in G.

Prove the following statements.

a. For every graph G, graphs H_1 and H_2 are isomorphic if and only if $G \leftarrow H_1$ and $G \leftarrow H_2$ are isomorphic.
b. $|E(G \leftarrow H)| = |EH||VG| + |VH|^2|EG|$.

1.4.23. Using the operation of graph extension defined in the Exercise 1.4.22,

(a) construct a selfcomplementary graph of order 5^n for every $n \geq 1$;

(b) prove that for any selfcomplementary graphs G, H the graph $G \leftarrow H$ is also selfcomplementary.

1.4.24. Prove that the n-cube Q_n may be defined as follows. The vertices are taken to be the $(0, 1)$-vectors of length n. Two vertices are adjacent if and only if the corresponding vectors differ exactly by one component.

1.4.25. Draw the 5-cube.

1.4.26. Find the numbers of vertices and edges of the n-cube Q_n.

1.4.27. Find the length of the longest induced simple cycle in Q_n for (a) $n = 4$, (b) $n = 5$.

1.4.28. Prove that Q_n is a regular bipartite graph.

1.5 Matrices Associated with Graphs

In what follows, the element of a matrix M at position (i, j) is denoted by M_{ij}. A matrix in which every element is either 0 or 1 is called a *binary matrix* or $(0,1)$-*matrix*.

Let G be a graph of order n whose vertices are labelled by the numbers $1, \ldots, n$, i.e., $VG = \{v_1, \ldots, v_n\}$. Let us define a binary $n \times n$-matrix $A = A(G)$ by setting $A_{ij} = 1$ if v_i, v_j are adjacent, and $A_{ij} = 0$ otherwise. The matrix $A(G)$ is called the *adjacency matrix* of the graph G.

The adjacency matrices of multi- and pseudographs are defined in a similar way: A_{ij} is equal to the number of edges connecting the vertices v_i, v_j assuming that a loop is counted twice.

Proposition 1.5.1 *Two graphs (multigraphs, pseudographs) are isomorphic if and only if the adjacency matrix of the first one is obtained from the adjacency matrix of the second one by the same permutations of rows and columns.*

Corollary 1.5.2 *The characteristic polynomials of adjacency matrices of isomorphic graphs are equal.*

Therefore we may define the *characteristic polynomial* of a graph to be the characteristic polynomial of its adjacency matrix.

Graphs with equal characteristic polynomials are called *cospectral.* Notice that there exist nonisomorphic cospectral graphs. This means that a graph is not defined by its characteristic polynomial. For example, Fig. 1.5.1 shows two pairs of nonisomorphic cospectral graphs.

Theorem 1.5.3 *The degree of a regular graph is a maximal-magnitude root of its characteristic polynomial.*

The *Kirchhoff matrix* $K(G)$ of a graph G is defined as follows:

$$K_{ij} = \begin{cases} -1, & \text{if } v_i \text{ and } v_j \text{ are adjacent}, \\ 0, & \text{if } i \neq j \text{ and } v_i \text{ and } v_j \text{ are nonadjacent}, \\ \deg v_i, & \text{if } i = j. \end{cases}$$

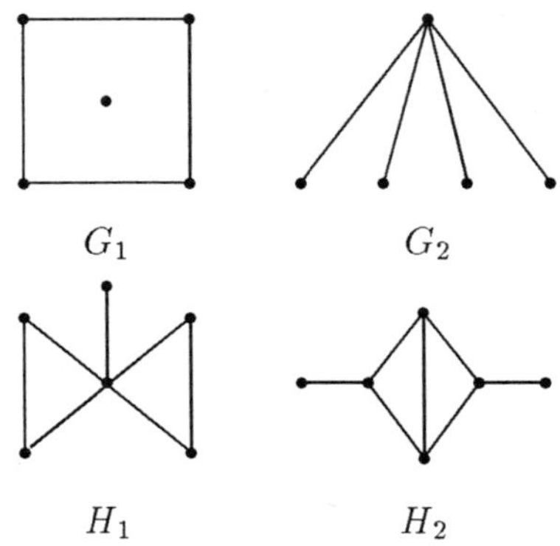

Figure 1.5.1: Nonisomorphic cospectral graphs

Proposition 1.5.1 remains valid if we take the Kirchhoff matrices instead of the adjacency ones.

Assume now that the edges of G are indexed, i.e., $EG = \{e_1, \ldots, e_m\}$.

Define the binary matrix $I(G)$ called the *incidence matrix* as follows:

$$I_{ij} = \begin{cases} 1, & \text{if the vertex } v_i \text{ and the edge } e_j \text{ are incident,} \\ 0, & \text{otherwise.} \end{cases}$$

Proposition 1.5.4 *Graphs are isomorphic if and only if their incidence matrices may be obtained from each other by permutations of rows and columns.*

* * *

1.5.1. Prove that the sum of the elements of a row of the adjacency matrix of a graph is equal to the degree of the corresponding vertex.

1.5.2. Prove that the sum of the elements of a row or of a column of the Kirchhoff matrix is zero.

1.5.3. The *trace* of a square matrix is the sum of its diagonal elements. Prove that the trace of the adjacency matrix of a pseudograph is twice the number of its loops, and the trace of the Kirchhoff matrix of an ordinary graph is twice the number of its edges.

1.5.4. Prove the following statements:

a) If graphs G, H are isomorphic then the following conditions hold:

1. $A(G)$ is obtained from $A(H)$ by the same permutations of rows and columns;
2. $K(G)$ is obtained from $K(H)$ by the same permutations of rows and columns;
3. $I(G)$ is obtained from $I(H)$ by permutations of rows and columns;

b) If one of the above conditions 1–3 holds then $G \cong H$.

1.5.5. Prove that by the same permutations of rows and columns the matrix $A(G)$ may be reduced to

a) the block-diagonal form:

$$\operatorname{diag}[A_1, A_2, \ldots, A_k], \; k > 1,$$

where A_i, $i = 1, 2, \ldots, k$, are square matrices, if and only if G is not connected.

b) the block form

$$\begin{bmatrix} O_k & B \\ C & O_m \end{bmatrix},$$

where O_p is the zero $(p \times p)$-matrix, if and only if G is a nontrivial bipartite graph.

1.5.6. Let s be a permutation acting on the set $\{1, 2, \ldots, n\}$. Define the *permutation matrix* M_s to be the binary $(n \times n)$-matrix such that

$$(M_s)_{ij} = \begin{cases} 1, & \text{if } s(j) = i, \\ 0, & \text{otherwise.} \end{cases}$$

Prove the following statements:

a) Each row and each column of M_s contains a single one and $\det M_s = \pm 1$.

b) $(M_s A)_{ij} = A_{s^{-1}(i)j}$, $i = 1, 2, \ldots, n$, $j = 1, 2, \ldots, m$, for any permutation s and any $(n \times m)$-matrix A.

c) $(BM_s)_{ij} = B_{is(j)}$, $i = 1, 2, \ldots, m$, $j = 1, 2, \ldots, n$, for any permutation s and any $(m \times n)$-matrix B.

d) Graphs G, H are isomorphic if and only if $A(H) = M_s A(G)(M_s)^{-1}$ for some permutation s.

e) Graphs G, H are isomorphic if and only if $I(H) = M_s I(G)(M_t)^{-1}$ for some permutations s and t.

1.5.7. Let G and H be graphs of order n with the same vertex set V. Prove that the solutions of the equation

$$A(H)X = XA(G), \tag{1.5.1}$$

where X is a permutation matrix of size $(n \times n)$, bijectively correspond to isomorphisms of G and H, i.e., a permutation $s : V \to V$ is an isomorphism of G and H if and only if $X = M_s$ is a solution of (1.5.1).

1.5.8. Let $f(x) = x^n + c_1 x^{n-1} + \ldots + c_{n-1} x + c_n$ be the characteristic polynomial of a graph G. Prove that

a) $c_1 = 0$;

b) $-c_2$ is the number of edges of G;

c) $-c_3$ is twice the number of triangles in G.

1.5.9. Prove that the graphs G_1, G_2 and H_1, H_2 from Fig. 1.5.1 are indeed pairs of nonisomorphic cospectral graphs.

1.5.10. Prove that the (i, j)-th element of the matrix $(A(G))^k$ is the number of (v_i, v_j)-walks of length k in G.

1.5.11. Prove that a graph G of order n is connected if and only if the matrix $(A(G) + E_n)^{n-1}$ has no zeroes. Here E_n denotes the identity matrix.

1.5.12. Prove that after a renumbering of the vertices and edges of a graph the rank of its incidence matrix remains unchanged.

1.5.13. Prove that the rank of the incidence matrix of a connected graph of order $n > 1$ is $n-1$, if the graph is bipartite; otherwise the rank is equal to n.

1.5.14. A matrix is called *totally unimodular* if all its minors have the values $0, \pm 1$. Prove that the incidence matrix of a bipartite graph is totally unimodular.

1.5.15. Let $s = (1, 2, \ldots, n)$ be a cyclic permutation; define

$$h = M_s = \begin{bmatrix} 0 & 0 & \ldots & 0 & 1 \\ 1 & 0 & \ldots & 0 & 0 \\ 0 & 1 & \ldots & 0 & 0 \\ & & \ldots & & \\ 0 & 0 & \ldots & 1 & 0 \end{bmatrix}.$$

Let further $f(x) = a_0 + a_1 x + \ldots + a_{n-1} x^{n-1}$ be a polynomial with real coefficients. We may define a matrix $f(h)$ as follows:

$$f(h) = a_0 + a_1 h + \ldots + a_{n-1} h^{n-1} = \begin{bmatrix} a_0 & a_{n-1} & a_{n-2} & \ldots & a_2 & a_1 \\ a_1 & a_0 & a_{n-1} & \ldots & a_3 & a_2 \\ a_2 & a_1 & a_0 & \ldots & a_4 & a_3 \\ & & \ldots & & & \\ a_{n-2} & a_{n-3} & a_{n-4} & \ldots & a_0 & a_{n-1} \\ a_{n-1} & a_{n-2} & a_{n-3} & \ldots & a_1 & a_0 \end{bmatrix}.$$

A matrix of such type is called a *circulant* of order n.

Prove that a graph G is isomorphic to a circulant graph (see Exercise 1.1.26) if and only if under an appropriate numbering of the vertices the adjacency matrix of G is a circulant.

1.5.16. Let n be a prime number and let $G_i = G_{n,U_i}$, $i = 1, 2$, be two circulant graphs (see Exercise 1.1.26). Prove that each of the conditions below is necessary and sufficient for G_1 and G_2 to be isomorphic:

a) The spectra of G_1 and G_2 are the same.

b) $U_2 = kU_1$ for some positive integer k not divisible by n (see Exercise 1.1.26d).

1.6 Automorphism Group of Graph

First we present some notions from the theory of permutation groups.

A bijective mapping $X \to X$ of a nonempty set X is called a *permutation* of X, or a *permutation acting* on X. The product fg of two permutations f, g is specified by the equality $fg(x) = f(g(x))$ for any $x \in X$. This product operation defines the structure of group over the set of all permutations acting on X. This group is called the *symmetric group* and is denoted by $S(X)$. The unit element of this group is the identity transformation

$$e : X \to X,\ x \mapsto x,\ x \in X.$$

If $|X| = n$, the group S(X) is denoted by S_n, and it is called the *symmetric group of degree* n.

Any subgroup F of $S(X)$ is called a *permutation group* over the set X.

If for any pair $x_1, x_2 \in X$ the group F contains a permutation f such that $f(x_1) = x_2$ then the group F is called *transitive*. Otherwise the group is called *intransitive*.

We may specify an equivalence relation $\sim$ over X by setting $x \sim y$ if and only if the group F has a permutation f such that $f(x) = y$. This relation partitions X into the classes of equivalent elements: any two elements from the same class are mapped into each other by an appropriate permutation from F, while elements from different classes cannot be mapped into each other by permutations from F. These classes are called the *orbits of the group* F. The orbits of the group F are F*-invariant subsets* of X, i.e., if Y is an orbit, then $F(Y) = \{f(y) : y \in Y, f \in F\} = Y$. Every F-invariant subset of X is the union of some orbits of F.

Let F be a permutation group over X and suppose that $a \in X$. Denote $F_a = \{f \in F : f(a) = a\}$. The set F_a is a subgroup of F. It is called the *stabilizer of the element* a.

Theorem 1.6.1 *Let F be a permutation group over X. Let us take $a \in X$, and let Y be an orbit of F that contains a. For every element $y \in Y$ let us pick $f_y \in F$ such that $f_y(a) = y$. Then $\{f_y : y \in Y\}$ is a complete system of left coset representatives of F modulo F_a, i.e.,*

$$F = \bigcup_{y \in Y} f_y F_a$$

is the decomposition of F into left cosets modulo F_a.

Let F be a permutation group over X and let H be a permutation group over Y. If there exists a bijection $\varphi : X \to Y$ such that

$$H = \varphi F \varphi^{-1} = \{\varphi f \varphi^{-1} : f \in F\}, \tag{1.6.1}$$

then it is said that the permutation groups F and H are *similar*. The relation (1.6.1) implies that the mapping

$$\psi : F \to H : f \mapsto \varphi f \varphi^{-1},\ f \in F$$

is an isomorphism of groups F and H. However similarity is a stronger relation among groups than isomorphism: groups may be isomorphic but not similar.

Consider one of such examples. In what follows, $\langle a, b, \ldots, c\rangle$ denotes the subgroup of a group generated by the elements $a, b, \ldots, c$.

Let $a = (1,2)$ and $b = (3,4)$ be transpositions and let $c = (1,2)(3,4)$ and $d = (1,3)(2,4)$ be products of two transpositions. Suppose further $F = \langle a, b\rangle$ and $H = \langle c, d\rangle$. Then F and H are isomorphic subgroups of S_4 which are not similar. The group H is called the *Klein four-group.*

Similar permutation groups are isomorphic, and what is more, they "act similarly", i.e., a permutation $f \in F$ acts on $x \in X$ in the same way as $\varphi f \varphi^{-1}$ acts on $\varphi(x) = y$, i.e., if $f(x) = x'$ then $\varphi f \varphi^{-1}(\varphi(x)) = \varphi(x')$.

Let us define two operations on the set of permutation groups. The first one is the direct product. If X_1, X_2 are nonempty disjoint sets and F_i is a permutation group over X_i, $i = 1, 2$, then the *direct product* $F_1 \times F_2$ is the set

$$F_1 \times F_2 = \{(f_1, f_2) : f_1 \in F_1,\ f_2 \in F_2\}$$

that acts on the union $X_1 \cup X_2$ as follows:

$$(f_1, f_2)(x) = \begin{cases} f_1(x), & x \in X_1, \\ f_2(x), & x \in X_2. \end{cases}$$

Clearly, $|F_1 \times F_2| = |F_1||F_2|$.

The second operation is the wreath product. Let F be a permutation group over X and let H be a permutation group over Y. Let further Π be the set of all mappings $Y \to F$. Define the operation "+" over Π by setting

$$(\pi_1 + \pi_2)(y) = \pi_1(y)\pi_2(y) \tag{1.6.2}$$

for any $\pi_1, \pi_2 \in \Pi, y \in Y$. Then the set Π turns out to be a group isomorphic to the direct product of $k = |Y|$ copies of F. The zero of the group Π is the mapping

$$0 : Y \to F : y \mapsto e_X,\ y \in Y,$$

where e_X is the unit element of F. For $\pi \in \Pi$, $(-\pi)(y) = (\pi(y))^{-1}$.

If $\pi \in \Pi, h \in H$, then $\pi h \in \Pi$.

Let F Wr H denote the set of all pairs (h, π), $\pi \in \Pi, h \in H$. Define the action of a pair (h, π) on the Cartesian product $Z = X \times Y$ as follows: if $(x, y) \in Z$, then

$$(h, \pi)(x, y) = (\pi(y)(x), h(y)). \tag{1.6.3}$$

The set F Wr H of all pairs (h, π) acting on Z according to (1.6.3) is called the *wreath product* of permutation groups F and H.

Theorem 1.6.2 *The wreath product F Wr H is a permutation group over the set $Z = X \times Y$ such that*

$$(h_1, \pi_1)(h_2, \pi_2) = (h_1 h_2, \pi_1 h_2 + \pi_2) \tag{1.6.4}$$

for any $(h_i, \pi_i) \in F$ Wr H, $i = 1, 2$.

The equality (1.6.4) implies that the set of all pairs of form (e_Y, π), where e_Y is the unit of H, constitutes a subgroup Π' isomorphic to Π, and the set of all pairs of form $(h, 0)$ constitutes a subgroup H' isomorphic to H. The group Π' is normal in F Wr H. In addition, the following equalities are true:

$$F \text{ Wr } H = H'\Pi', \quad |F \text{ Wr } H| = |H||F|^{|Y|}.$$

Proposition 1.6.3 *F Wr H is transitive if and only if both groups F and H are transitive.*

A permutation φ that acts on the set VG of the vertices of a graph G and preserves the adjacency relation (i.e., such that the images $\varphi(u)$ and $\varphi(v)$ of vertices u, v are adjacent if and only if u, v are adjacent) is called an *automorphism* of the graph G. In other words, an automorphism of a graph is an isomorphism of the graph onto itself.

Theorem 1.6.4 *The set of all automorphisms of a graph is a permutation group over the set of its vertices.*

It is called the *automorphism group* of the graph G and is denoted by Aut G.

Theorem 1.6.5 *If a graph G is the disjoint union of graphs F, H such that no component of F is isomorphic to a component of H then* Aut $G =$ Aut $F \times$ Aut H.

Theorem 1.6.6 *If all components of an n-component graph G are isomorphic to each other then* Aut G *is similar to the wreath product* (Aut F) *Wr S_k, where F is a component of G and S_k is the symmetric group of degree k.*

The latter two theorems imply that finding the automorphism group of a graph amounts to two problems for connected graphs: recognition of isomorphism of connected graphs and finding the automorphism group of a connected graph.

* * *

1.6.1. Let F be a permutation group over X and let Y be an F-invariant subset of X. Prove that Y is the union of some orbits of F.

1.6.2. Prove that isomorphic subgroups of the symmetric group may turn out to be dissimilar.

1.6.3. Prove that $|F \text{ Wr } H| = |F|^k |H|$ if H is a subgroup of the symmetric group S_k.

1.6.4. Prove a) the Theorem 1.6.2; b) the Proposition 1.6.3.

1.6.5. Prove the Theorem 1.6.4.

1.6.6. Prove that Aut $\overline{G} =$ Aut G for any graph G.

1.6.7. Prove that the automorphism group of the graph from Fig. 1.6.1 is the identity group: Aut $G = \langle e \rangle$, where e is the identity permutation. Find other graphs with such automorphism group.

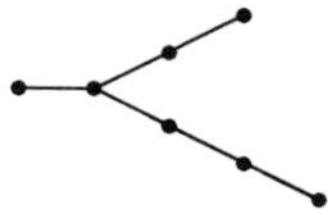

Figure 1.6.1: To Exercise 1.6.7

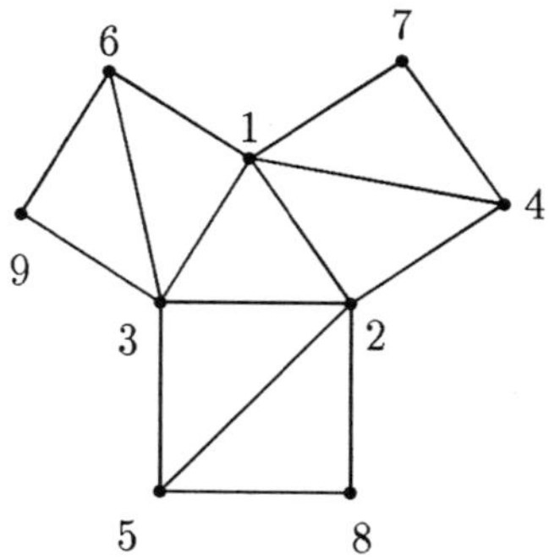

Figure 1.6.2: To Exercise 1.6.10

1.6.8. a)Prove that if G is a graph of order $n \leq 5$ then $\text{Aut } G \neq \langle e \rangle$.

b) Find all six-vertex graphs with the identity automorphism group.

1.6.9. Prove that if graphs are isomorphic then their automorphism groups are similar.

1.6.10. Find the automorphism groups for a) O_n; b) K_n; c) P_n; d) C_n; e) the graph from Fig. 1.6.2.

1.6.11. A permutation of the finite set is called *even* if it is a product of even number of transpositions. Prove that

a) the set A_n of all even permutations is the subgroup of the symmetric group S_n (it is called the *alternate group*).

b) there are no graphs of order $n > 3$ with $\text{Aut } G = A_n$.

1.6.12. Prove a) the Theorem 1.6.5; b) the Theorem 1.6.6.

1.6.13. Prove that the automorphism group of a disconnected graph is transitive if and only if all components of the graph are isomorphic and their automorphism groups are transitive.

1.6.14. Find Aut G if G is a graph that has exactly

a) four components: C_4, C_4, C_4, P_3;

b) two components: K_m, K_n, $m \neq n$.

c) two components: K_n, K_n.

1.6.15. Solve the Exercise 1.6.14b after the addition of an edge making the graph connected.

1.6.16. Let G be a graph. For $f \in \text{Aut } G$, let $U \subseteq VG$ be a subset invariant with respect to f and let f' be the *restriction of the mapping* f onto U, i.e.,

$$f' : U \to U,\ f'(x) = f(x),\ x \in U.$$

Prove that $f' \in \text{Aut } (G(U))$, where $G(U)$ is the subgraph induced by the set U.

1.6.17. Let f be an automorphism of a graph G. Define a mapping $\overline{f}$ of the set of maximal cliques of G by setting $\overline{f}(X) = \{f(x) : x \in X\}$ for any maximal clique X. Prove that $\overline{f}$ is an automorphism of the clique graph $Q(G)$.

1.6.18. Let G be a nonempty graph, $f \in \text{Aut } G$. Define a mapping $f_E : EG \to EG$ by setting $f_E(e) = f(a)f(b)$ for $e = ab$. Further, define

$$\text{Aut}_E G = \{f_E : f \in \text{Aut } G\}.$$

It is called the *edge group of graph* G. Prove the following statements:

a) $\text{Aut}_E G$ is a subgroup of the group $\text{Aut } (L(G))$.

b) $\text{Aut}_E G \cong \text{Aut } G$ if and only if the following two conditions hold:
(1) G has at most one isolated vertex;
(2) no component of G is isomorphic to K_2.

c) If one of the conditions
(1) G has no triangles and $\delta(G) > 1$;
(2) $\delta(G) > 3$
holds, then $\text{Aut } (L(G)) = \text{Aut}_E(G) \cong \text{Aut } G$.

1.6.19. Find $\text{Aut } (L(G))$ for a) $G = K_4$; b) $G = K_5$.

1.6.20. Find the automorphism group of the Petersen graph.

1.6.21. Find an edge-symmetric graph which is not vertex-symmetric. (This means that $\text{Aut}_E G$ is symmetric but $\text{Aut } G$ is not.)

1.6.22. Find the edge group for the graph from Fig. 1.6.3.

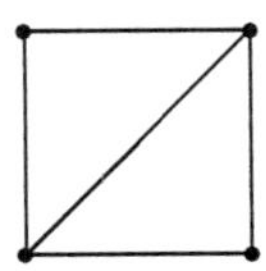

Figure 1.6.3: To Exercise 1.6.22

1.6.23. Prove that the number of pairwise distinct labelled graphs isomorphic to a graph G of order n is $n!/|\text{ Aut } G|$.

1.6.24. Let G be a graph of order n. Demonstrate that

$$\text{Aut } G = \{s \in S_n : M_s A(G) = A(G) M_s\},$$

where M_s is the permutation matrix corresponding to s (see Exercise 1.5.6), $A(G)$ is the adjacency matrix of G.

1.6.25. Prove that a graph G of order n is isomorphic to a circulant graph (see Exercise 1.1.26) if and only if Aut G contains a cyclic permutation of length n.

1.6.26. Let G, F be connected graphs such that $\Delta(G) = \Delta(H)$. Define a new graph $H = G \cup F + ab$ by adding an edge ab to the disjoint union $G \cup F$. Prove that for any vertex a of maximal degree in G there exists a vertex b of maximal degree in F such that the following statement is valid: $G \cong F$ if and only if there exists an automorphism φ of H such that $\varphi(a) \neq a$.

Chapter 2

Trees

2.1 Trees: Basic Notions

A *tree* is a connected graph without cycles. An arbitrary graph without cycles is called a *forest* (or *acyclic graph*). Therefore the components of a forest are trees.

Theorem 2.1.1 *The following statements are equivalent for an (n, m)-graph G :*

1. *G is a tree.*
2. *G is a connected graph and $m = n - 1$.*
3. *G is a forest and $m = n - 1$.*
4. *Any two vertices of G are connected by a single simple path;*
5. *G is a forest with the following property: if we connect any pair of nonadjacent vertices by an edge, a single cycle will be created in the graph.*

* * *

2.1.1. Prove that an acyclic graph with n vertices and k components has exactly $n - k$ edges.

2.1.2. Are there isomorphic graphs among the trees shown at Fig. 2.1.1?

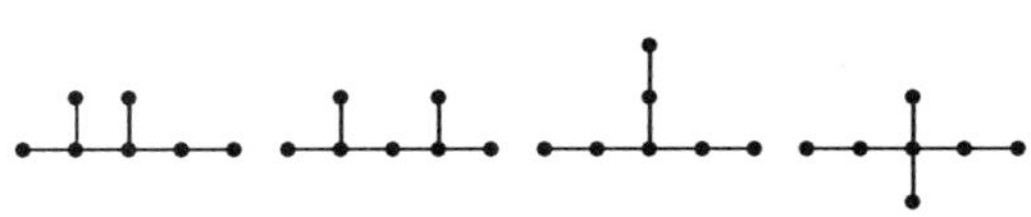

Figure 2.1.1: To Exercise 2.1.2

2.1.3. Draw all pairwise distinct labelled trees of order 4.

2.1.4. Draw all pairwise nonisomorphic trees of order n for a) $n = 5$; b) $n = 6$; c) $n = 7$; d) $n = 8$.

2.1.5. Let G be a connected (n, m)-graph. Prove that if G is not a tree then $m > n - 1$.

2.1.6. Prove that each tree is a bipartite graph. Which complete bipartite graphs are trees?

2.1.7. Prove that every connected subgraph of a tree is the induced one.

2.1.8. What is the maximal possible degree $\Delta(G)$ of the vertices of a tree G of order $n \geq 3$?

2.1.9. Find all trees that are selfcomplementary graphs.

2.1.10. Which trees are regular graphs?

2.1.11. Prove that the following statements are equivalent for a connected graph G:

- G has a single cycle;
- G is an (n, n)-graph for some n;
- G has an edge e such that $G - e$ is a tree.

2.1.12. Let $\mathcal{T}_k$ be the set of all trees without vertices of degree 2 and with k pendant vertices (and with any number of all vertices). Prove that $\mathcal{T}_k$ is finite for any k.

2.1.13. The *mean degree* (of vertices) of a graph of order n is the value $d_{mean} = \frac{1}{n}\sum_{i=1}^{n} \deg v$. Express the number of vertices of a tree in terms of the mean degree.

2.1.14. A vertex i of a tree is called *branching* if $d_i = \deg i \geq 3$. Express the number of the pendant vertices of a tree in terms of the degrees d_i of the branching ones.

2.1.15. Prove that any nontrivial tree

a) has at least two pendant vertices;

b) has exactly two pendant vertices if and only if the tree is a path.

2.1.16. Prove the following statements for a tree:

a) For any simple path P starting at a vertex v there exists a pendant vertex w such that P is a subpath of the simple (v, w)-path.

b) If a vertex v is the most distant one from a vertex u then v is a pendant vertex.

c) A vertex of maximal excentricity is pendant.

2.1.17. Consider a tree T with exactly k pendant vertices. If we number these vertices by $1, 2, \ldots, k$, we may define the $k \times k$-matrix D of *distances between the pendant vertices* of T as follows: $D_{ij} = d(i, j)$.

a) How will D change if we change the numbering of the pendant vertices of T?

b) Prove that

$$D_{ij} + D_{jk} - D_{ik} > 0;$$

$$D_{ij} + D_{jk} - D_{ik} \equiv 0 \pmod 2$$

for any three different indices i, j, k.

c) Prove that a tree is specified by its matrix D up to isomorphism.

d) Can you draw the trees with the following matrices D?

$$D_1 = \begin{bmatrix} 0 & 2 & 1 & 2 \\ 2 & 0 & 1 & 2 \\ 1 & 1 & 0 & 1 \\ 2 & 2 & 1 & 0 \end{bmatrix}, \quad D_2 = \begin{bmatrix} 0 & 2 & 3 & 3 \\ 2 & 0 & 3 & 3 \\ 3 & 3 & 0 & 2 \\ 3 & 3 & 2 & 0 \end{bmatrix}.$$

e) Design an algorithm for reconstructing a tree from its matrix D.

2.1.18. Find a formula for the number of pendant vertices for a tree of order n whose non-pendant vertices are of the same degree.

2.1.19. Prove that for any tree T of order $n \geq 3$ the number of pendant vertices is at most

$$\frac{n(\Delta(T) - 2) + 2}{\Delta(T) - 1},$$

where $\Delta(T)$ is the maximal vertex degree in T.

2.1.20. Let a tree of order $n \geq 2$ have k pendant vertices and no vertices of degree 2.

a) Prove that $k \geq (n + 2)/2$.

b) Characterize the trees that attain this bound (i.e., the trees with $n = 2k - 2$.)

2.1.21. Prove the following statements:

a) Every central vertex of a tree belongs to every its diametral path.

b) The center of a tree coincides with the center of any its diametral path.

c) The center of a tree consists of either one or two vertices (*Jordan's theorem*).

d) The center of a tree T may be found by means of the following *Jordan's algorithm.* Delete all pendant vertices of T to obtain a tree T'. Delete all pendant vertices of T' to obtain T''.... Repeat this operation until only one or two vertices remain, which constitute the center of the initial tree T.

e) In a tree of odd diameter any two diametral paths have a common edge.

2.1.22. Using Jordan's algorithm, find the center of the tree shown at Fig. 2.1.2.

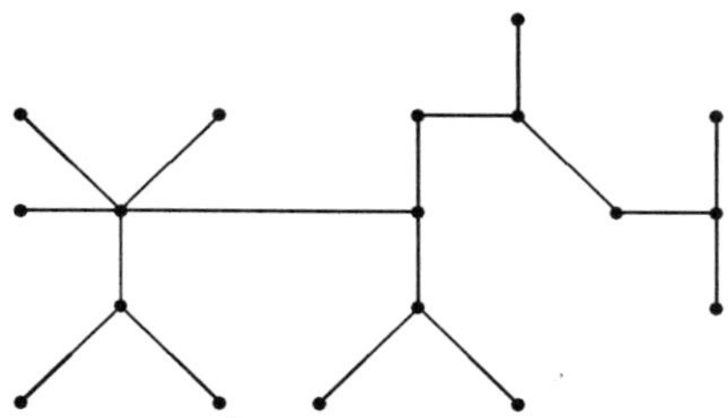

Figure 2.1.2: To Exercise 2.1.22

2.1.23. Prove that a tree is of diameter 2 if and only if it is a star.

2.1.24. Draw all pairwise nonisomorphic trees of order n with one and with two central vertices for $n = 4, 5, 6, 7, 8$ (cf. Exercise 2.1.4).

2.1.25. Characterize all connected graphs in which every maximal under inclusion simple path is diametral.

2.1.26. Prove that the radius $r(T)$ and the diameter $d(T)$ of a tree T are related by

$$r(T) = \lceil d(T)/2 \rceil .$$

2.1.27. Find all trees with the diameter equal to the radius.

2.1.28. What values can the diameter assume for the graph complementary to a nontrivial tree that is not a star?

2.1.29. Prove that a tree of order n with diameter at least $2k-3$ has at least $n-k$ different paths of length k.

2.1.30. Centroid of Tree. A *branch of a tree* T at vertex v is a maximal subtree of T in which v is a pendant vertex. Thus the number of the branches at a vertex v is $\deg v$. Define *weight of a vertex* $v \in VT$ to be the number

$$w(v) = \max\{|EH| : H \in T_v\},$$

where T_v is the number of branches of T at v. Clearly, the weight of a pendant vertex is $|ET|$. A vertex v_0 is called *centroidal* for a tree T if

$$w(v_0) = \min\{w(v) : v \in VT\}.$$

The number $w(v_0)$ is called the *weight* of T and is denoted by $w(T)$. The *centroid of a tree* is the set of its centroidal vertices.

a) Draw a tree with 1) one central and one centroidal vertices; 2) one central and two centroidal vertices; 3) two central and one centroidal vertices; 4) two central and two centroidal vertices;

b) Construct a tree that has disjoint center and centroid, both consisting of two adjacent vertices.

c) Prove that for every tree the centroid consists either of one vertex or two adjacent vertices.

d) Prove that a tree T has a single centroidal vertex if and only if

$$w(T) \leq \frac{1}{2}|ET|,$$

and it has two centroidal vertices if and only if

$$w(T) = \frac{1}{2}(|ET| + 1).$$

e) Prove that the centroid of a tree may consist of two vertices only if the tree is of even order.

2.1.31. A *caterpillar* is a tree that turns into a simple path after the deletion of all pendant vertices. Prove that:

a) a tree is a caterpillar if and only if it contains no subgraph isomorphic to the graph from Fig. 2.1.3.

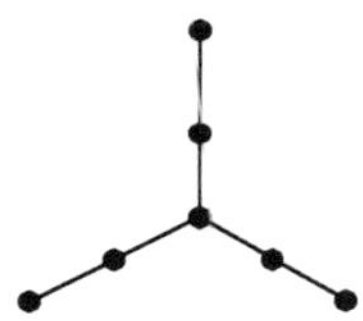

Figure 2.1.3: To Exercise 2.1.31

b) A tree T of order n is a caterpillar if and only if there exists a bijection $\varphi : ET \to \{1, 2, \ldots, n-1\}$ such that the following condition holds: for every vertex $v \in VT$ the set of edges $\{e_i\}$ incident to v may be indexed in such a way that $\varphi(e_{i+1}) = \varphi(e_i) + 1$, $i = 1, 2, \ldots, \deg v - 1$. A tree that satisfies the latter condition is called an *interval tree.*

2.1.32. Find a tree of minimal order $n \geq 2$ with the identity automorphism group.

2.1.33. Find all trees with transitive automorphism group.

2.1.34. Prove that every tree of order n is isomorphic to a subgraph of every graph with minimal vertex degree at least $n - 1$.

2.1.35. Characterize the classes of trees specified by the following *forbidden subgraphs,* i.e., the classes of graphs that do not contain subgraphs isomorphic to: a) $K_{1,3}$; b) P_3; c) P_k, $k \geq 4$; d) the graph from Fig. 2.1.4; e) the graph from Fig. 2.1.5.

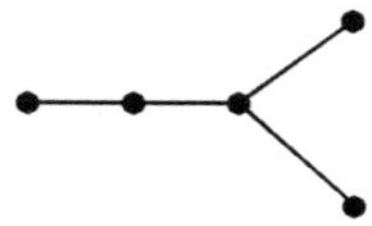

Figure 2.1.4: To Exercise 2.1.35d

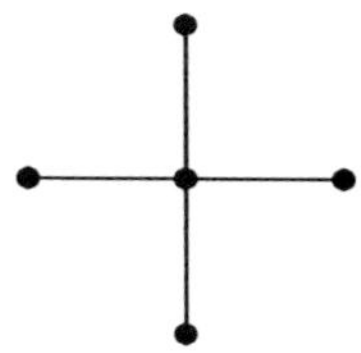

Figure 2.1.5: To Exercise 2.1.35e

2.1.36. Let A be the incidence matrix of a tree of order $n \geq 2$. Prove that any $n-1$ rows of A are linearly independent over the field $\{0,1\}$.

2.1.37. Prüfer Codes. These codes are intended for coding the trees. Consider a tree T of order n with the vertex set $\{1,2,\ldots,n\}$. The Prüfer code of T is constructed as follows. Let b_1 be the smallest number which is a pendant vertex of T. Let $e_1 = b_1a_1$ denote the corresponding pendant edge. Deleting e_1 and b_1 from T we obtain another tree T_1. We repeat this operation for T_1 : we find the smallest number b_2 which is a pendant vertex, and $e_2 = b_2a_2$ denotes the corresponding pendant edge. This operation is repeated until we are left with a single edge $e_{n-1} = b_{n-1}a_{n-1}$. The resulting sequence $(a_1,\ldots,a_{n-2})$ is called the *Prüfer code of the tree T*.

Construct the Prüfer codes for the trees from Fig. 2.1.6.

2.1.38. Let $(a_1,a_2,\ldots,a_{n-2})$ be the Prüfer code of a tree T with the vertex set $\{1,2,\ldots,n\}$. Prove that

$$\{1,2,\ldots,n\}\backslash\{a_1,a_2,\ldots,a_{n-2}\}$$

is the set of the pendant vertices of T.

2.1.39. a) Design an algorithm for reconstructing a tree from its Prüfer code.

b) Reconstruct the trees from their Prüfer codes:
(1) $(3,3,5,5,6,6)$; (2) $(1,5,1,5,9,8,2)$; (3) $(1,5,2,2,1,5,5)$.

2.1.40. Prove the following statements:

a) The Prüfer codes of two different labelled trees are different.

b) Every sequence of length $n-2$ of numbers from the set $\{1,2,\ldots,n\}$ is the Prüfer code of some labelled tree.

2.1.41. What tree has the Prüfer code $(i,i,\ldots,i)$?

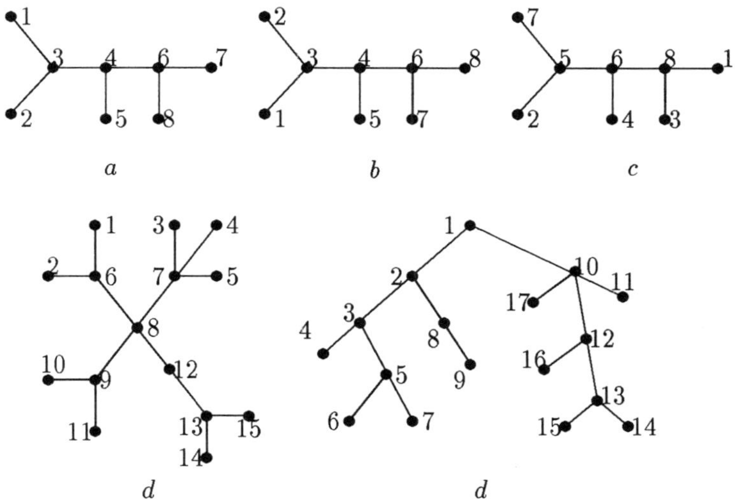

Figure 2.1.6: To Exercise 2.1.37

2.1.42. For what trees of order $n \geq 4$ all components of the Prüfer code are different?

2.1.43. Using Prüfer codes, prove the *Cayley Theorem:* the number of labelled trees of order n is n^{n-2}.

2.1.44. Suppose that the collection of graphs $P(G) = (G_1, G_2, \ldots, G_n)$ is the deck of a graph of order $n \geq 3$ (see Sect. 1.3). The graphs G_i are unlabelled. Prove that the collection $P(G)$ allows to recognize whether G is a tree.

2.1.45. Let $T_1, T_2, \ldots, T_k$, $k > 1$, be subtrees of a tree T such that any two of them intersect each other. Prove that

a) there exists a vertex of T common to all T_i;

b) the subgraph induced by $\bigcap_{i=1}^{n} VT_i$ is a tree.

2.1.46. Consider the following function defined over the vertex set of a tree T :

$$e(x) = \max_{v \in VT} d(x, y), \ x \in VT$$

(the *excentricity of* x). Prove that this function has the *convexity* property:

$$2e(x) \leq e(y) + e(z),$$

where y and z are vertices adjacent to x.

2.1.47. Consider the following function defined over the vertex set of a tree T :

$$\sigma(x) = \sum_{v \in VT} d(x, v), \ x \in VT.$$

Prove that it has the *strict convexity* property:

$$2\sigma(x) < \sigma(y) + \sigma(z),$$

where y and z are vertices adjacent to x.

2.1.48. Barycenter of Tree. The vertex v_0 of a tree T is called *barycentral* if

$$\sigma(v_0) = \min\{\sigma(v) : v \in VT\},$$

where σ function is defined in Exercise 2.1.47.

The *barycenter of a tree* is the set of its barycentral vertices.

a) Prove that each tree has a barycenter consisting of one or two adjacent vertices.

b) Give an example of a tree whose barycentral and central vertices are at distance 100 apart.

2.1.49. Find trees of a given order for which $\sum_{(x,y)} d(x,y)$, where the summation is over all unordered pairs of vertices, is a) minimal; b) maximal.

2.1.50. Find a recurrence relation for the number $t(n,k)$ of labelled trees of order n with k pendant vertices.

2.1.51. Prove that in an odd-order tree every automorphism has a stationary vertex.

2.1.52. Prove that the excentricity of a vertex v of a tree T may be computed by the formula

$$e(v) = d(v, Z(T)) + r(T),$$

where $Z(T)$ is the center of T and $r(T)$ is the radius.

2.1.53. The *subtree intersection graph* $SI(T)$ of a tree T is defined as follows. The vertices of $SI(T)$ bijectively correspond to the subtrees of T. Two vertices of $SI(T)$ are adjacent if and only if the corresponding subtrees of T have a common vertex.

Prove that for any tree T the graph $SI(T)$ has no induced subgraphs isomorphic to C_k, $k \geq 4$.

2.2 Skeletons and Spanning Trees

Let H be a spanning subgraph of a graph G. If H induces a tree on every connected domain of G then H is called a *skeleton of the graph* G. Thus, a skeleton of a graph is a forest (*spanning forest*). For a connected graph, any spanning subgraph which is a tree (*spanning tree*) is a skeleton. Clearly, every graph has a skeleton: one can obtain it successively destructing cycles by deletions of edges.

Theorem 2.2.1 *Let G be an (n, m)-graph with k components. The number of edges that must be deleted from G to obtain a skeleton is independent on the order of their deletion and is equal to $m - n + k$.*

The number $m - n + k$ is called the *cyclomatic number* or *cyclic rank* of the graph and is denoted by $\nu(G)$. The number $n - k$ of edges of its skeleton is called *cocyclic rank* of the graph and is denoted by $\nu^*(G)$.

Theorem 2.2.2 *Every acyclic subgraph of a graph G is contained in some skeleton of G.*

Theorem 2.2.3 (Kirchhoff Theorem) *The number of skeletons in a connected labelled graph G of order $n \geq 2$ is equal to the cofactor of any element of the Kirchhoff matrix $K(G)$.*

Theorem 2.2.4 (Cayley Theorem) *The number of labelled trees of order n is n^{n-2}.*

A graph G is called *weighted* if a *weight function* $w : EG \to \mathbf{R}^+$ is specified that assigns a *weight* $w(e)$ to every edge $e \in EG$. The *weight* $w(H)$ of a subgraph H of G is the sum of the weights of edges of H :

$$w(H) = \sum_{e \in EH} w(e).$$

A weighted graph G with the weight function w is denoted by (G, w).

Minimal Spanning Tree Problem: Find a spanning tree of minimal weight in a weighted connected graph (G, w).

We describe the algorithms due to Kruskal and Prim for solving this problem.

Kruskal's Algorithm

1. Construct a graph $T_1 = O_n + e_1$ by adding an edge $e_1 \in EG$ of minimal weight to the empty graph with the vertex set VG.

2. Having T_i, $i < n - 1$, constructed, we construct $T_{i+1} = T_i + e_{i+1}$, where e_{i+1} is an edge of G with minimal weight among the edges which are not in T_i and do not produce a cycle when added to T_i.

T_{n-1} is the minimum weight spanning tree for (G, w).

Prim's Algorithm

1. Pick an edge $e_1 = ab$ of minimal weight and construct the tree T_1 by setting

$$VT_1 = \{a, b\},\ ET_1 = \{e_1\}.$$

2. Having T_i of order $i + 1$, $i < n - 1$, constructed, among the edges connecting it with the vertices of $G \backslash T_i$ we pick an edge e_{i+1} of minimal weight. T_{i+1} is obtained from T_i by adding e_{i+1} together with the endpoint which is not in T_i.

T_{n-1} is the minimum weight spanning tree for (G, w).

The problem of finding a spanning tree of maximal weight may also be solved by the algorithms of Kruskal and Prim; one must only pick edges of maximal weight rather than of minimal.

* * *

2.2.1. Prove that the cyclic rank of a graph cannot be negative.

2.2.2. Consider the following operations with a graph:

1. if a vertex v is incident only to two edges $e_1 = vu_1$ and $e_2 = vu_2$, then v is deleted (together with e_1 and e_2) and the edge u_1u_2 is added;
2. an edge u_1u_2 is replaced by two edges vu_1, vu_2, v being a new vertex.

Prove that the cyclic rank of a graph is invariant with respect to these operations.

2.2.3. Prove that a graph is a forest if and only if its cyclic rank is zero.

2.2.4. Find all skeletons of the following labelled graphs: a) K_4, b) P_6, c) C_7, d) $K_{1,3}$, e) $K_{2,3}$.

2.2.5. Find the numbers of nonisomorphic skeletons of the following graphs: a) K_4, b) K_5, c) K_6, d) C_7, e) a 6-vertex wheel graph W_5, f) $K_{2,3}$, g) $K_{3,3}$.

2.2.6. Using the Kirchhoff theorem, find the number of skeletons in graphs from Fig. 2.2.1 (the vertices of graphs must be indexed first).

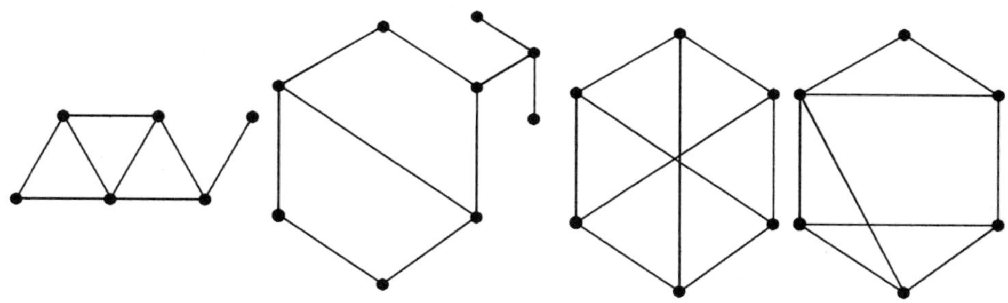

a *b* *c* *d*

Figure 2.2.1: To Exercise 2.2.6

2.2.7. In what cases a graph has a single skeleton?

2.2.8. Prove that any nontrivial connected graph has a skeleton such that if we delete all its edges, the graph becomes disconnected.

2.2.9. Demonstrate how one can find a skeleton of a graph using the breadth-first search.

2.2.10. Let H be a subgraph of a connected graph G of order n. Consider the following conditions:

a) H has n vertices;

b) H has $n-1$ edges;

c) H is connected;

d) H is a forest.

Prove that:

1. Any three of these conditions taken together imply that H is a spanning tree of G.
2. The conditions b), d) together imply this;
3. No other pair of conditions a)-d) imply this.

2.2.11. Prove that a subgraph H is a skeleton of a graph G if and only if the following equalities hold:

$$|H| = |G|; \; m(H) = \nu^*(G); \; k(H) = k(G).$$

2.2.12. Find spanning trees of maximal (minimal) weights for graphs from Fig. 2.2.2 using algorithms of Kruskal and Prim.

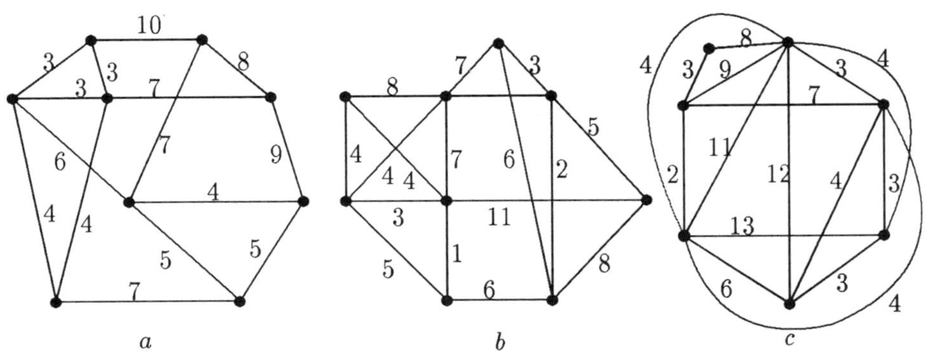

Figure 2.2.2: To Exercise 2.2.12

2.2.13. Prove that every pendant vertex of a graph is contained in every skeleton of the graph.

2.2.14. Prove that every edge of a graph belongs to some skeleton of the graph.

2.2.15. Prove that a subgraph of a graph G is its skeleton if and only if it is a maximal subgraph of G without cycles.

2.2.16. Let G be obtained by "gluing together" two connected graphs G', G'' at a single vertex. Express the number of spanning trees of G in terms of the numbers of spanning trees of G' and G''.

2.2.17. Prove that if a graph has a simple cycle of length k then the graph has at least k skeletons.

2.2.18. Does there exist a graph with exactly two skeletons?

2.2.19. Which graphs are such that every skeleton has exactly two pendant vertices?

2.2.20. Draw a graph that has two skeletons with disjoint edge sets.

2.2.21. Prove the following two statements for connected graphs.

a) For any vertex v of a graph G there exists a spanning tree T of G such that $d_G(v, w) = d_T(v, w)$ for any vertex $w \in VG$. Here $d_H(v, w)$ denotes the distance between v and w in the graph H.

b) The center of a graph is contained in the union of the centers of its skeletons.

2.2.22. Let T be a spanning tree of a connected graph G. Prove that $d(G) \leq d(T)$, where $d(X)$ denotes the diameter of the graph X.

2.2.23. Is it true that in a connected graph of diameter $d \geq 2$

a) the diameter of any spanning tree is at least d?

b) a spanning tree of the same diameter exists?

2.2.24. Prove or disprove the statements:

a) If the diameter of a graph is 2 then the graph has a skeleton which is a star.

b) If a graph has a skeleton which is a star then the diameter of G is 2.

2.2.25. How many skeletons are there in the labelled complete graph of order n.

2.2.26. Using the Kirchhoff theorem, find the number of skeletons in the labelled complete bipartite graph $K_{m,n}$.

2.2.27. Prove that the number of labelled trees which are labelled bipartite graphs with sizes of parts m and n, is $m^{n-1}n^{m-1}$.

2.2.28. In a random labelled tree of order n a random vertex is picked. Assuming that all outcomes are equiprobable, prove that

$$\lim_{n \to \infty} P_n = \frac{1}{e},$$

where P_n denotes the probability of the event that the picked vertex is pendant.

2.2.29. Let T_1, T_2 be skeletons of a graph G.

a) Prove that for any edge $e_1 \in ET_1$ there exists an edge $e_2 \in ET_2$ such that $(T_1 - e_1) \cup e_2$ is also a skeleton of G. This replacement of e_1 by e_2 will be called *switch*.

b) Prove that any skeleton may be obtained from any other one by a sequence of switches (producing a skeleton at each switch).

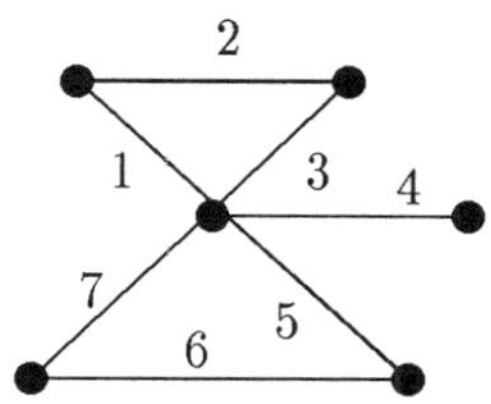

Figure 2.2.3: To Exercise 2.2.30a

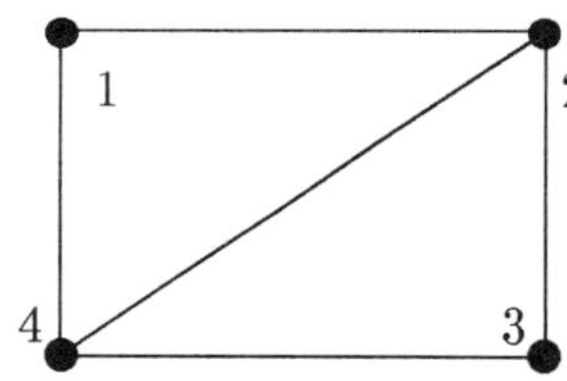

Figure 2.2.4: To Exercise 2.2.31

2.2.30. A *graph of skeletons* $S(G)$ of a graph G is the graph whose vertex set is the set $T_1, T_2, \dots, T_r$ of skeletons of G and two its vertices T_i and T_j are adjacent if and only if the union of skeletons $T_i \cup T_j$ in G is a graph with a single cycle.

a) Construct a graph of skeletons for the graph from Fig. 2.2.3.

b) Prove that the skeletons T_i and T_j of G are adjacent in $S(G)$ if and only if there exist edges $e_1 \in ET_i$ and $e_2 \in ET_j$ such that $(T_i - e_1) \cup e_2 = T_j$.

c) Prove that $S(G)$ is connected.

d) Prove that for skeletons T_i and T_j of a connected graph G the distance between the corresponding vertices in $S(G)$ is equal to $|ET_i \backslash ET_j|$.

e) Prove that the diameter of $S(G)$ is at most $|G| - 1$.

f) Characterize nontrivial graphs for which the diameter $d(S(G))$ of its graph of skeletons is $|G| - 1$.

g) Prove that $|G| - d(S(G))$ may be arbitrarily large.

h) What is the maximal possible diameter of $S(G)$ if $k(G) = k$?

i) Characterize graphs G with the mentioned above diameter of $S(G)$.

2.2.31. Construct all skeletons of the labelled graph G from Fig. 2.2.4 and draw the $S(G)$.

2.2.32. Prove that every minimal spanning tree of a connected graph may be constructed by Kruskal's algorithm.

2.2.33. Let T be a minimal spanning tree of a connected weighted graph (G, w) with weights $c_1 \leq c_2 \leq \ldots \leq c_m$ of the edges of T. Let further T' be a spanning tree of the same graph with weights $d_1 \leq d_2 \leq \ldots \leq d_m$ of the edges of T'. Prove that $c_i \leq d_i$, $i = 1, 2, \ldots, .$

Chapter 3

Independence and Coverings

3.1 Independent Vertex Sets and Cliques

A set of vertices of a graph is called *independent* if no two vertices in it are adjacent. In other words, a set $S \subseteq VG$ is independent in G if the induced subgraph $S(G)$ is empty. Clearly, in this case any subset $S' \subseteq S$ is also independent.

An independent vertex set is called *maximal* if it is not a proper subset of any other independent set. An independent set with maximal number of elements is called a *maximum cardinality independent set.* The number of vertices in a maximum cardinality independent set of a graph G is called the *independence number* of G and is denoted by $\alpha_0(G)$.

Below are some estimates for the independence number.

$$\alpha_0(G) \geq \sum_{v \in VG} (1 + \deg v)^{-1}. \tag{3.1.1}$$

$$\alpha_0(G) \geq \frac{|G|}{1+d}, \tag{3.1.2}$$

$$\alpha_0(G) \geq \frac{|G|}{1+\Delta(G)}, \tag{3.1.3}$$

where $d = (\sum_{v \in VG} \deg v)/|G|$.

The notion of the clique is antipodal to the notion of the independent set. A nonempty subset of vertices of a graph is called a *clique* if it induces a complete subgraph. A clique is called *maximal* if it is not a proper subset of any other clique. A clique with maximal number of elements is called a *maximum cardinality clique.* The number of vertices in the maximum cardinality clique of a graph G is called the *toughness* of the graph G and is denoted by $\varphi(G)$.

Clearly, a subset of vertices of a graph G is a clique if and only if it is an independent set in the complementary graph $\overline{G}$. Therefore $\varphi(G) = \alpha_0(\overline{G})$, and the inequalities (3.1.1)–(3.1.3) provide bounds for $\varphi(G)$ as well.

All cliques, as well as all maximal cliques, constitute a covering of the vertex set of a graph. The minimal number of cliques of G covering VG is denoted by $c(G)$. Clearly, $c(G) \geq \alpha_0(G)$ for any G.

The minimal number of complete graphs whose union is G is denoted by $cc(G)$. This number is also related to cliques, since cliques induce complete subgraphs.

The intersection graph for the set of maximal cliques of G is called the *clique graph* and is denoted by $Q(G)$. Thus the vertices of $Q(G)$ bijectively correspond to the maximal cliques of G and two vertices of $Q(G)$ are adjacent if and only if the corresponding cliques intersect.

* * *

3.1.1. Define the independence numbers for the graphs from Fig. 1.2.4.

3.1.2. Define the independence numbers for the following graphs: a) K_n; b) $K_{1,n}$; c) $K_{m,n}$; d) P_n; e) C_n; f) Petersen graph; g) n-cube Q_n.

3.1.3. Define the independence numbers for the graphs of Platonic solids (Fig. 1.1.2).

3.1.4. Characterize the graphs for which

a) every vertex subset is independent;

b) no independent set has more than one vertex.

3.1.5. Is it true that every maximal independent set of vertices of a graph is the maximum cardinality independent set?

3.1.6. How may the independence number of a graph change

a) after the deletion of one vertex;

b) after the deletion of one edge;

c) after the addition of one edge.

3.1.7. Let $G(A, B; E)$ be a bipartite graph. Is it true that $\alpha_0(G) = \max\{|A|, |B|\}$?

3.1.8. Prove that G is bipartite if and only if $\alpha_0(G) \geq |H|/2$ for every subgraph H of G.

3.1.9. Prove that if G is a tree then $\alpha_0(G) \geq |G|/2$.

3.1.10. Prove that every connected graph other than K_2 has a maximal independent set of vertices which contains all pendant vertices of the graph.

3.1.11. Propose an algorithm for the construction of a maximal independent set of vertices of a tree.

3.1.12. Prove that if the distance between any two pendant vertices of a tree is even then the tree has a single maximum cardinality independent vertex set. Is the converse true?

3.1.13. Let $\alpha_0'(G)$ denote the minimal cardinality of maximal independent sets of vertices of a graph G. Find $\alpha_0'(G)$ for a) K_n; b) $K_{1,n}$; c) $K_{m,n}$; d) P_n; e) C_n; f) Petersen graph; g) n-cube Q_n.

3.1.14. For any positive integer a, construct a graph G such that

$$\alpha_0(G) - \alpha_0'(G) = a.$$

3.1.15. Prove that for graphs G, H

a) $\alpha_0(G \cup H) = \alpha_0(G) + \alpha_0(H)$, if $VG \cap VH = \emptyset$.

b) $\alpha_0(G + H) = \max(\alpha_0(G), \alpha_0(H))$;

c) $\alpha_0(G \times H) \geq \alpha_0(G)\alpha_0(H)$.

See Section 1.4 for the operation with graphs.

3.1.16. a) Prove the formula (3.1.1).

b) Prove that the bound (3.1.1) is attainable for a nontrivial graph.

c) Design an algorithm for finding an independent set of cardinality at least $\sum_{v \in VG}(1 + \deg V)^{-1}$ in a graph G.

3.1.17. Prove the formula (3.1.2).

3.1.18. Let I be a maximal independent vertex set of a graph G and let $\rho(I)$ denote the number of edges connecting the vertices of I with the vertices of $VG \backslash I$. Prove that $\rho(I) + |I| \geq |G|$.

3.1.19. Prove the formula (3.1.3) without making use of (3.1.2).

3.1.20. Let $l(G)$ be the minimal number of paths of a graph whose union contains all vertices of G. Prove that $l(G) \leq \alpha_0(G)$.

3.1.21. Prove that $\alpha_0(G) \leq |EG|/\delta(G)$ for all graphs G without isolated vertices.

3.1.22. Prove that $\alpha_0(G) \leq p^0 + (p^+ + p^-)/2$ for all graphs G, where p^0, p^+, p^- are the numbers of zero, positive and negative eigenvalues of the adjacency matrix $A(G)$ respectively.

3.1.23. Let $e_i = uv_i$, $i = 1, 2, \ldots, \deg u$, be the edges of a graph G incident to a vertex u. Consider the graph G' obtained from G by the addition of a new vertex u' and new edges uu' and $u'v_i$, see Fig. 3.1.1. (This operation producing G' from G is called *vertex splitting*.) Prove that $\alpha_0(G') = \alpha_0(G)$.

3.1.24. Turan graph: Let G be a graph satisfying the following conditions:

(1) $|G| = n$;

(2) $\alpha_0(G) \leq \alpha$;

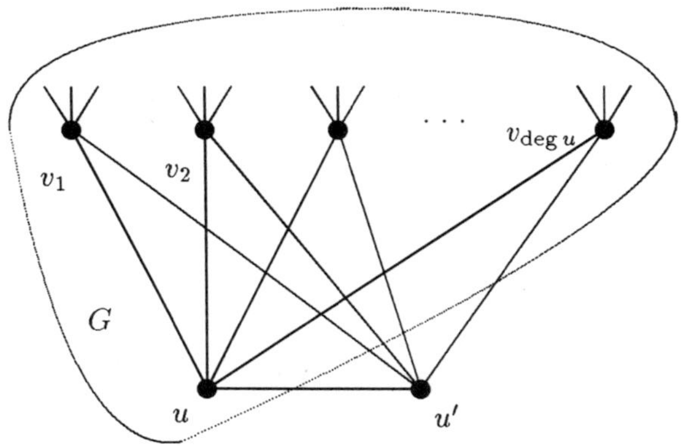

Figure 3.1.1: To Exercise 3.1.23

(3) among all graphs satisfying the conditions (1), (2), G has the minimal number of edges.

Such graph is called *the Turan graph* and is denoted by $G^m_{n,\alpha}$, where m is the number of its edges.

Prove the following properties of the Turan graph:

a) any two adjacent vertices are of the same degree;

b) the degrees of any two nonadjacent vertices differ by at most 1;

c) every component of $G^m_{n,\alpha}$ is a complete graph;

d) the components of $G^m_{n,\alpha}$ are q graphs K_{p+1} and $\alpha - q$ graphs K_p, where p is the quotient and q is the remainder under the division of n by α ($n = \alpha p + q$, $0 \leq q \leq n-1$.)

3.1.25. Express the number m of edges of Turan's graph $G^m_{n,\alpha}$ in terms of p, α, q (see the previous exercise).

3.1.26. Prove that if $\alpha_0(G) = \alpha$ for an (n,m)-graph G then

$$m \geq p(n - \frac{\alpha}{2} - \frac{\alpha p}{2}),$$

where $p = \lfloor n/\alpha \rfloor$.

3.1.27. Prove that

$$\alpha_0(G) > \frac{n^2}{2(m+n)}$$

for any (n,m)-graph G.

3.1.28. a) Prove that $\alpha_0(G) = \alpha_0(G^p)$, $p \geq 3$, (see Sect. 1.4) if and only if every component of G is a complete graph.

b) Construct a graph for which the above statement fails for $p = 2$.

3.1.29. A graph is called *edge α_0-critical* if after the deletion of any edge the independence number increases. Give examples of edge α_0-critical graphs and of graphs not of this class.

3.1.30. Prove that

a) every nonempty graph has a subgraph which is an edge α_0-critical graph;

b) every graph is an induced subgraph of an edge α_0-critical graph.

3.1.31. Which Platonic solids (see Fig. 1.1.2) are edge α_0-critical graphs?

3.1.32. Construct the clique graphs $Q(G)$ for a) O_n; b) K_n; c) $K_{1,n}$; d) P_n; e) C_n; f) $K_{m,n}$; g) Petersen graph.

3.1.33. Prove that if $\varphi(G) \geq p$ then G has at least p vertices with degrees at least $p-1$.

3.1.34. Find a graph with maximal number of edges for a given number of vertices n and toughness φ. How many edges does it have?

3.1.35. Prove that a graph G is isomorphically embeddable into G' if and only if $\varphi(G \diamond G') \geq |G|$ (see Sect. 1.4).

3.1.36. Prove that a graph G has a subgraph isomorphic to G' if and only if $\varphi(G \dot{\diamond} G') \geq |G|$ (see Exercise 1.4.21).

3.1.37. Threshold Graph: Let G be a graph with vertex set $VG = \{v_1, v_2, \ldots, v_n\}$. For a subset $U \subseteq VG$ the characteristic vector $x_U = (x_1, x_2, \ldots, x_n)$ is defined as follows:

$$x_i = \begin{cases} 1, & \text{if } v_i \in U; \\ 0, & \text{otherwise.} \end{cases}$$

A graph G is called a *threshold graph* if there exist nonnegative reals

$$\alpha_1, \alpha_2, \ldots, \alpha_n, \beta$$

such that a subset $U \subseteq VG$ is independent if and only if the vector x_U satisfies the inequality

$$\alpha_1 x_1 + \alpha_2 x_2 + \ldots + \alpha_n x_n \leq \beta,$$

which is called a *separating inequality* for G. (The empty set is defined to be independent.)

Prove the following statements:

a) The mapping $U \mapsto x_U$ is a bijection of the set of subsets of VG onto the set of strings of length n consisting of zeroes and ones.

b) O_n is a threshold graph.

c) K_n is a threshold graph.

d) A graph obtained from a threshold graph by adding an isolated vertex is a threshold one.

e) Every threshold graph has a separating inequality with positive coefficients.

f) A graph obtained from a threshold graph by adding a dominating vertex is a threshold one.

g) the graph at Fig. 3.1.2 is a threshold graph. (Find a separating inequality for it.)

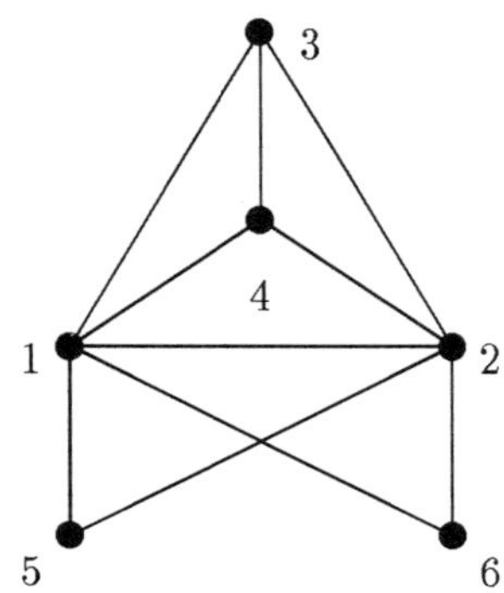

Figure 3.1.2: To Exercise 3.1.37g

h) Every induced subgraph of a threshold graph is a threshold one.

i) Graphs $2K_2$, P_4, C_4 are not threshold.

j) Every threshold graph has either an isolated vertex or a dominating vertex.

k) A graph is a threshold graph if and only if it may be obtained from K_1 by a sequence of additions of isolated and dominating vertices.

3.1.38. Draw all pairwise nonisomorphic four-vertex threshold graphs.

3.1.39. a) Find $cc(G)$ for the graphs from Fig. 1.2.4.

b) Prove than $c(G) \leq cc(G)$ for all graphs G.

c) Find a graph with $c(G) = cc(G)$; find a graph with $c(G) < cc(G)$.

3.1.40. Column Intersection Graph: Let A be a binary $(m \times n)$-matrix without zero columns. Let $\mathbf{1}$ denote the column of ones of height m and let $\mathbf{0}$ denote the row of zeroes of height n. The set

$$P(A) = \{x : Ax \leq \mathbf{1},\ x \geq \mathbf{0}\},$$

i.e., the set of all real-valued solutions $x = (x_1, x_2, \ldots, x_n)$ with nonnegative components x_i of the system of inequalities $Ax \leq \mathbf{1}$ will be called a *polytope*.

We define the *column intersection graph* G_A for the matrix A to be the graph with $VG_A = \{1, 2, \ldots, n\}$ in which two vertices i, j are adjacent if and only if at least one row of A contains ones at positions i and j.

Prove the following statements:

a) A subset U of vertices of a graph G_A is independent if and only if the characteristic vector x_U (see Exercise 3.1.37) belongs to the polytope $P(A)$.

b) For every graph G there exists a matrix A such that $G \cong G_A$.

3.1.41. Let $G_1, G_2, \ldots, G_p$ be a covering of an n-vertex graph G by complete subgraphs. A binary matrix A is called the *clique covering matrix of the graph* G if $A_{ij} = 1$ if and only if the vertex $v_j \in VG$ belongs to the clique G_i.

Reconstruct graphs from the following clique covering matrices:

$$\text{a) } A = \begin{pmatrix} 1 & 1 & 1 & 0 & 0 \\ 0 & 1 & 1 & 1 & 0 \\ 0 & 0 & 0 & 1 & 1 \end{pmatrix}, \quad \text{b) } A = \begin{pmatrix} 1 & 1 & 0 & 1 & 1 & 0 & 0 & 0 \\ 0 & 1 & 1 & 0 & 1 & 1 & 0 & 0 \\ 0 & 1 & 1 & 0 & 0 & 0 & 1 & 0 \\ 0 & 0 & 0 & 1 & 1 & 0 & 0 & 1 \end{pmatrix}.$$

3.1.42. Prove that a graph G is isomorphic to a column intersection graph G_A if and only if the matrix A is the clique covering matrix for G.

3.1.43. An integer point of the polytope $P(A)$ (see Exercise 3.1.40) is a vector $x \in P(A)$ all whose components are integer. Let A, B be binary matrices without zero columns.

a) Find conditions necessary and sufficient for the coincidence of the sets of integer points of polytopes $P(A), P(B)$.

b) Does the coincidence of the polytopes $P(A), P(B)$ follow from the coincidence of their sets of integer points?

3.1.44. Let G be a graph. Consider all binary matrices A such that $G = G_A$ (see Exercises 3.1.40 and 3.1.42). Prove that the minimal number of rows for such matrices is $cc(G)$.

3.1.45. Prove that a graph G is isomorphic to the line graph $L(H)$ of a graph H if and only if there exists a covering $\{G_1, G_2, \ldots, G_p\}$ of G by complete graphs satisfying the conditions:

a) every vertex of G is in at most two G_i;

b) every edge of G is in exactly one G_i.

3.2 Coverings

We shall say that a vertex and an edge *cover each other* if they are incident to each other. Thus, an edge $e = ab$ covers a and b and each of these vertices covers e. A set $V' \subseteq VG$ is called a *covering* (*vertex covering*) of a graph G if each edge from EG is incident to at least one vertex from V'. A covering of a graph is called *minimal* if it does not contain a covering with less vertices and it is called *minimum cardinality vertex covering* if it has minimal number of elements among the coverings of G. This minimal number is called the *covering number*, or the *vertex covering number* of the graph and is denoted by $\beta_0(G)$. The independence number and the covering number are related with each other as follows:

$$\alpha_0(G) + \beta_0(G) = |G|. \tag{3.2.1}$$

An *edge covering of a graph* G is a subset of edges $E' \subseteq EG$ that covers all vertices of a graph G, i.e., such that each vertex from VG is incident to at least one edge from E'. An edge covering of a graph is called *minimal* if it does not contain a covering with less edges, and it is called a *minimum cardinality edge covering* if it has minimal number of elements among the edge coverings of G. This minimal number is called the *edge covering number* of the graph and is denoted by $\beta_1(G)$.

* * *

3.2.1. Define the values of $\beta_0(G)$ and $\beta_1(G)$ for graphs from Fig. 1.2.4.

3.2.2. Define the values of $\beta_0(G)$ and $\beta_1(G)$ for the following graphs: a) P_n; b) C_n; c) K_n; d) $K_{1,n}$; e) $K_{m,n}$; f) the Petersen graph.

3.2.3. Give an example of a bipartite graph with $\beta_0 < \min(m, n)$, where m and n are the cardinalities of the parts.

3.2.4. Does there exist a connected graph whose minimal covering contains all vertices?

3.2.5. Characterize connected graphs for which a minimal edge covering contains all edges.

3.2.6. Prove that there exists a minimum cardinality vertex covering of a tree other than K_2 that contains all vertices adjacent to pendant vertices.

3.2.7. Give an algorithm for the construction of a minimum cardinality vertex covering of a tree.

3.2.8. Find a graph whose minimal vertex covering is not a minimum cardinality one.

3.2.9. Find a graph whose minimal edge covering is not a minimum cardinality one.

3.2.10. Prove that for a graph G without isolated vertices the following inequalities hold:

$$|VG|/2 \le \beta_1(G) \le |VG| - 1.$$

For any positive integer n, find a graph of order n for which one of the inequalities turns into the equality.

3.2.11. a) Prove that every edge covering of a graph contains all pendant edges.

b) Design an algorithm for the construction of a minimum cardinality edge covering of a tree.

3.2.12. Prove that a set U of vertices of a graph G is a (minimal, minimum cardinality) covering if and only if $VG \backslash U$ is a (maximal, maximum cardinality) independent set.

3.2.13. Prove the equality (3.2.1).

3.2.14. Prove that for a bipartite graph G with q edges the relation $\alpha_0(G)\beta_0(G) \ge q$ holds, and the equality takes place only for complete bipartite graphs.

3.2.15. Prove that $\beta_0(G) \ge \delta(G)$ for any graph G. Find a graph for which the equality holds.

3.2.16. For the complete n-partite graph G prove

a) $\beta_0(G) = |VG| - N$, where N is the maximum of the cardinalities of the parts.

b) $\beta_0(G) = \delta(G)$.

3.2.17. Prove that an edge covering of a graph is minimal if and only if every component of a graph induced by the edges of the covering is a star.

3.2.18. A vertex v of a graph G is called *critical* if $\beta_0(G - v) < \beta_0(G)$.

Prove that a vertex v is critical if and only if it belongs to some minimum cardinality vertex covering.

3.2.19. An edge e of a graph G is called *critical* if $\beta_0(G - e) < \beta_0(G)$.

Prove that if an edge is critical then its ends are critical vertices.

3.2.20. Find a graph with critical vertices and without critical edges.

3.2.21. A graph is called *vertex-critical* if every its vertex is critical. A graph is called *edge-critical* if every its edge is critical.

a) Give examples of vertex-critical and edge-critical graphs.

b) Prove that every edge-critical graph without isolated vertices is vertex-critical.

c) Give an example of a vertex-critical graph which is not edge-critical.

3.2.22. Prove or disprove the statement: an edge of a graph is critical if and only if it belongs to a minimum cardinality edge covering of the graph.

3.3 Dominating Sets

A set D of vertices of a graph G is called a *dominating set* if every vertex from $VG\backslash D$ is adjacent to a vertex from D, i.e., every vertex of G is at distance at most 1 from some vertex of a dominating set. A dominating set is called *minimal* if it does not contain a dominating set with less vertices; and it is called a *minimum cardinality dominating set* if it has minimal number of elements among the dominating sets of G. This minimal number is called the *dominating number* and is denoted by $\gamma(G)$.

A subset of vertices of a graph that is both independent and dominating is called a *kernel* of the graph.

* * *

3.3.1. Evaluate $\gamma(G)$ for the graphs shown at Fig. 1.2.4.

3.3.2. Evaluate $\gamma(G)$ for the graphs a) P_n; b) C_n; c) K_n; d) $K_{1,n}$; e) $K_{m,n}$; f) the Petersen graph.

3.3.3. Find a graph for which a minimal dominating set is not a minimum cardinality dominating set.

3.3.4. Find a graph for which a minimal dominating set is not independent.

3.3.5. Find a graph for which $\alpha_0(G) = \gamma(G) = 3$ and every maximum cardinality independent set is a minimum cardinality dominating set.

3.3.6. Prove that

a) an independent set of vertices of a graph is maximal if and only if it is dominating.

b) every maximal independent vertex set is a kernel.

3.3.7. Prove that $\alpha_0(G) \geq \gamma(G)$ for every graph G.

3.3.8. Prove the inequalities $|VG|/(\Delta(G)+1) \leq \gamma(G) \leq |VG| - \Delta(G)$ for all graphs G.

Give examples of graphs G_1, G_2 such that

$$\gamma(G_1) = |VG_1|/(\Delta(G_1)+1),\ \gamma(G_2) = |VG_2| - \Delta(G_2).$$

3.3.9. Prove that a dominating set D is minimal if and only if for every $v \in D$ one of the following conditions holds:

(1) the graph has no edges vu with $u \in D$;

(2) if the graph has an edge vu with $u \in D$ then it has an edge uw with $w \notin D$, where u is the unique vertex from D adjacent to w.

3.3.10. Prove that a graph G without isolated vertices has a dominating set D such that $VG\backslash D$ is also dominating.

3.3.11. Prove that if a graph G has no isolated vertices then $\gamma(G) \leq |G|/2$. For every even n, find a graph of order n with $\gamma(G) = |G|/2$.

3.3.12. Prove that if D is a minimal dominating set of a graph G without isolated vertices then $VG \backslash D$ is also a dominating set.

3.3.13. Prove that for every minimal dominating set of a graph without isolated vertices there exists another minimal dominating set disjoint from the former one.

3.3.14. Prove that every graph has a minimal dominating set that contains all vertices adjacent to pendant vertices.

3.3.15. Give an algorithm for the construction of a minimum cardinality dominating set of a tree.

3.3.16. Prove that for every graph G there exists a graph H containing a minimum cardinality dominating set D such that $H(D) \cong G$.

3.3.17. A kernel D of a graph G is called a *minmax kernel* if $|D| = \alpha_0(G) = \gamma(G)$. Find a graph with a minmax kernel.

3.3.18. The *domatic number* $d(G)$ is the maximal number of dominating sets the vertex set of G may be partitioned into.

Prove the inequality $d(G) \leq \Delta(G) + 1$.

3.3.19. Let $\gamma'(G)$ denote the minimal number of vertices in a dominating set of vertices of a graph G, which is independent.

a) Prove that $\gamma(G) \leq \gamma'(G) \leq \alpha_0(G)$ for any graph G.

b) Give examples of graphs for which the last relation holds as a strict inequality and as an equality.

c) Find a graph G of the minimal order such that $\gamma(G) < \gamma'(G)$.

d) Prove that $\gamma'(G) - \gamma(G)$ may be arbitrarily large.

3.3.20. Compute the numbers $\gamma'(G)$ for graphs shown at Fig. 1.2.4.

3.3.21. Compute the numbers $\gamma'(G)$ for a) P_n; b) C_n; c) K_n; d) $K_{1,n}$; e) $K_{m,n}$; f) Petersen graph.

3.3.22. Prove that $\gamma(G) \leq \beta_0(G)$. Find a graph with $\gamma(G) < \beta_0(G)$.

3.3.23. Find a graph with (a) $\gamma'(G) > \beta_0(G)$. (b) $\gamma'(G) < \beta_0(G)$.

3.4 Matchings

A set of pairwise nonadjacent edges of a graph is called a *matching,* or an *independent set of edges.* A matching is called a *perfect matching,* or a *1-factor* if every vertex of a graph is incident to an edge of the matching. Clearly, a perfect matching is a minimum cardinality edge covering of a graph.

A graph is called *1-factorizable* if it may be represented as a union of edge-disjoint 1-factors.

Sometimes, when the set of edges of a graph is the (only) matching, the graph itself is called a matching. We have already met such kind of graph in the exercises, namely, $2K_2$.

A matching in a graph G is called *maximal* if it is not contained in a larger matching; it is called a *maximum cardinality matching* if it contains the maximal number of edges among the matchings of the graph. This maximal number is called the *matching number* of the graph and is denoted by $\alpha_1(G)$.

The independent edge sets of a graph G bijectively correspond to the independent vertex sets of the line graph $L(G)$, hence, $\alpha_1(G) = \alpha_0(L(G))$.

For a graph without isolated vertices, the matching number is related with the edge covering number as follows:

$$\alpha_1(G) + \beta_1(G) = |G|. \tag{3.4.1}$$

The number of edges in any matching of a graph does not exceed the number of vertices in any vertex cover; in particular,

$$\beta_0(G) \geq \alpha_1(G). \tag{3.4.2}$$

Let M be a matching in a graph G. A path of G whose edges are alternately in M and in $EG \backslash M$ is called an *alternating path with respect to M.* By definition, a path of length 1 is alternating. The vertices of G incident to the edges of M are called *saturated (by M),* the remaining ones are called *unsaturated (by M).* A path alternating with respect to M that connects two unsaturated vertices is called an *augmenting path with respect to M.*

Theorem 3.4.1 (Berge Theorem) *A matching M in a graph is of maximal cardinality if and only if the graph has no augmenting paths with respect to M.*

* * *

3.4.1. Evaluate $\alpha_1(G)$ for the graphs shown at Fig. 1.2.4. Which of these graphs have a perfect matching?

3.4.2. Evaluate $\alpha_1(G)$ for the graphs a) P_n; b) C_n; c) K_n; d) $K_{1,n}$; e) $K_{m,n}$; f) Petersen graph.

3.4.3. Give an example when a maximal matching is not of maximal cardinality.

3.4.4. Find a graph such that

a) $\beta_0(G) = \alpha_1(G)$;

b) $\beta_0(G) > \alpha_1(G)$.

3.4.5. Prove that the graphs Q_n and K_{2n} are 1-factorizable and the Petersen graph is not.

3.4.6. Prove the relation (3.4.1).

3.4.7. Prove that $\beta_1(G) = \alpha_1(G)$ if and only if the graph G has a perfect matching.

3.4.8. Prove that if a connected graph of even order has no induced subgraphs isomorphic to $K_{1,3}$ then it has a perfect matching.

3.4.9. Let M_1, M_2 be matchings in a graph G and let G' be a graph induced by the set of edges

$$M_1 \oplus M_2 = (M_1 \backslash M_2) \cup (M_2 \backslash M_1).$$

Prove that the components of G' may be only of two types:

(1) cycles of even length with edges alternately in M_1 and M_2;

(2) paths whose edges are alternately in M_1 and M_2 and whose terminal vertices are unsaturated in one of the matchings.

3.4.10. Prove the Berge Theorem.

3.4.11. Let M, N be disjoint matchings in a graph G, $|M| > |N|$. Prove that there exist disjoint matchings M', N' such that $|M'| = |M| - 1$, $|N'| = |N| + 1$, $M' \cup N' = M \cup N$.

3.4.12. Prove that a tree T has a perfect matching if and only if $T - v$ has a single component of odd order for any $v \in VT$.

3.4.13. Prove that every tree has at most one perfect matching.

3.4.14. Prove that $\alpha_1(G) = \min\{\delta(G), \lfloor |G|/2 \rfloor\}$ if G is a complete multipartite graph.

3.4.15. Prove that the following inequalities hold for a maximal matching M of a graph G :

a) $\lfloor \delta(G)/2 \rfloor \leq |M|$;

b) $\alpha_1(G) \leq 2|M|$.

3.4.16. We shall say that an edge *weakly covers* itself and the adjacent edges. The minimal numbers of edges and independent edges necessary to weakly cover EG are denoted by $\beta_{11}(G)$ and $\beta'_{11}(G)$ respectively.

Evaluate $\beta_{11}(G)$ and $\beta'_{11}(G)$ for the graphs shown at Fig. 1.2.4.

3.4.17. Evaluate $\beta_{11}(G)$ and $\beta'_{11}(G)$ for the graphs a) P_n; b) C_n; c) K_n; d) $K_{1,n}$; e) $K_{m,n}$; f) Petersen graph.

3.4.18. Prove that every component of a graph induced by a minimal number of edges weakly covering the edges of a graph G is a star $K_{1,p}$, $p \geq 1$.

3.4.19. Prove that $\beta_{11}(G) = \beta'_{11}(G)$ for every graph G.

3.5 Matchings in Bipartite Graphs

We shall consider a bipartite graph $G = (X, Y; E)$ with fixed parts X and Y. Let M be a matching in this graph. Clearly, M may be interpreted as an injective mapping from some subset $A \subseteq X$ into Y : if an edge xy belongs to M then the vertex y is the image of x. In such situation M will be called a *matching from A into Y* or simply a *matching into Y*. Similarly, we may speak about a *matching from A into B, $B \subseteq Y$*.

For $A \subseteq VG$ we denote

$$N_G(A) = N(A) = \bigcup_{v \in A} N(v) \backslash A.$$

Theorem 3.5.1 *A bipartite graph $G(X, Y; E)$ has a matching from X into Y if and only if*

$$|A| \leq |N(A)| \tag{3.5.1}$$

for every $A \subseteq X$.

Corollary 3.5.2 *A regular bipartite graph of nonzero degree is 1-factorizable.*

Define

$$\delta(A) = \begin{cases} |A| - |N(A)|, & A \neq \emptyset, \\ 0, & \text{otherwise}. \end{cases}$$

The value

$$\delta_0(X, Y; E) = \max_{A \subseteq X} \delta(A)$$

is called the *deficiency* of the bipartite graph $G = (X, Y; E)$.

Theorem 3.5.3 *For a bipartite graph G with nonempty part X, Y,*

$$\alpha_1(G) = |X| - \delta_0(X, Y; E).$$

Clearly, for a graph G the number of vertices in a covering is not less than the number of edges in any matching; in particular, $\beta_0(G) \geq \alpha_1(G)$.

Theorem 3.5.4 *For a bipartite graph G, $\beta_0(G) = \alpha_1(G)$.*

Corollary 3.5.5 *For a bipartite graph G,*

$$\alpha_0(G) + \alpha_1(G) = |G|, \quad \beta_1(G) = \alpha_0(G).$$

* * *

3.5.1. Let $G = (X, Y; E)$ be a bipartite graph and let M_i, $i = 1, 2$, be perfect matchings from X_i into Y_i, $X_i \subseteq X$, $Y_i \subseteq Y$, respectively. Prove that there exists a matching $M' \subseteq M_1 \cup M_2$ such that the vertices of X_1 and Y_2 are saturated.

3.5.2. Prove that a nonempty bipartite graph has a matching such that all vertices of maximal degree are saturated.

3.5.3. Prove that the edge set of a bipartite graph G may be partitioned into $\Delta(G)$ matchings.

3.5.4. Let $G = (X, Y; E)$ be a bipartite graph and let t be a positive integer, $t \leq |X|$. Prove that G has a matching of cardinality t if and only if

$$|N_G(A)| \geq |A| + t - |X|$$

for every $A \subseteq X$.

3.5.5. Prove that

$$|X| - \delta_0(X, Y; E) = |Y| - \delta_0(Y, X; E)$$

for every bipartite graph $G = (X, Y; E)$.

3.5.6. Prove that G is bipartite if and only if $\alpha_0(H) = \beta_1(H)$ for every subgraph H without isolated vertices.

3.5.7. A *line of a matrix* is either its row or its column. Two elements of a matrix are called *independent* if they do not share a line.

Prove the König theorem: the maximal number of pairwise independent ones of a binary matrix is equal to the minimal number of its lines containing all ones of this matrix.

3.5.8. Prove that a binary matrix in which every line contains exactly k '1's may be represented as a sum of k binary matrices in which every line contains exactly one '1'.

3.5.9. Let $P = \{p_1, \ldots, p_n\}$ be a finite nonempty set and let $S = (S_1, \ldots, S_m)$ be an m-element collection of its subsets. A subset $T \subseteq P$ is called a *transversal* (or a *system of different representatives*) of S if there exists a bijective mapping $\varphi : T \to \{1, 2, \ldots, m\}$ such that $t \in S_{\varphi(t)}$ for every $t \in T$. This means that we may index the elements of T in such a way that $t_i \in S_i$, $i = 1, 2, \ldots, m$.

Prove the following Hall theorem: S has a transversal if and only if for any $k \in \{1, \ldots, m\}$ the union

$$S_{i_1} \cup S_{i_2} \cup \ldots \cup S_{i_k}$$

of any k subsets S_i contains at least k different elements.

3.5.10. A subset $T \subseteq P$ (see Exercise 3.5.9) is called a *partial transversal* of a collection S if there exists an injective mapping $\varphi : T \to \{1, 2, \ldots, m\}$ such that $t \in S_{\varphi(t)}$ for every $t \in T$.

Prove that a collection S has a partial transversal of cardinality $p \le m$ if and only if for every $1 \le k \le m$ the union of any k subsets S_i contains at least $k + p - m$ different elements.

3.5.11. A collection $(P_1, P_2, \ldots, P_m)$ of subsets of P (see Exercise 3.5.9) is called a
$(k_1, k_2, \ldots, k_m)$-*transversal* of a collection S, where $(k_1, k_2, \ldots, k_m)$ is a sequence of positive integers, if

(1) $P_i \subseteq S_i$, $i = 1, 2, \ldots, m$;

(2) $|P_i| = k_i$, $i = 1, 2, \ldots, m$;

(3) $P_i \cap P_j = \emptyset$, $i \ne j$.

Prove that a collection $S = (S_1, \ldots, S_m)$ has a $(k_1, k_2, \ldots, k_m)$-transversal if and only if

$$|\bigcup_{i \in I} S_i| \ge \sum_{i \in I} k_i$$

for every $I \subseteq \{1, 2, \ldots, m\}$.

Chapter 4

Connectivity

4.1 Biconnected Graphs and Biconnected Components

A vertex v of a graph G is called a *cutpoint* if $G - v$ has more components than G. In particular, if G is connected and v is a cutpoint then $G - v$ is disconnected. Similarly, an edge is called a *bridge* if its deletion increases the number of components in the graph.

A connected graph is called *biconnected* (*2-connected*) if it is not K_2 and has no cutpoints.

Theorem 4.1.1 *For a connected graph G, $|G| > 2$, the following statements are equivalent:*

1) G is biconnected;

2) any two vertices of G are on a common simple cycle;

3) any vertex and any edge of G are on a common simple cycle;

4) any two edges of G are on a common simple cycle;

5) for any two vertices a, b and any edge e, the graph G has a simple (a, b)-path containing e;

6) for any three vertices a, b, c, the graph G has a simple (a, b)-path containing c.

A maximal under inclusion connected subgraph without cutpoints is called *block*. A biconnected graph G is called *critically biconnected* if $G - e$ is not biconnected for every edge e.

* * *

4.1.1. Let G be a biconnected graph. Prove the following statements without any usage of the Theorem 4.1.1:

a) $\delta(G) \geq 2$;

b) if P is a simple path with both ends belonging to G then $G \cup P$ is also biconnected;

c) for any three different vertices a, b, c, the graph G has an (a, b)-path not containing c.

4.1.2. Prove that an edge of a graph is not a bridge if and only if it belongs to a cycle.

4.1.3. Let $cp(G)$ denote the number of cutpoints of a graph G. Prove the following statements:

a) if e is an edge that belongs to a cycle of G then $cp(G) \leq cp(G-e)$;

b) $cp(G) \leq cp(T)$ for any skeleton T of G;

c) $cp(G) \leq |G| - 2$;

d) if $cp(G) = |G| - 2$ then G is a path graph.

4.1.4. Prove that if $\delta(G) \leq 3$ and $G \neq K_2$ then G has a cutpoint if and only if G has a bridge. Show that this statement fails for $\delta(G) \geq 4$.

4.1.5. Prove that every biconnected graph G has at least $n = |G|$ edges; moreover, the unique biconnected graph with n edges is C_n.

4.1.6. State necessary and sufficient conditions for a line graph $L(G)$ to have a cutpoint.

4.1.7. Let G be a connected graph with $\delta(G) \geq 3$. Prove that $L(G)$ has no bridges.

4.1.8. Prove that $d(G) \leq \lfloor |G|/2 \rfloor$ for a biconnected graph G, where $d(G)$ is the diameter of G.

4.1.9. Construct biconnected n-vertex graphs G with $d(G) = 2$, $|EG| = 2n - 5$ for all $n \geq 5$.

4.1.10. Let G be a biconnected graph. For $V_1 \subseteq VG$ and $v \in VG \backslash V_1$ prove that G has a simple (a, b)-path H such that $VH \cap V_1 = \{a, b\}$ and H contains v.

4.1.11. Let $B = \{B_i\}$ and $C = \{c_j\}$ be the block set and the cutpoint set of a graph G respectively. Let $bc(G)$ denote the graph with the vertex set $B \cup C$ and the edge set $\{B_i c_j : B_i \in B,\ c_j \in C,\ c_j \in VB_i\}$.

a) For a graph G shown at Fig. 4.1.1 construct the graph $bc(G)$.

b) Prove that if G is connected then $bc(G)$ is a tree.

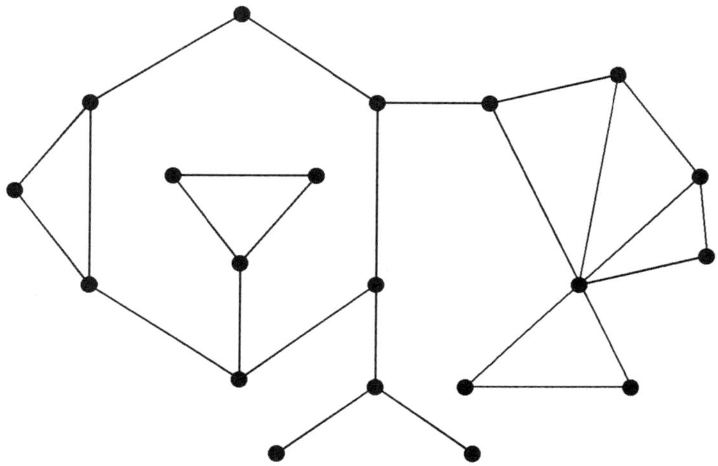

Figure 4.1.1: To Exercise 4.1.11

4.1.12. Let G be a connected graph and let $b(v)$ denote the number of blocks of G containing a vertex v. Prove that the number of blocks of G is

$$1 + \sum_{v \in VG} (b(v) - 1).$$

4.1.13. Let G be a biconnected graph with $\delta(G) \geq 3$. Prove that there exists a vertex $v \in VG$ such that $G - v$ is also biconnected.

4.1.14. Prove that if every two blocks of a graph have a common vertex then all blocks have a common vertex.

4.1.15. Let x be vertex of graph G and let an edge e be incident to x. Prove that if x is not a cutpoint of G then x is not a cutpoint of $G - e$.

4.1.16. Let G be a biconnected graph. Prove that if $G - e$ has a cutpoint, then the ends of e are in different blocks of $G - e$.

4.1.17. Prove that for every simple cycle of a graph there exists a block of the graph that contains this cycle.

4.1.18. Prove that any two blocks of a graph have at most one common vertex.

4.1.19. Prove that every edge of a graph belongs to a single block.

4.1.20. Prove that if a block B and a simple cycle C of a graph are such that $|VB \cap VC| \geq 2$ then $VC \subseteq VB$.

4.1.21. An edge ab is called a *diagonal* of a cycle C if $a, b \in VC$ and $ab \notin EC$. Prove that an edge e of a biconnected graph G is a diagonal of a simple cycle in G if and only if $G - e$ is biconnected.

4.1.22. Prove that a critically biconnected graph with more than 3 vertices has no triangles.

4.1.23. Let G be a critically biconnected graph. Prove that every block of $G - e$, $e \in EG$, not isomorphic to K_2 is also a critically biconnected graph.

4.1.24. Prove that a critically biconnected graph has a vertex of degree 2.

4.1.25. Prove that $|EG| \leq 2|VG| - 4$ for a critically biconnected graph G, $|G| \geq 4$.

4.1.26. Is it true that for a critically biconnected graph every vertex of degree greater than 2 is adjacent to a vertex of degree 2?

4.1.27. Let v be a vertex of a connected graph G such that $G(N(v))$ is connected. Prove that $G - v$ is connected.

4.1.28. Let v be a cutpoint of G. Prove that v is not a cutpoint of $\overline{G}$.

4.1.29. A block is called *pendant* if it contains only one cutpoint of the graph. Prove that every graph with cutpoints has at least two pendant blocks.

4.1.30. Let G be a biconnected graph such that $G - a - b$ is disconnected for every two its nonadjacent vertices a, b. Prove that G is a simple cycle.

4.1.31. Let $\{B_1, B_2, \ldots, B_t\}$ be the block set of a graph G. The *block graph* $B(G)$ for the graph G is the graph with the vertex set $\{v_1, v_2, \ldots, v_t\}$ such that $v_i v_j \in EB(G)$ if and only if $B_i \cap B_j \neq \emptyset$. Prove the following statements:

a) If G is connected and a block B_k is not pendant then the graph $\bigcup_{i=1, i\neq k}^{t} B_i$ is disconnected.

b) If $B(G)$ is biconnected then it is a complete graph.

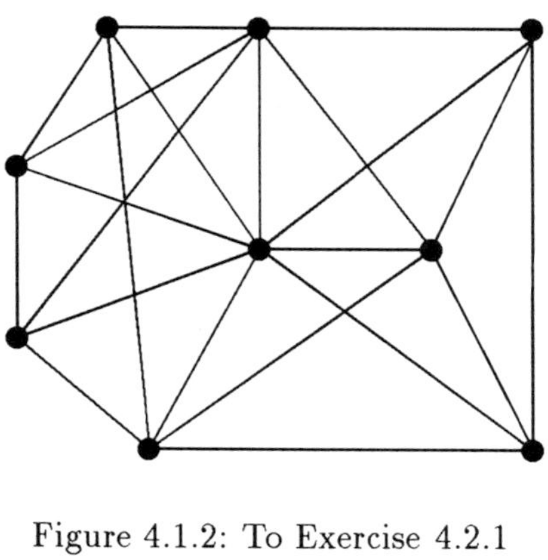

Figure 4.1.2: To Exercise 4.2.1

4.2 k-connectivity

The *vertex connectivity number* $\kappa(G)$ of a graph G is the minimal number of vertices whose removal makes G disconnected or single-vertex.

The *edge connectivity number* $\lambda(G)$ of a nontrivial graph G is the minimal number of edges whose removal makes G disconnected. The edge connectivity of a single-vertex graph is defined to be zero.

A graph is called *k-connected* if $\kappa(G) \geq k$; it is called *k-edge connected* if $\lambda(G) \geq k$.

A maximal k-connected subgraph of G is called its *k-component.*

Theorem 4.2.1 *Any two different k-components of a graph have at most $k-1$ common vertices.*

Theorem 4.2.2 *For any graph G,*

$$\kappa(G) \leq \lambda(G) \leq \delta(G).$$

* * *

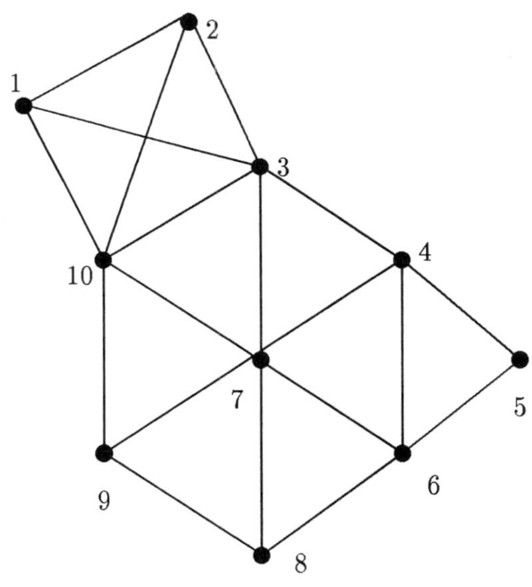

Figure 4.2.1: To Exercise 4.2.2

4.2.1. Find $\kappa(G)$, $\lambda(G)$ for the graph G shown at Fig. 4.1.2.

4.2.2. Find a 3-component in the graph G shown at Fig. 4.2.1.

4.2.3. How many edges must be added to the graph shown at Fig. 4.2.2 to make it triconnected?

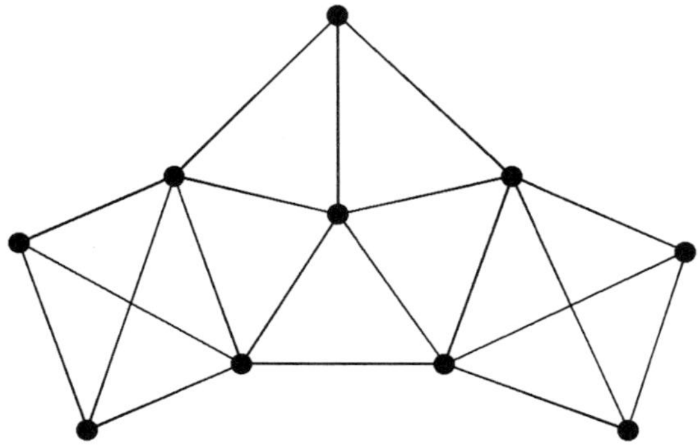

Figure 4.2.2: To Exercise 4.2.3

4.2.4. Prove that $|EG| \geq k|G|/2$ for a k-edge connected graph.

4.2.5. Prove that if $\delta(G) > (|G| + k - 2)/2$ then G is k-connected.

4.2.6. Find a graph G with $\kappa(G) < k$ and $\delta(G) = \lfloor(|G| + k - 3)/2\rfloor$.

4.2.7. Prove that if $\delta(G) \geq |G| - 2$ then $\kappa(G) = \delta(G)$.

4.2.8. For every $n = 4, 5, \ldots,$ construct an n-vertex graph with $\delta(G) = |G| - 3$ and $\kappa(G) < \delta(G)$.

4.2.9. Prove that if an edge e is a bridge of a connected graph G then $G - e$ has exactly two components.

4.2.10. Let S be a subset of edges of a connected graph G of size $|S| = \lambda(G)$. Prove that if $G - S$ is disconnected then it has exactly two components.

4.2.11. Prove that if $\delta(G) \geq |G|/2$ then $\lambda(G) = \delta(G)$.

4.2.12. Find an n-vertex connected graph G with $\lambda(G) < \delta(G) = \lfloor|G|/2\rfloor - 1$.

4.2.13. Prove that if G is not a complete graph then $\kappa(G) \geq 2\delta(G) - |G| + 2$.

4.2.14. Prove that $\lambda(G) = \kappa(G)$ for a cubic graph G.

4.2.15. Prove that $|EG| \geq \kappa(G)|G|/2$.

4.2.16. Prove that if G is an r-regular graph with a cutpoint then $\lambda(G) \leq \lfloor r/2\rfloor$.

4.2.17. Prove that if G is of diameter less than 3 then $\lambda(G) = \delta(G)$.

4.2.18. Prove that $\kappa(G) = 1 + \min_{v \in VG} \kappa(G - v)$ for a connected graph G.

4.2.19. Prove that there are no triconnected graphs with 7 edges.

4.2.20. How many are there triconnected graphs with 8 edges?

4.2.21. For which k there exists a k-connected graph with $(k^2 + k + 2)/2$ edges?

4.2.22. Is it true that in a triconnected graph any three edges not incident to the same vertex belong to a common simple cycle?

4.2.23. Prove that in a triconnected graph any three vertices belong to a common simple cycle.

4.2.24. Is it true that in a triconnected graph any four vertices belong to a common simple cycle?

4.2.25. Let G be a bipartite graph. Prove that if $\delta(G) \geq |G|/4 + 1$ then $\lambda(G) = \delta(G)$.

4.2.26. Prove that if G is an r-regular graph with $|G| = 2r + 1$ then $\lambda(G) = r$.

4.2.27. A connected graph is called *rigid* if $|X| \geq k(G - X)$ for any $X \subseteq VG$, $X \neq \emptyset$,where $k(H)$ is the number of components of the graph H. Is it true that in a rigid graph any three vertices belong to a common simple cycle?

4.2.28. For each $n > 2$ find a rigid graph (see Exercise 4.2.27) which is not triconnected.

4.2.29. For each $n > 1$ find an n-connected graph which is not rigid (see Exercise 4.2.27).

4.2.30. A graph is called *cyclically k-connected* if there is no $S \subseteq EG$, $|S| < k$, such that $G - S$ is disconnected and every component of $G - S$ contains a cycle. Find a cyclically 4-connected graph which is not 4-connected.

4.2.31. Find a graph in which every 3 vertices are on a common simple cycle but there are 4 vertices that are not on a common simple cycle.

4.3 Cycles and Cuts

The length of a shortest cycle of a graph G is called the *girth* of G and is denoted by $g(G)$.

A subset $U \subseteq EG$ is called a *separator set* of G if the number of components of $G - U$ is greater than that of G. A minimal under inclusion separator set is called a *cut*. The notion of cut is also applicable to pseudographs.

* * *

4.3.1. Find a graph G with $|G| = 7$, $\delta(G) \geq 3$, $g(G) = 4$.

4.3.2. Prove that if $\delta(G) \geq 3$ and $|G| \geq 8$ then $g(G) \leq |G|/2$.

4.3.3. Prove that if G is biconnected and has no cycles of even length then G is a cycle of odd length.

4.3.4. Let $h(G)$ denote the length of the longest simple cycle in G. Prove that if $\delta(G) \geq 2$ then

a) $h(G) \geq \delta(G) + 1$;

b) $h(G) \geq 2|EG|/(|G| - 1)$.

4.3.5. Prove that if X is a cut of a connected graph G then $G - X$ has exactly two components.

4.3.6. Suppose that every vertex of a simple cycle C is colored in one of two colors. Prove that the number of edges of the cycle with ends colored differently is even.

4.3.7. Prove that any simple cycle and any cut in a graph have even number of common edges.

4.3.8. Prove that for any cut X of a graph G there exists a partition of the vertex set $VG = V_1 \cup V_2$ such that $X = \{ab : a \in V_1, b \in V_2, ab \in EG\}$.

4.3.9. How many cuts has a graph shown at Fig. 4.3.1? Find them.

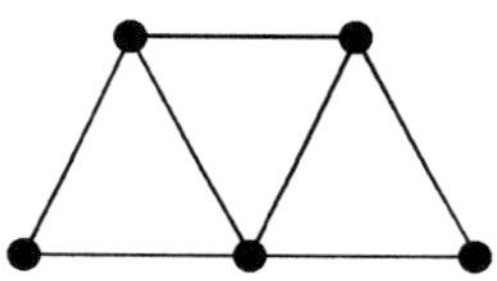

Figure 4.3.1: To Exercise 4.3.9

4.3.10. Prove that a cut of a graph induces a bipartite subgraph.

4.3.11. Prove that every cut of a graph contains at least one edge of every skeleton of the graph.

4.3.12. For a vertex v of a graph G, let X be a set of edges incident to v belonging to the same block of G. Prove that X is a cut of G.

4.3.13. Prove that every edge of a graph is in some cut of the graph.

4.3.14. Is it true that every pair of edges of a graph is in some cut of the graph?

4.3.15. Prove that every separator set of a graph has a subset which is a separator set for some block of the graph.

4.3.16. Prove that every cut of a graph belongs to some block of the graph.

4.3.17. Let $VG = V_1 \cup V_2$ be a partition of the vertex set of a connected graph G. Prove that the set of edges $(V_1, V_2) = \{ab : a \in V_1, b \in V_2, ab \in EG\}$ is a cut of G if and only if $G - (V_1, V_2)$ has exactly two components.

4.3.18. Let C be a simple cycle of a graph G and let e_1, e_2 be any edges of the cycle. Prove that G has a cut S such that $S \cap EC = \{e_1, e_2\}$.

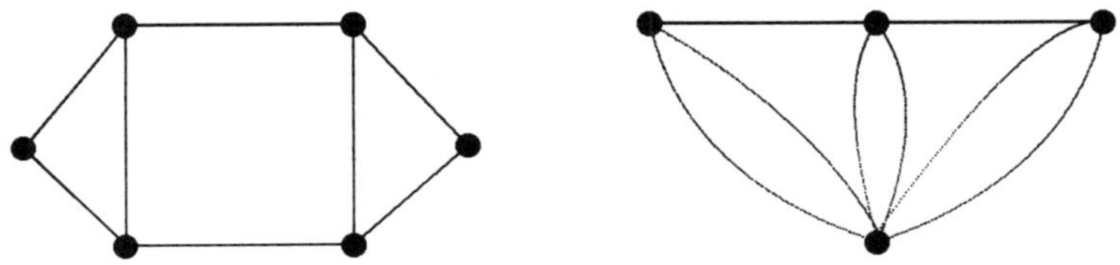

G *H*

Figure 4.3.2: To Exercise 4.3.19

4.3.19. A pseudograph G^* is called *abstract dual* for a pseudograph G if there exists a bijection $EG \to EG^*$ such that the subset of edges of EG constitutes a simple cycle if and only if the corresponding subset of edges of EG^* constitutes a cut. Prove the following statements:

a) For graphs G and H shown at Fig. 4.3.2, $H \cong G^*$.

b) Graphs K_5 and $K_{3,3}$ have no abstract duals.

Chapter 5

Matroids

5.1 Independence Systems

Let E be a finite nonempty set and let I be a set of its subsets satisfying the following two conditions, called *independence axioms:*

I.1. $\emptyset \in I$.

I.2. If $X \in I$ and $Y \subseteq X$ then $Y \in I$.

The pair $S = (E, I)$ is called the *independence system* with *support* E and *set of independent sets* I. The number $|E|$ is called the *order* of the system.

Maximal under inclusion elements of I are called *bases* of the independence system S. A subset $X \subseteq E$ is called *dependent* if it is not independent, i.e., if $X \notin I$. Minimal under inclusion dependent sets are called *cycles* of S.

The set of independent sets of an independence system S, the set of its bases and the set of its cycles will be denoted $\mathcal{I}(S)$, $\mathcal{B}(S)$, and $\mathcal{C}(S)$, respectively. Each of these sets together with the support E uniquely specifies the remaining two ones and hence the system S.

It is possible that $\mathcal{C}(S) = \emptyset$. In this case the system S is called *free.* Each subset of a free independence system is independent. At the other extremity, when $\mathcal{B}(S) = \{\emptyset\}$, the independence system is called *trivial.*

For any independence system S the set of bases $\mathcal{B}(S)$ is an *antichain* with respect to set inclusion, i.e., if $B_1, B_2 \in \mathcal{B}(S)$ and $B_1 \subseteq B_2$ then $B_1 = B_2$. The set $\mathcal{C}(S)$ is also an antichain.

Let $S_k = (E_k, I_k)$, $k = 1, 2$, be two independence systems and let $\varphi : E_1 \to E_2$ be a bijection. Then if $\varphi(X) \in I_2$ if and only if $X \in I_1$ for every $X \subseteq E_1$, then the mapping φ is called *isomorphism* of the independence systems S_1, S_2, and these systems are called *isomorphic* ($S_1 \cong S_2$).

If $S_k = (E, I_k)$, $k = 1, 2$, are two independence systems with the same support and $I = I_1 \cap I_2$, then the pair (E, I) is also an independence system; it is called the *intersection* of systems S_1, S_2 and is denoted by $S_1 \cap S_2$.

* * *

5.1.1. Let E be $\{1,2,3,4\}$. Check whether the pair $S=(E,I)$ is an independence system for

a) $I=\{\{1\},\{2\},\{3\},\{1,2\},\{1,2,3,4\}\}$;

b) $I=\{\{1\},\{2\},\{3\},\{1,2\},\{1,3\}\}$;

c) $I=\{\{1\},\{2\},\{3\},\{1,2\},\{1,3\},\emptyset\}$.

If yes then find $\mathcal{B}(S)$ and $\mathcal{C}(S)$.

5.1.2. Let E be a nonempty finite set and let $\mathcal{B}$ be a nonempty set of its subsets that is an antichain. Prove that there exists an independence system S with support E such that

a) $\mathcal{B}(S)=\mathcal{B}$;

b) $\mathcal{C}(S)=\mathcal{B}$;

5.1.3. Let $S_k=(E_k,I_k)$, $k=1,2$, be two independence systems and let $\varphi: E_1 \to E_2$ be their isomorphism. Prove that for any $X \subseteq E_1$,

a) $\varphi(X) \in \mathcal{B}(S_2) \iff X \in \mathcal{B}(S_1)$;

b) $\varphi(X) \in \mathcal{C}(S_2) \iff X \in \mathcal{C}(S_1)$.

5.1.4. Let $S_k=(E_k,I_k)$, $k=1,2$, be two independence systems and let $\varphi: E_1 \to E_2$ be a bijection. Prove that if any of the claims of the previous exercise holds then φ is an isomorphism of the independence systems.

5.1.5. Prove that the intersection of independence systems is also an independence system. How does the set of cycles of the intersection look like?

5.1.6. Let G be a graph. Let IG denote the set of independent subsets of vertices of G plus the empty set.

a) Prove that the pair (VG, IG) is an independence system. It is called *graphical independence system.*

b) Which subsets of vertices of G are bases of the graphical independence system (VG, IG)?

c) Describe the cycles of the graphical independence system.

5.1.7. Prove that an independence system S is isomorphic to a graphical independence system if and only if one of the following conditions holds:

a) S is a free independence system;

b) the number of elements in every cycle of S is 2.

5.1.8. Find the number of pairwise nonisomorphic independence systems of order 3. How many of them are graphical independence systems? Draw the corresponding graphs.

5.1.9. Let f be a monotonic Boolean function in n variables, $f \not\equiv 1$. For $E = \{1, 2, \ldots, n\}$, let I denote the set of subsets $U \subseteq E$ whose characteristic vectors (see Exercise 3.1.37) are zeroes of f.

a) Prove that $S = (E, I)$ is an independence system. We shall say that it is *specified by the function f*.

b) Characterize the cycles of S.

c) Prove that every independence system is isomorphic to the independence system specified by some monotonic Boolean function.

d) Which monotonic Boolean functions specify graphical independence systems?

5.1.10. Find the cycles and the bases of the independence system specified by the Boolean function:

a) $f(x_1, x_2, \ldots, x_n) = x_1 \vee x_2 \vee \ldots \vee x_n$;

b) $f(x_1, x_2, x_3, x_4) = x_1 \vee x_2x_3 \vee x_1x_2 \vee x_4$.

5.1.11. For $E = \{1, 2, \ldots, n\}$, let I denote the set of subsets $U \subseteq E$ whose characteristic vectors (see Exercise 3.1.37) satisfy the system of inequalities

$$\begin{aligned} a_{11}x_1 + a_{12}x_2 + \ldots + a_{1n}x_n &\leq b_1, \\ a_{21}x_1 + a_{22}x_2 + \ldots + a_{2n}x_n &\leq b_2, \\ &\ldots \\ a_{m1}x_1 + a_{m2}x_2 + \ldots + a_{mn}x_n &\leq b_m, \end{aligned} \tag{5.1.1}$$

where all a_{ij}, b_i are nonnegative reals. Prove the following statements:

a) The pair (E, I) is an independence system. It is said that it is *specified by the system of inequalities* (5.1.1).

b) Prove that every independence system is isomorphic to the independence system specified by some system of inequalities.

c) Which system of inequalities specify graphical independence systems?

5.1.12. For a nonempty graph G let IG denote the set of matchings of G plus the empty set.

a) Prove that the pair (EG, IG) is an independence system.

b) Find the set of cycles of this system.

c) Characterize the bases of the system.

5.2 Matroids

A *matroid with support* E is a pair (E, I), where E is a nonempty finite set and I is a set of subsets of E satisfying the following three conditions (*independence axioms of matroid*):

I.1. $\emptyset \in I$.

I.2. If $X \in I$ and $Y \subseteq X$ then $Y \in I$.

I.3. If $X, Y \in I$ and $|X| < |Y|$ then there exists $y \in Y \backslash X$ such that $X \cup \{y\} \in I$.

Thus, a matroid is an independence system satisfying the additional axiom **I.3**. Therefore bases and cycles of such system (matroid) have additional properties.

Theorem 5.2.1 *The set $\mathcal{B}(M)$ of bases of a matroid M satisfies the following two conditions:*

B.1. *No base is contained in another base.*

B.2. *If B_1, B_2 are bases then for any element $b \in B_1 \backslash B_2$ there exists an element $c \in B_2 \backslash B_1$ such that $(B_1 \backslash \{b\}) \cup \{c\} \in \mathcal{B}(M)$.*

Theorem 5.2.2 *Let E be a nonempty finite set and let $\mathcal{B}$ be a nonempty set of its subsets satisfying both axioms of bases. Then if I denotes the set of all subsets of elements of $\mathcal{B}$ ($I = \{X \subseteq B : B \in \mathcal{B}\}$), then the pair (E, I) is the matroid with the support E, the set of independent sets I and the set of bases $\mathcal{B}$.*

All bases of a matroid are of the same cardinality; this number is called the *rank* of the matroid and is denoted by $\rho(M)$.

Theorem 5.2.3 *The set $\mathcal{C}(M)$ of cycles of a matroid satisfies the following two conditions:*

C.1. *No cycle is contained in another cycle.*

C.2. *If C_1, C_2 are different cycles then for any element $c \in C_1 \cap C_2$ the set $(C_1 \cup C_2) \backslash \{c\}$ contains a cycle.*

The conditions of the above theorem are called the *axioms of cycles*.
A single-element cycle is called a *loop*.

Theorem 5.2.4 *Let E be a nonempty finite set and let C be a set of its subsets satisfying both axioms of cycles. Let I be the set of subsets $X \subseteq E$ such that no subset of X is in C. Then the pair (E, I) is the matroid with support E, set of independent sets I, and set of cycles C.*

Proposition 5.2.5 *For any base B of a matroid and for any element $e \notin B$ of the matroid the set $B \cup \{e\}$ contains a unique cycle of the matroid.*

The *rank function* ρ of a matroid M

$$\rho(A) = \max\{|X| : X \subseteq A, \ X \in \mathcal{I}(M)\}$$

assigns a number $\rho(A)$ to each subset $A \subseteq E$, which is called the *rank* of A.

In what follows, $(E, \mathcal{B})$, (E, C), and (E, ρ) will denote, respectively, the matroid with the set of bases $\mathcal{B}$, with the set of cycles C and with the rank function ρ.

Proposition 5.2.6 *A matroid (E, ρ) has k pairwise nonintersecting bases if and only if $k\rho(A) + |\overline{A}| \geq k\rho(E)$ for any $A \subseteq E$.*

Let $M = (E, \mathcal{B})$ be a matroid. For $B \in \mathcal{B}$, the set $\overline{B} = E \backslash B$ is called a *cobase* of the matroid M. The set of all cobases of M will be denoted by $\mathcal{B}^*(M)$.

Proposition 5.2.7 *The pair $(E, \mathcal{B}^*(M))$ is a matroid with the set of bases $\mathcal{B}^*(M)$.*

The matroid $\mathcal{B}^*(M)$ is called *dual* to M and is denoted by M^*. An independent (dependent) set of elements of M^* is called *coindependent* (*codependent*) set of M. A cycle of M^* is called a *cocycle* of M. The rank function of M^* is called the *corank function* of M and is denoted by $\rho^*(M)$. For every statement about matroid in terms of cycles, bases, etc., there exists the dual one about the dual matroid in terms of cocycles, cobases, etc. If any of the two mutually dual statements is valid for all matroids then the second one is also valid.

Proposition 5.2.8 *For every nonempty independent set X of a matroid there exists a cocycle Y such that $|X \cap Y| = 1$.*

Proposition 5.2.9 *If matroids M_1, M_2 are isomorphic then M_1^*, M_2^* are isomorphic too.*

Consider some examples of matroids. Let L be a linear space over a field F. If $E \subseteq L$ is a finite nonempty set and I is the set of all linearly independent over F sets of vectors in L plus the empty set, then (E, I) is a matroid. It is called the *vector matroid induced by the vector set E*. A matroid isomorphic to the vector one is also called vector matroid.

Let G be a nonempty graph. We may define the matroid $M(G)$ with the support $E = EG$ whose bases are the sets of edges of the skeletons of G. This matroid is called the *cyclic matroid*, or the *matroid of cycles of the graph G*. A matroid isomorphic to cyclic matroid is called *graphic matroid.*

Notice a terminological whim: a graphic matroid is not necessarily a graphical independence system.

The dual matroid $(M(G))^* = M^*(G)$ is called the *matroid of cuts* of the graph G. A matroid isomorphic to the matroid of cuts of a graph is called *cographic.*

Let E be a finite nonempty set and let $S = \{E_i : i = 1, 2, \ldots, k, E_i \neq \emptyset\}$ be a covering of E, i.e., $E = E_1 \cup E_2 \cup \ldots \cup E_k$. Let I denote the set of the partial transversals of the collection S (see Exercise 3.5.9) plus the empty set. Then the pair (E, I) is a matroid. A matroid isomorphic to the one described above is called a *transversal matroid.*

Let $M_k = (E_k, I_k)$, $k = 1, 2$, be two matroids specified by the sets of independent sets I_1, I_2. If $E = E_1 \cup E_2$ and $I = \{X \cup Y : X \in I_1, Y \in I_2\}$ then the pair (E, I) is a matroid with the system of independent sets I. It is called the *union of the matroids M_1, M_2* and is denoted by $M_1 \cup M_2$.

Consider the following problem of discrete optimization. Let E be a nonempty finite set and let $w : E \rightarrow \mathbf{R}^+$ be a function providing every $e \in E$ with a nonnegative real *weight* $w(e)$. For $X \subseteq E$, the weight of X is defined as the sum of weights of all $x \in X$. Let further I be a set of subsets of E. It is necessary to find an element in I of maximal weight. Let us describe a simple algorithm for solving such problems.

Greedy Algorithm

Step 1: Find an element $e_1 \in E$ such that

$$\{e_1\} \in I, \ w(e_1) = \max\{w(e) : \ \{e\} \in I\}.$$

Step k : Find $e_k \in E \backslash \{e_1, \ldots, e_{k-1}\}$ such that $\{e_1, \ldots, e_{k-1}, e_k\} \in I$ and

$$w(e_k) = \max\{w(e) : \ e \neq e_i, \ i = 1, \ldots, k-1, \ \{e_1, \ldots, e_{k-1}, e\} \in I\}.$$

If there is no such e_k then terminate.

The greedy algorithm always produces an element of I maximal under inclusion but not necessarily of maximal weight. But if I is the set of independent sets of a matroid with support E then the greedy algorithm always constructs an independent set of maximal weight (this set turns out to be a base).

One may easily modify the greedy algorithm to make it construct a minimum weight independent set of a matroid.

* * *

5.2.1. Prove that an independence system whose all cycles are single-element is a matroid.

5.2.2. Prove that an independence system of order 2 is a matroid, but not all independence systems of order 3 are matroids.

5.2.3. List all pairwise nonisomorphic matroids of order 3. For each of them, find its rank, all bases, and all cycles. How many of them are graphical independence systems? And graphic matroids?

5.2.4. Let $M = (E, I)$ be a matroid, $\emptyset \neq A \subseteq E$,

$$I_1 = \{X \in I : \ X \cap A = \emptyset\}, \ \ I_2 = \{X \cap A : \ X \in I\}.$$

Prove that the pairs (E, I_k), $k = 1, 2$, are the matroids with the sets of independent subsets I_k.

5.2.5. Prove that for a matroid M of order n

a) the number of bases is at most $C_n^{\rho(M)}$;

b) the number of cycles is at most $C_n^{\rho(M)+1}$.

5.2.6. Let (E, C) be a matroid with the set of cycles C. For a positive integer k let C' denote the set obtained from C by the deletion of all cycles with more than k elements. Prove the following statements.

a) If $k \leq 2$ then (E, C') is the matroid with the cycle set C'.

b) If $k = 3$ then the statement from a) is not necessarily valid.

5.2.7. a) Prove that the vector matroid is matroid indeed. What are its bases and cycles? What is its rank?

b) Find all bases, cycles, cobases, cocycles, rank and corank of the vector matroid induced by the columns of the matrix

$$A = \begin{bmatrix} 1 & 2 & 3 & 1 \\ 0 & 1 & 1 & 2 \\ 1 & 1 & 0 & 1 \end{bmatrix}.$$

5.2.8. Let E be a finite nonempty set and let $\mathcal{B}_k$ be the set of k-element subsets of E. Prove that $(E, \mathcal{B}_k)$ is the matroid with the set of bases $\mathcal{B}_k$. It is called the *uniform matroid (of rank k)*.

5.2.9. Prove that the union of matroids is matroid.

5.2.10. Let E be a finite nonempty set and let

$$\{E_i,\ i = 1, \ldots, k\} \tag{5.2.1}$$

be a covering of E, i.e., $E = \bigcup_{i=1}^{k} E_i$. Let further $m = (m_1, \ldots, m_k)$ be an integer-valued vector with $0 \le m_i \le |E_i|$, $i = 1, \ldots, k$. Set

$$I = \{X \subseteq E : |X \cap E_i| \le m_i,\ i = 1, \ldots, k\}.$$

Prove the following statements.

a) The pair $S = (E, I)$ is the independence system with the support E and the set of independent sets I.

b) Every independence system is isomorphic to some independence system S defined in a).

c) If the covering (5.2.1) is a partition, i.e., $E_i \cap E_j = \emptyset$ for all i, j, $i \neq j$, then $S = (E, I)$ is a matroid. It is called the *partition matroid specified by the vector m*.

d) Every independence system of order 2 is a partition matroid.

e) Every matroid of order 3 is a partition matroid, but this is not so for matroids of larger orders.

5.2.11. Prove the following statements.

a) The transversal matroids and only such ones may be written as unions of uniform matroids of rank 1 (see Exercise 5.2.8).

b) The uniform matroids of rank k and only such ones may be written as the union of k copies of a uniform matroid of rank 1.

c) The partition matroids (see Exercise 5.2.10) and only such ones may be written as unions uniform matroids with pairwise disjoint supports.

d) A uniform matroid is a partition matroid, but there are partition matroids that are not uniform ones.

e) A partition matroid is a transversal matroid, but there are transversal matroids that are not partition ones.

5.2.12. Prove that the number of pairwise nonisomorphic

a) matroids of order n is at most 2^{2^n};

b) transversal matroids of order n is at most 2^{n^2}.

5.2.13. Prove that every matroid dual to a uniform one is uniform, dual to a partition one is a partition matroid, but there are transversal matroids whose duals are not transversal.

5.2.14. A graph is called an *M-graph* if every its component is a complete graph. Prove the equivalence of the following statements.

a) A graph G is an M-graph.

b) The graphical independence system (VG, IG) (see the Exercise 5.1.6) is a matroid.

c) The path graph P_3 is not an induced subgraph of G.

5.2.15. Prove the following statements.

a) Every independence system is an intersection of matroids considered as independence systems.

b) The minimal number of matroids whose intersection is the graphical independence system (VG, IG) (see the Exercise 5.1.6) is equal to the minimal number of M-graphs (see the Exercise 5.2.14) whose union is the graph G.

5.2.16. Let (EG, IG) be the independence system specified in the Exercise 5.1.12. Prove the equivalence of the following statements.

a) The independence system (EG, IG) is the matroid with the set of independent sets IG.

b) Every nontrivial component of G is either the triangle K_3 or the star $K_{1,n}$, $n \geq 1$.

c) P_4 is not a subgraph of G.

5.2.17. Let G be a nonempty graph.

a) Prove that the cyclic matroid $M(G)$ is matroid indeed. What are its bases and cycles, cobases and cocycles? What are its rank and corank?

b) Find the bases, cycles, cobases and cocycles, rank and corank for $M(G)$, where G is shown at Fig. 5.2.1.

5.2.18. Does the isomorphism of the cyclic matroids $M(G)$and $M(H)$ imply the isomorphism of the graphs G and H?

5.2.19. Characterize the graphs whose cyclic matroids are uniform (the Exercise 5.2.8).

5.2.20. Prove that a nonempty graph G has s skeletons with pairwise disjoint edge sets if and only if every its spanning subgraph H satisfies the inequality $m(G)-m(H) \geq s(k(H)-k(G))$, where k denotes the number of components, m denotes the number of edges.

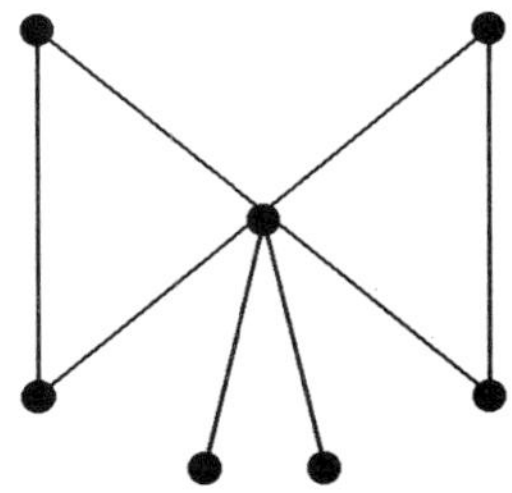

Figure 5.2.1: To Exercise 5.2.17

5.2.21. Prove the following statements.

a) If D_1, D_2 are two different cuts of a graph G and $e \in D_1 \cap D_2$ then the set $(D_1 \cup D_2) - e$ is a separator set.

b) For every nonempty acyclic subgraph F of a graph G there exists a cut in G with exactly one edge in common with F.

5.2.22. Prove that if a graph G^* is an abstract dual for a graph G (the Exercise 4.3.19) then G is an abstract dual for G^*.

5.2.23. Prove that the matroids of cycles $M(K_5)$ and $M(K_{3,3})$ are not cographic and the matroids of cuts $M^*(K_5)$ and $M^*(K_{3,3})$ are not graphic.

5.2.24. For the matrix

$$\begin{bmatrix} 1 & 0 & 2 & 0 & 3 \\ 0 & 3 & 1 & 0 & 1 \\ 2 & 2 & 3 & 3 & 6 \\ 1 & 2 & 0 & 2 & 1 \\ 0 & 0 & 1 & 1 & 0 \\ 1 & 2 & 1 & 3 & 1 \end{bmatrix}$$

find

a) a linearly independent set of rows with maximal sum of elements;

b) a maximal linearly independent set of rows with minimal sum of elements.

5.2.25. A set of tasks must be executed on a single computer. The jobs have their deadlines. All tasks have the same execution time.

a) Prove that the set of all subsets of tasks that can be performed within the deadlines constitute the set of independent sets of some matroid.

b) Assume that to every task that fails to meet its deadline a penalty is assigned, all penalties being of the same value. Which order of the execution of the tasks minimizes the total penalty?

5.3 Binary Matroids

Let M be a matroid with support $E = \{e_1, \ldots, e_m\}$ and let F^n be the linear space of columns of height n over a field F. A mapping $\varphi : E \to F^n$ is called a *representation of the matroid M over the field F* if it satisfies the following condition: for any subset

$$X = \{e_{i_1}, e_{i_2}, \ldots, e_{i_k}\} \subseteq E$$

the system of vectors $\varphi(e_{i_1}), \varphi(e_{i_2}), \ldots, \varphi(e_{i_k})$ is linearly independent over F if and only if $X \in I(M)$. If φ is a representation then the matrix

$$\varphi(E) = [\varphi(e_1)\varphi(e_2)\ldots\varphi(e_m)]$$

whose columns $\varphi(e_i)$ are the images of the elements of the support E under the representation φ is also called a *representation of the matroid M*. A matroid is called *representable over the field F* if there exists a representation of the matroid over F.

Theorem 5.3.1 *If a matroid M is representable over a field F then the dual matroid M^* is also representable over F.*

In what follows, $\mathbf{Z}_2$ is the two-element field. A matroid representable over $\mathbf{Z}_2$ is called *binary matroid.*

Let M be a binary matroid and let L be the set whose elements are the empty set, all cycles of M and the unions of pairwise disjoint cycles. Let us define the operations of *addition* modulo 2 and *multiplication* by 0 and 1 on L :

$$X + Y = (X \cup Y)\backslash(X \cap Y),\ 0 \cdot X = \emptyset,\ 1 \cdot X = X.$$

Let us select a base B of M. For any $e \in E\backslash B$, exactly one cycle of M is a subset of $B \cup \{e\}$. Let us denote this cycle by C_e.

Theorem 5.3.2 *The set L supplied with the above operations is a linear space over $\mathbf{Z}_2$ of dimension $\rho^*(M)$. The set of cycles*

$$\{C_e : e \notin B\} \tag{5.3.1}$$

is its basis.

The space L is called the *cycle space* and the basis (5.3.1) is called the *cycle basis (with respect to base B)* of the matroid M.

The *cocycle space* L^* is the object dual to L. The set of cocycles $\{C_e^* : e \in B\}$ is its basis, where B is some fixed base of M and C_e^* is the unique cocycle which is a subset of $(E\backslash B) \cup \{e\}$. This basis is called the *cocycle basis (with respect to base B)* of the matroid M.

The cycle and cocycle bases with respect to the same base are easily expressed in terms of each other; namely, the following theorem is valid.

Theorem 5.3.3 *If $f \in B$ then*

$$C_f^* = \{f, e_1, e_2, \ldots, e_k\},$$

where $\{C_{e_1}, C_{e_2}, \ldots, C_{e_k}\}$ is the set of cycles from the basis (5.3.1) *containing f.*

The cyclic matroid $M(G)$ for every nonempty graph G is binary, since the following theorem is valid.

Theorem 5.3.4 *The incidence matrix $I(G)$ is a representation of the cyclic matroid $M(G)$ over $\mathbf{Z}_2$.*

Corollary 5.3.5 *The cyclic matroid $M(G)$ is isomorphic to the vector matroid over $\mathbf{Z}_2$ induced by the columns of the incidence matrix $I(G)$.*

The cycle basis of $M(G)$ is called the *cycle basis* of G and the cocycle basis $M(G)$ is called the *cut basis* of G.

* * *

5.3.1. Prove the following two statements.

a) A matroid is a vector matroid if and only if there exists a field over which it is representable.

b) If a matroid has more than one loop then it cannot be a vector one.

5.3.2. a) Prove that the uniform matroid of order 4 and rank 2 is not graphic.

b) Is it a vector matroid?

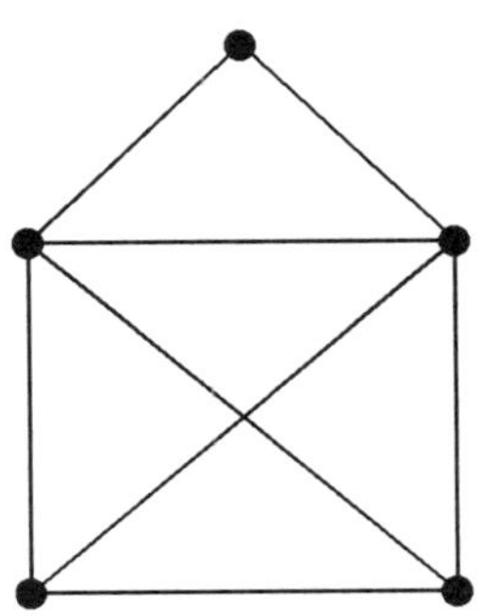

Figure 5.3.1: To Exercise 5.3.5

5.3.3. a) Prove that the matroid $(E, \mathcal{B})$, where $E = \{1, 2, 3, 4\}$ and the set of bases $\mathcal{B}$ is the set of all three-element subsets of E containing 1, is binary.

b) Find all cycles of this matroid.

c) Is this matroid graphic?

5.3.4. Prove that

a) the dimension of the cycle space of a nonempty graph G is equal to the cyclic rank $\nu(G)$;

a) the dimension of the cocycle space of a nonempty graph G is equal to the cocyclic rank $\nu^*(G)$;

c) the number of cycles of a graph G is at most $2^{\nu(G)} - 1$, and the number of cuts is at most $2^{\nu^*(G)} - 1$ (both these bounds are attainable).

5.3.5. Find a cycle basis, all simple cycles, a cut basis, and all cuts of the graph shown at Fig. 5.3.1.

5.3.6. Find a cycle basis and a cut basis of the graph with the incidence matrix

$$I(G) = \begin{bmatrix} 1 & 0 & 1 & 0 & 0 & 0 \\ 1 & 1 & 0 & 1 & 0 & 0 \\ 0 & 1 & 0 & 0 & 1 & 0 \\ 0 & 0 & 0 & 1 & 1 & 1 \\ 0 & 0 & 1 & 0 & 0 & 1 \end{bmatrix}.$$

Chapter 6

Planarity

6.1 Embeddings of Graphs. Euler Formula

Recall that a continuous rectifiable curve without self-intersections is called a *Jordan curve.*

It is said that a graph G is *embeddable into a space* L if there exists a mapping of its vertices and edges into different points and Jordan curves of L respectively such that

– the ends of the edges are mapped into the endpoints of the corresponding curves;

- these curves intersect only at their common endpoints.

Such mapping (and its image) is called an *embedding of the graph* G *into the space* L.

If a graph is embeddable onto the plane, it is called *planar* and the (image of the) embedding is called a *plane graph.* In this case it is said that the graph *has a plane embedding.*

Note. Rigorously speaking, a plane graph is not a graph; it is rather a representation of the graph which in addition to all graph-theoretical properties possesses some specific ones.

It is assumed that a graph is *isomorphic* to all its embeddings and all embeddings of the same graph are isomorphic to each other. In particular, a planar graph is isomorphic to the plane graphs that are its plane embeddings.

The graphs of Platonic solids, Fig. 1.1.2, are examples of planar graphs.

It is evident that

a) a subgraph of a planar graph is planar;

b) a graph is planar if and only if every its component is a planar graph.

Theorem 6.1.1 (Wagner Theorem) *For every planar graph there exists an isomorphic plane graph whose edges are straight line segments.*

Theorem 6.1.2 *Every graph is embeddable into the three-dimensional space* E^3.

Theorem 6.1.3 *A graph is embeddable onto a sphere if and only if it is planar.*

A *face* of a plane graph G is a maximal under inclusion set of points of the plane such that any two points of it may be connected by a Jordan curve that does not intersect the edges of the graph, with a possible exception of the endpoints of the curve.

The set of vertices and edges of a plane graph that belong to a face is called the *boundary of the face.* A face of a plane graph bounded by a 3-cycle, or triangle, will also be called *triangle.*

A plane graph has exactly one face of infinite area (face 3 at Fig. 6.1.1). Such face is called *external,* or *outer,* the remaining ones are called *internal* (faces 1, 2 at Fig. 6.1.1).

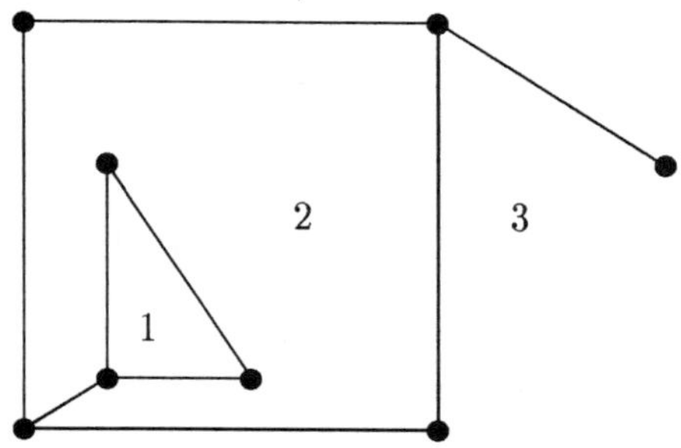

Figure 6.1.1: Plane graph

The notions of plane embedding and face are naturally extended to pseudo- and multigraphs.

A connected graph is called *outerplanar* if it has a plane embedding such that all its vertices belong to the boundary of the same face (it is convenient for this face to be the outer one). An outerplanar graph is called *maximal outerplanar* if it ceases to be outerplanar after the addition of any new edge.

In what follows, the numbers $n = n(G)$, $m = m(G)$, $f = f(G)$ will denote the numbers of the vertices, edges and faces of a plane graph respectively.

Euler Formula. For a connected plane graph,

$$n - m + f = 2.$$

Corollary 6.1.4 *The number of faces of every embedding of a planar connected* (n, m)*-graph is* $m - n + 2$.

Corollary 6.1.5 *For a connected plane* (n, m)*-graph with* $n \geq 3$,

$$m \leq 3n - 6.$$

Corollary 6.1.6 *If the boundary of every face of a connected plane* (n, m)*-graph is an* r*-cycle,* $r \geq 3$, *then* $m(r - 2) \geq r(n - 2)$.

Theorem 6.1.7 *A plane graph is biconnected if and only if the boundaries of all its faces are simple cycles.*

Theorem 6.1.8 *A connected graph is planar if and only if every its block is planar.*

* * *

6.1.1. Prove that every internal face of a plane graph G may be transformed into the external face of a plane graph isomorphic to G using the stereographic projection.

6.1.2. Redraw the plane graph shown at Fig. 6.1.2 making the face a) 1, b) 2, c) 3 external.

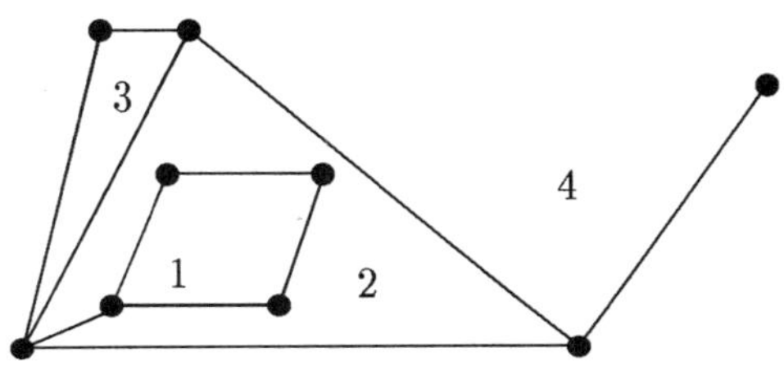

Figure 6.1.2: To Exercise 6.1.2

6.1.3. Using the Corollary 6.1.5, prove that K_5 is nonplanar.

6.1.4. Three Houses and Three Wells Problem: There are three houses and three wells. Each family uses each of the wells. After a quarrel, they decided to lay noncrossing paths from every house to every well. Is this possible?

6.1.5. Construct plane embeddings of the graphs a) $K_5 - e$, b) $K_{3,3} - e$, c) $K_{2,n}$ such that the edges are presented by straight line segments.

6.1.6. Verify the Euler formula for a) the wheel W_n, b) graphs of Platonic solids, c) the graph defined by the crossings and edges on the $m \times n$-chessboard, d) $K_{2,n}$.

6.1.7. Prove that the Euler formula may be generalized for the case of a disconnected plane (n, m)-graph with k components: $n - m + f = k + 1$.

6.1.8. To how many faces of a plane graph may a point of an edge that is not a vertex belong?

6.1.9. To how many faces of a plane graph may a vertex of degree $d \geq 1$ belong?

6.1.10. Let G be a plane biconnected graph. Prove that the set of all simple cycles that are the boundaries of the internal faces of G is the cycle basis of G (see Sect. 5.3).

6.1.11. Let the boundaries of all faces of a plane biconnected graph contain even numbers of edges. Prove that the lengths of simple cycles of the graph are even.

6.1.12. Prove that $m(h-2) \leq h(n-2)$ for a connected planar (n, m)-graph with $n \geq 3$ of girth $h \geq 3$. Using this relation, prove that the Petersen graph is nonplanar.

6.1.13. Prove that $m \leq 2n-4$ for a connected planar (n, m)-graph with $n \geq 3$ and without triangular faces.

6.1.14. Prove that $m \leq 2(p+q-2)$ for a planar bipartite m--edge graph with the sizes p, q of the parts at least 2.

6.1.15. Prove that $|EH_n| \leq 2^{n+1} - 4$, where H_n is a planar spanning subgraph of Q_n, $n \geq 2$.

6.1.16. Prove that every planar graph has a vertex of degree at most 5.

6.1.17. Prove that there are no planar 6-connected graphs.

6.1.18. Prove that every planar 5-connected graph has at least 12 vertices. Construct one of them.

6.1.19. For which r there exists a planar r-regular graph?

6.1.20. Prove that a planar connected cubic graph whose every face has at least 5 vertices is of order at least 20.

6.1.21. Let G be a planar connected cubic graph. Prove that

$$\sum_{i \geq 3} (6-i) f_i = 12,$$

where f_i is the number of faces of G each of which is bounded by i edges. (Here and in the sequel, an edge that belongs to a single face is counted twice in f_i.) Prove that G has at least one face with at most 5 edges.

6.1.22. Let G be a plane graph with the number of faces less than 12. Prove that if the degrees of the vertices of G are at least 3 then G has a face bounded by at most 4 edges.

6.1.23. Prove that a plane graph of order $n \geq 4$ without triangular faces has the maximal number of edges $2n-4$ if and only if every its face (including the external one) is bounded by a 4-cycle.

6.1.24. Prove that the maximal number of edges in a planar spanning subgraph of $K_{p,q}$, $p, q \geq 2$, is $2(p+q-2)$.

6.1.25. Prove that a planar spanning subgraph of $K_{p,q}$, $p, q \geq 2$, has $2(p+q-2)$ edges if and only if every its face (including the external one) is bounded by a 4-cycle.

6.1.26. Solve problem 6.1.15 using problem 6.1.23.

6.1.27. Prove that a planar spanning subgraph of Q_n, $n \geq 2$, has $2^{n+1} - 4$ edges if and only if every its face (including the external one) is bounded by a 4-cycle.

6.1.28. Prove that

a) if the faces of a planar graph are bounded by pentagons and hexagons, then the number of pentagonal faces is even;

b) if in addition this graph is cubic, then the number of pentagonal faces is 12 and it has no hexagonal faces.

6.1.29. For which n the square of the cycle C_n is planar?

6.1.30. Prove that the following statements are equivalent.

1) a graph G is outerplanar;
2) a graph G' obtained from G by adding a new vertex and connecting it with all other vertices of G by new edges is planar;
3) every block of G is outerplanar.

6.1.31. Prove that K_4 and $K_{2,3}$ are not outerplanar.

6.1.32. Prove that a connected subgraph of an outerplanar graph is outerplanar.

6.1.33. Which graphs shown at Fig. 6.1.3 are outerplanar?

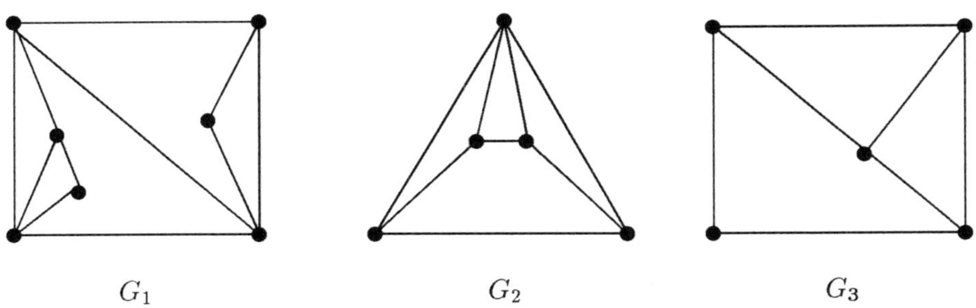

Figure 6.1.3: To Exercise 6.1.33

6.1.34. For which $n \geq 3$ the wheel W_n is outerplanar?

6.1.35. Prove that (a) $m \leq 2n - 3$ and (b) $3(f-1) + n \leq 2m$ for every outerplanar (n, m)-graph, $n \geq 2$.

6.1.36. For which $n \geq 3$ the square of the cycle C_n is outerplanar?

6.1.37. Let every connected subgraph F of G ($F \neq G$) be outerplanar. Is G necessarily outerplanar?

6.1.38. Prove that if G contains a subgraph isomorphic to $K_{1,3}$ then G^2 is not outerplanar.

6.1.39. Prove that the square of a connected graph G is outerplanar if and only if G is either K_3 or a path graph.

6.1.40. Prove that $m \leq (3n-4)/2$ for an outerplanar (n,m)-graph without triangles.

6.1.41. Let a maximal outerplanar graph G have $n \geq 3$ vertices. Prove that G has a) a vertex of degree 2; b) a plane embedding with $n-2$ internal faces.

6.1.42. Draw all pairwise nonisomorphic maximal outerplanar graphs of order 6.

6.1.43. Prove that every maximal outerplanar graph G of order $n \geq 3$ has a) $2n-3$ edges; b) at least two vertices of degree 2; c) $\kappa(G) = 2$; d) at least 3 vertices of degree at most 3.

6.1.44. Prove that every graph may be embedded into the three-dimensional space in such a way that all its edges are represented by straight line segments.

6.2 Plane Triangulation

A connected plane graph is called a *plane triangulation* if all its faces are triangles.

A *maximal planar graph* is a graph of order $n \geq 3$ that ceases to be planar after the addition of any edge. A *maximal plane graph* is a plane graph isomorphic to a maximal planar graph.

Theorem 6.2.1 *A graph is maximal plane graph if and only if it is a plane triangulation.*

Corollary 6.2.2 *Every plane graph is a spanning subgraph of a plane triangulation.*

Corollary 6.2.3 *For any maximal planar (n,m)-graph, $m = 3n-6$.*

Proposition 6.2.4 *If the number of vertices of a plane triangulation is at least* 4 *then the degree of every vertex is at least* 3.

* * *

6.2.1. Prove that a plane triangulation has no cutpoints.

6.2.2. How many faces has an n-vertex plane triangulation?

6.2.3. Prove that a planar graph of order $n \geq 4$ has at least four vertices of degree at most 5.

6.2.4. Prove that the single plane triangulation which is a 5-regular graph is the icosahedron graph, see Fig. 1.1.2.

6.2.5. What is the maximal number of edges in a planar graph of order $n \geq 3$?

6.2.6. Draw a plane graph of order $n > 5$ that has exactly four vertices of degree at most 5.

6.2.7. Does there exist a plane triangulation that has exactly two vertices of odd degree and these vertices are adjacent?

6.2.8. The vertices of a plane triangulation are colored in three colors. A face is said to be tricolored if its vertices are colored in different colors. Prove that the number of tricolored faces is even.

6.3 Planarity Criteria

The *edge subdivision* operation for an edge $e = ab$ is the deletion of e from the graph and the addition of two edges av and vb, where v is a new vertex.

Two graphs are called *homeomorphic* if they may be obtained from the same graph by edge subdivisions. By the definition, every graph is homeomorphic to itself.

Kuratowski Criterion: *A graph is planar if and only if it has no subgraphs homeomorphic to* K_5 *or* $K_{3,3}$.

A graph G is said to be *contractible to a graph* H if H may be obtained from G by edge contractions. By the definition, every graph is contractible to itself.

Wagner Criterion: *A graph is planar if and only if it has no subgraphs contractible to* K_5 *or* $K_{3,3}$.

* * *

6.3.1. Using the Kuratowski criterion, prove that the Petersen graph is nonplanar.

6.3.2. Prove that the n-transposition graph (see Exercise 1.1.27 for the definition) is nonplanar for $n \geq 3$.

6.3.3. For which n the graphs G_n defined in the Exercise 1.3.4 are planar?

6.3.4. How many are there nonisomorphic nonplanar graphs of sixth order?

6.3.5. Which of graphs shown at Fig. 6.3.1 are nonplanar?

6.3.6. For which n the graphs of order $2n$ shown at Fig. 6.3.2 are planar?

6.3.7. Let $H_n = (V_1, V_2; E)$ be a bipartite graph defined as follows.

$$V_1 = \{a_1, \ldots, a_n\};\ V_2 = \{b_1, \ldots, b_n\};$$

$$E = \{x_1, \ldots, x_n, y_1, \ldots, y_n, z_1, \ldots, z_n\};$$

$$x_i = a_i b_i,\ i = 1, \ldots, n;\ y_i = a_i b_{i+1},\ i = 1, \ldots, n-1;\ y_n = a_n b_1;$$

$$z_i = a_i b_{i+2},\ i = 1, \ldots, n-2;\ z_{n-1} = a_{n-1} b_1;\ z_n = a_n b_2.$$

For which $n \geq 3$ the graphs H_n are planar?

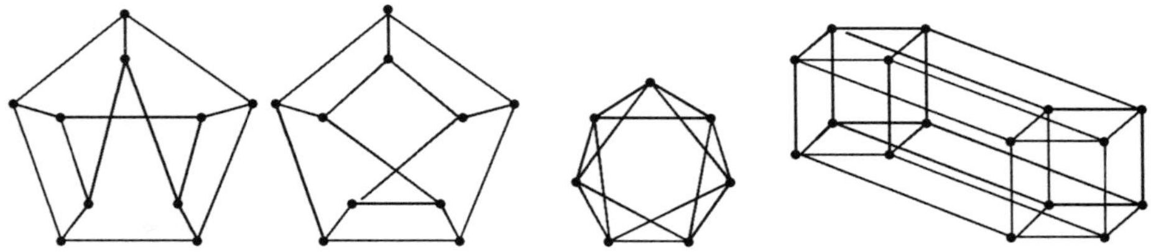

a *b* *c* *d*

Figure 6.3.1: To Exercise 6.3.5

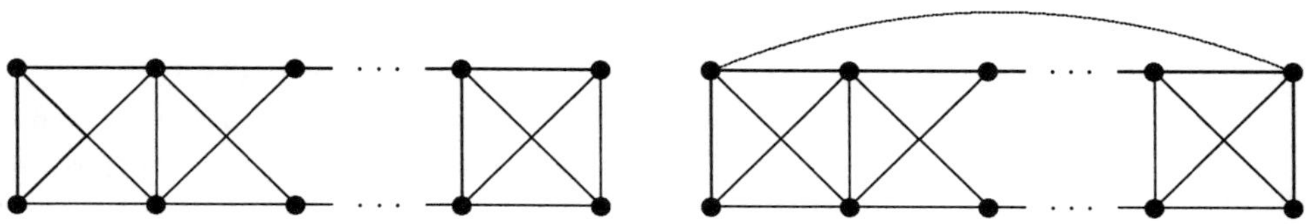

a *b*

Figure 6.3.2: To Exercise 6.3.6

6.3.8. Let G be a graph of order $n \leq 7$. Prove that either G or $\overline{G}$ is planar.

6.3.9. What is the minimal number n such that there exists a nonplanar n-vertex graph with nonplanar complement.

6.3.10. Find a graph G of order 8 such that

a) both G and $\overline{G}$ are planar;

b) both G and $\overline{G}$ are nonplanar;

c) one of G and $\overline{G}$ is planar and the other is not.

6.3.11. Is it true that for every 8-vertex triangulation the complementary graph is planar?

6.3.12. Let G be a graph of order $n \geq 11$. Prove that G and $\overline{G}$ cannot be both planar.

6.3.13. Prove that if an (n_1, m_1)-graph and an (n_2, m_2)-graph are homeomorphic then
$n_1 + m_2 = n_2 + m_1$.

6.3.14. Give examples of homeomorphic graphs with homeomorphic line graphs.

6.3.15. Is it true that for homeomorphic graphs their line graphs are homeomorphic?

6.3.16. Using the Wagner criterion, prove that the Petersen graph is nonplanar.

6.3.17. Let G_1 be contractible to G_2. Is it true that $L(G_1)$ is always contractible to $L(G_2)$?

6.3.18. Is it true that if a graph is contractible to K_5 then it has a subgraph homeomorphic to K_5?

6.3.19. Prove that a graph obtained by a contraction of an edge of a planar graph is planar.

6.3.20. Prove that if a graph is contractible to a nonplanar graph then it is nonplanar.

6.3.21. Prove or disprove: every connected nonplanar graph is contractible either to K_5 or to $K_{3,3}$.

6.3.22. Verify that the line graphs of K_5 and $K_{3,3}$ are nonplanar.

6.3.23. Let a graph G_1 be homeomorphic to a graph G_2 that has no vertices of degree two. Prove that $L(G_1)$ is contractible to $L(G_2)$.

6.3.24. Prove that if a graph is nonplanar then its line graph is also nonplanar.

6.3.25. Is it true that the line graph of a planar graph is planar?

6.3.26. For which $r \geq 1$ the line graph of an r-regular graph is planar?

6.3.27. Let the line graph of a graph G be planar. Prove that a) G is planar; b) $\Delta(G) \leq 4$; c) every vertex of G of degree 4 is a cutpoint.

6.3.28. Prove that if the square of a graph G is planar then $\Delta(G) \leq 3$.

6.3.29. Prove that a graph is outerplanar if and only if it has no subgraphs homeomorphic to K_4 or $K_{2,3}$.

6.3.30. A connected nonplanar graph is called *s-critical* if the deletion of any its edge produces a planar graph. Describe all s-critical graphs.

6.3.31. A connected nonplanar graph is called *t-critical* if the contraction of any its edge produces a planar graph. Describe all t-critical graphs.

6.4 Duality and Planarity

Note. In this section, the term "graph" will mean "pseudograph".

For a plane graph G we may construct another plane graph G^* called *geometric dual* for G as follows. Pick a point v_i^* inside every face F_i of G. These points will be the vertices of G^*. To every edge $e \in EG$ we assign a Jordan curve e^* that crosses only this edge and connects the points v_i^*, v_j^* lying in the faces whose boundaries contain e (there may be one or two such faces; in the former case the curve connects a point v_i^* with itself, i.e., it is a loop). The curves e^* will be the edges of G^*. Clearly, one may choose the edges e^* to be pairwise nonintersecting, with the exception of their common endvertices. At Fig. 6.4.1, the graph G is drawn in straight lines and the geometric dual graph G^* is drawn curved. As the figure shows, this construction not always gives an ordinary graph, hence the note above.

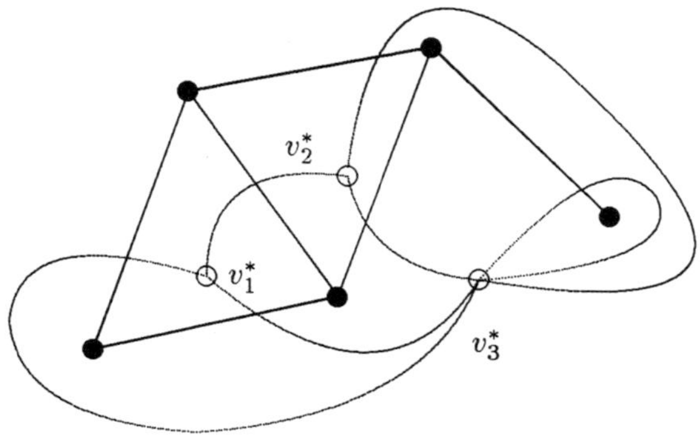

Figure 6.4.1: A graph and its geometric dual

One may easily verify that a plane graph G^* geometric dual to a plane graph G is unique up to isomorphism.

Proposition 6.4.1 *If G is a connected plane (n, m)-graph with f faces and G^* is an (n^*, m^*)-*
graph with f^* faces geometric dual to G, then $n^* = f$, $m^* = m$, $f^* = n$.

Since G^* is a plane graph, one may construct a geometric dual to it, naturally denoted by G^{**}.

Theorem 6.4.2 *If G is a plane connected graph then G^{**} is isomorphic to G.*

A graph geometrically dual to a planar graph is defined as follows: construct any plane embedding of the graph and take its geometric dual.

Theorem 6.4.3 *Let G be a planar graph with geometric dual G^*. A subset of edges in G constitutes a simple cycle if and only if the corresponding subset of edges of G^* constitutes a cut.*

This theorem naturally leads to the following combinatorial definition of duality for graphs that generalizes the geometric duality.

A graph G^+ is called *abstract dual to G* if there exists a bijection between the sets EG and EG^+ such that a subset of edges in G constitutes a simple cycle if and only if the corresponding subset of edges of G^+ constitutes a cut.

Fig. 6.4.2 shows a graph and its abstract dual. Fig. 6.4.3 shows another example of graph and its abstract dual. Notice that here G^+ is not a geometric dual to G, but G is a geometric dual to G^+.

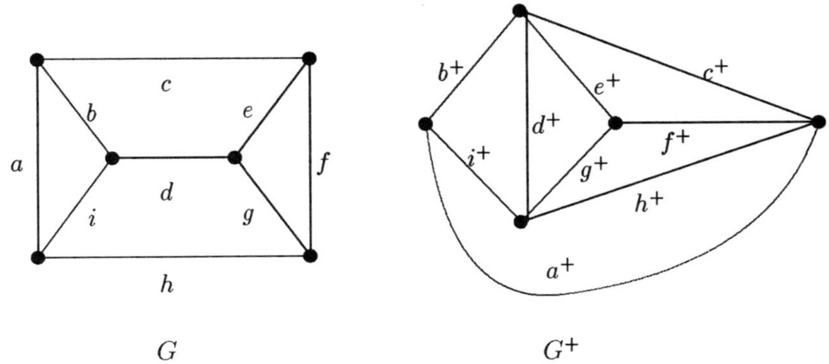

Figure 6.4.2: A graph and its abstract dual

Proposition 6.4.4 *A graph geometrically dual for a plane graph is also an abstract dual for it.*

Proposition 6.4.5 *A graph H is an abstract dual for a graph G if and only if the matroid $M(G)$ of cycles of G is isomorphic to the matroid $M^*(H)$ of cuts of H.*

Theorem 6.4.6 *If G^+ is an abstract dual for a graph G then G is an abstract dual for a graph G^+.*

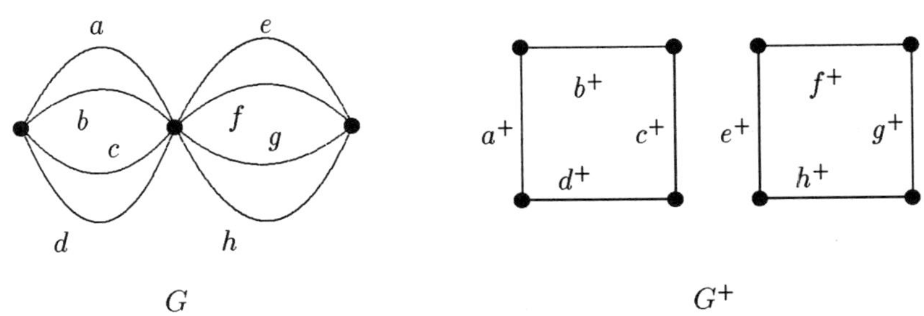

Figure 6.4.3: Another graph and its abstract dual

Notice that in contrast to Theorem 6.4.2, G is not required to be connected. In view of Theorem 6.4.6, graphs G and G^+ may be called *dual to each other*.

Theorem 6.4.7 *If a graph has an abstract dual then every its subgraph has an abstract dual.*

Theorem 6.4.8 *If a graph G has an abstract dual then every graph homeomorphic to G has an abstract dual.*

Whitney Criterion: *A graph is planar if and only if it has an abstract dual.*

The Whitney criterion implies that K_5 and $K_{3,3}$ have no abstract duals.

* * *

6.4.1. Construct geometric duals for the graphs from Fig. 6.4.4.

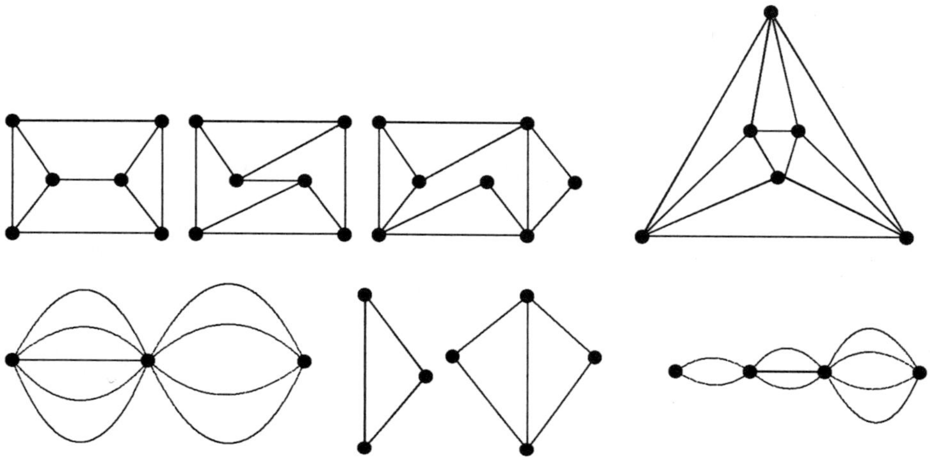

G

Figure 6.4.4: To Exercise 6.4.1

6.4.2. Construct geometric duals for the graphs of Platonic solids, Fig. 1.1.2.

6.4.3. Prove that a graph geometrically dual to a planar graph is connected.

6.4.4. Prove that if a plane graph has a cutpoint then its geometric dual also has a cutpoint.

6.4.5. Prove that a plane graph is biconnected if and only if its geometric dual is biconnected.

6.4.6. Prove that a graph geometric dual to a plane triangulation of order $n \geq 4$ is a biconnected cubic plane graph.

6.4.7. Verify the isomorphism of graphs shown at Fig. 6.4.5. Are their geometric duals isomorphic?

6.4.8. Let a plane graph G be disconnected. Prove that the iterated geometric dual G^{**} is not isomorphic to G.

6.4.9. Show that a geometric dual to a wheel is a wheel.

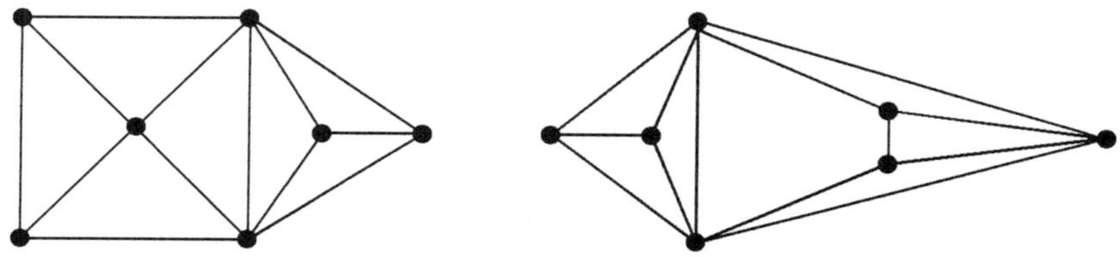

Figure 6.4.5: To Exercise 6.4.7

6.4.10. Under which conditions a graph geometric dual to an ordinary plane graph is an ordinary graph?

6.4.11. A plane graph is called *selfdual* if it is isomorphic to its geometric dual. Prove that $m = 2n - 2$ for a selfdual (n, m)-graph.

6.4.12. Show that a wheel W_n, $n \geq 3$, is selfdual. Give other examples of selfdual graphs.

6.4.13. How many are there pairwise nonisomorphic selfdual graphs of 6th order?

6.4.14. Prove that there are no plane graphs with five faces such that any two faces have a common edge.

6.4.15. Let G be a connected ordinary plane graph such that its geometric dual G^* is also an ordinary graph (cf. Exercise 6.4.10). Prove that G and G^* have the same number of spanning trees.

6.4.16. Prove that if the degrees of the vertices of a biconnected plane graph are even then the geometric dual is bipartite.

6.4.17. Find two nonisomorphic graphs abstractly dual to the graph shown at Fig. 6.4.6.

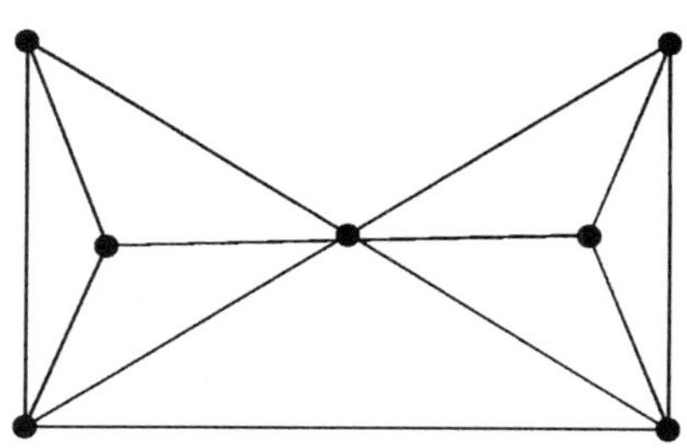

Figure 6.4.6: To Exercise 6.4.17

6.4.18. Do there exist nonisomorphic graphs abstractly dual to the graph from Fig. 6.4.7?

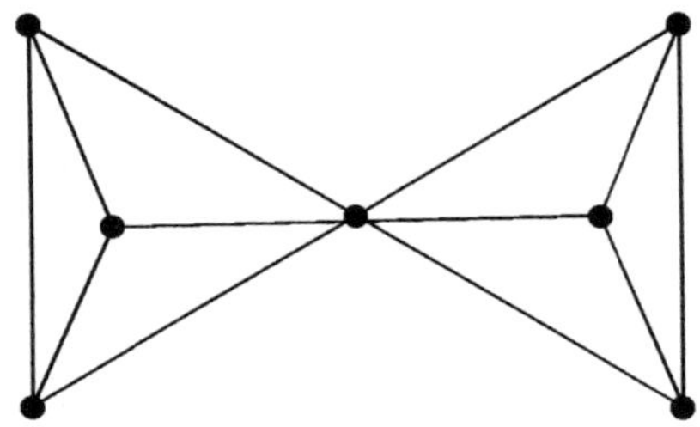

Figure 6.4.7: To Exercise 6.4.18

6.4.19. Prove that if G and G^+ are dual to each other then $n(G) - k(G) = \nu(G^+)$ and $n(G^+) - k(G^+) = \nu(G)$, where $n(G)$ is the number of vertices, $k(G)$ is the number of components, $\nu(G)$ is the cyclic rank of a graph G.

6.4.20. Which pairs of graphs from Fig. 6.4.8 are dual to each other?

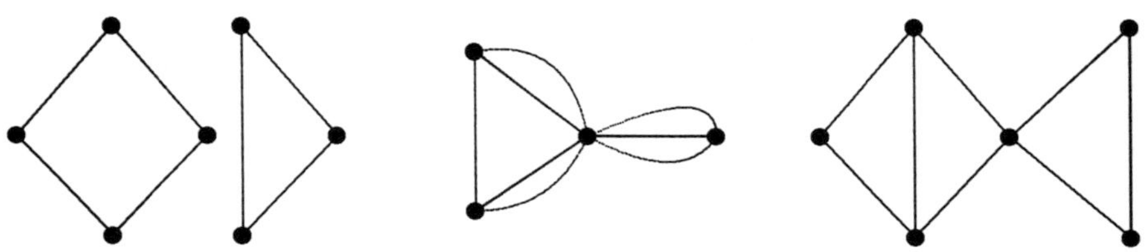

G H Q

Figure 6.4.8: To Exercise 6.4.20

6.4.21. Let G^+ be an abstract dual to a planar graph G and let G^{++} be an abstract dual to G^+. Is it always $G^{++} \cong G$?

6.4.22. a) Prove that the following statements are equivalent:

1. G is a planar graph.
2. The cyclic matroid $M(G)$ is cographic.
3. The matroid of cuts $M^*(G)$ is graphic.

b) Prove that K_5 and $K_{3,3}$ are nonplanar using this criterion.

6.4.23. A graph G^* is called a *generalized dual* to a graph G if there exists a bijection between the sets EG and EG^* such that for every spanning subgraph H of G

$$\nu(H) + \nu^*(\tilde{H^*}) = \nu^*(G^*),$$

where H^* is a subgraph of G^* that corresponds to H, $\tilde{H^*}$ is the graph obtained from G^* by the deletion of the edges of H^*, $\nu(G)$ denotes the cyclic rank, $\nu^*(G)$ denotes the cocyclic rank of a graph G.

Prove the following equalities.

a)$\nu(G) = \nu^*(G^*)$;

b)$\nu(G^*) = \nu^*(G)$;

c) $\nu^*(H) + \nu(\tilde{H}^*) = \nu^*(G)$.

6.4.24. Prove that if G^* is a generalized dual to a graph G then G is a generalized dual to G^*.

6.4.25. Prove that a geometrical dual to a connected plane graph is a generalized dual to this graph.

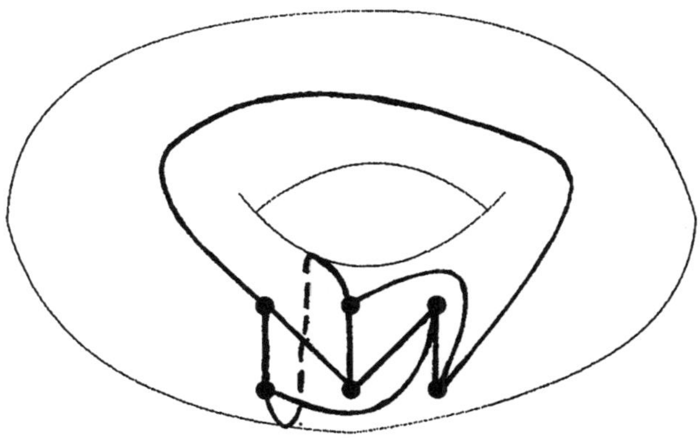

Figure 6.4.9: Embedding of $K_{3,3}$ onto the torus

6.5 Measures of Displanarity

The *genus* of a graph G is defined to be the minimal number $\gamma = \gamma(G)$ of handles necessary to attach to the sphere so that G is embeddable onto the resulting surface. Such surface is said to be *of genus* γ. The *graph of genus* γ is a graph embeddable onto the surface of genus γ but not onto the surface of genus $\gamma - 1$.

Clearly, the planar graphs are the graphs of genus 0.

A sphere with one handle (a genus 1 surface) is a *torus*. The graphs of genus 1, i.e., the ones that cannot be embeddable onto a plane (hence onto a sphere) but that can be embedded onto the torus are called *toroidal*. Fig. 6.4.9 shows an embedding of $K_{3,3}$ onto the torus.

An embedding of a graph onto the torus is convenient to sketch using the rectangle with opposite sides pairwise identified as shown at Fig. 6.5.1. At Fig. 6.5.2, an embedding of $K_{3,3}$ onto the torus is shown in such a way.

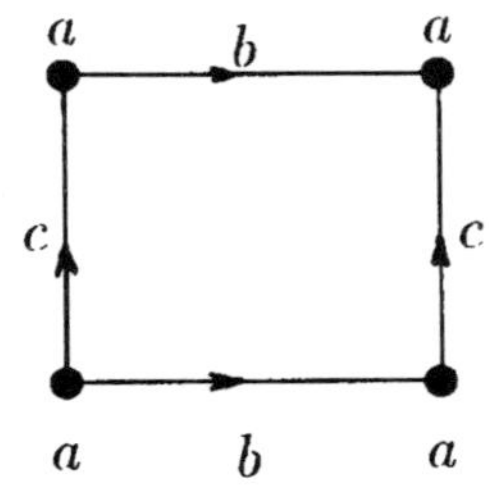

Figure 6.5.1: A representation of the torus

The following theorem generalizes the Euler formula (Sect.6.1). In this generalization a face of a genus-γ graph is naturally defined in terms of Jordan curves on the surface of genus γ.

Theorem 6.5.1 *For a connected graph of genus γ with n vertices, m edges and f faces, $n - m + f = 2 - 2\gamma$.*

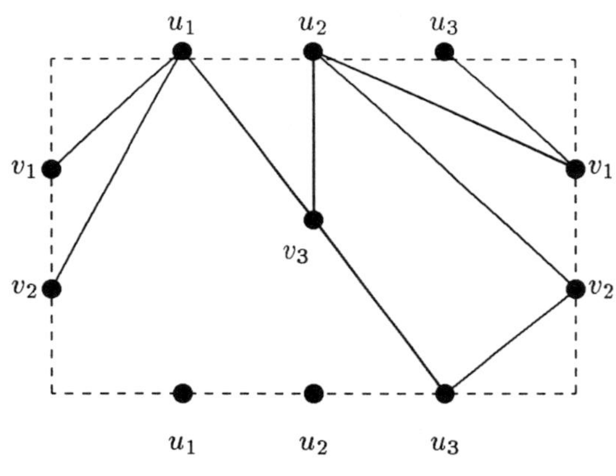

Figure 6.5.2: Embedding of $K_{3,3}$ onto the torus

The following formulae are known for the genus of some graphs:

$$\gamma(K_n) = \lceil (n-3)(n-4)/12 \rceil, \; n \geq 3. \tag{6.5.1}$$

$$\gamma(K_{p,q}) = \lceil (p-2)(q-2)/4 \rceil. \tag{6.5.2}$$

$$\gamma(Q_n) = (n-4)2^{n-3} + 1. \tag{6.5.3}$$

The *crossing number* $cr(G)$ of a graph G is the minimal number of pairwise intersections of edges possible when drawing a graph on the plane. Clearly, $cr(G) = 0$ if and only if G is a planar graph.

The *thickness* $t(G)$ of a graph G is the minimal number of its planar subgraphs whose union is the graph G. Clearly, the thickness of a planar graph is 1.

The *skewness* $sk(G)$ of a graph G is the minimal number of edges whose deletion produces a planar graph. Clearly, $sk(G) = 0$ if and only if G is a planar graph.

The *arboricity* $\zeta(G)$ of a graph G is the minimal number of edge-disjoint acyclic spanning subgraphs whose union is G.

Theorem 6.5.2 (Nash-Williams Formula) *If G is a nontrivial (n, m)-graph and m_p denotes the maximal number of edges in a subgraph of G with p vertices then*

$$\zeta(G) = \max_{p=2,\ldots,n} \lceil m_p/(p-1) \rceil .$$

* * *

6.5.1. Prove that a) $\gamma(K_5) = 1$; a) $\gamma(K_7) = 1$; a) $\gamma(K_{4,4}) = 1$.

6.5.2. Find the genus of the Petersen graph.

6.5.3. Embed Q_4 onto the torus.

6.5.4. Give an example of a graph of genus 2.

6.5.5. Prove that there are no complete graphs of genus 7. What is the next integer which is cannot be the value of the genus of a complete graph?

6.5.6. Using the Theorem 6.5.1 prove that

a) $\gamma(G) \geq \lceil (m - 3n)/6 \rceil + 1$ for a connected (n, m)-graph G;

b) $\gamma(G) \geq \lceil (m - 2n)/4 \rceil + 1$ for a connected triangle-free (n, m)-graph G;

c) $\gamma(K_n) \geq \lceil (n-3)(n-4)/12 \rceil$;

d) $\gamma(Q_n) \geq (n-4)2^{n-3} + 1$;

e) $\gamma(K_{p,q}) \geq \lceil (p-2)(q-2)/4 \rceil$.

6.5.7. Find a lower bound for the genus of an (n, m)-graph of girth g. Deduce from it that $\gamma(K_{p,q}) = \lceil (p-2)(q-2)/4 \rceil$.

6.5.8. What are the crossing numbers of $K_{3,3}$, K_5 and the Petersen graph?

6.5.9. Prove that if graphs G_1, G_2 are homeomorphic then $cr(G_1) = cr(G_2)$.

6.5.10. Prove that if p, q are even integers then $cr(K_{p,q}) \leq pq(p-2)(q-2)/16$. What is the upper bound for the crossing number of $K_{p,q}$ when p, q are not necessarily even?

6.5.11. Prove or disprove the statement: $cr(G) = 1$ for a nonplanar graph G if and only if it has an edge e such that $cr(G - e) = 0$.

6.5.12. Prove that $\gamma(G) \le cr(G)$. Give examples of graphs such that a) $\gamma(G) < cr(G)$;
b) $\gamma(G) = cr(G)$.

6.5.13. Find the thickness for $K_{3,3}$, K_5, and the Petersen graph.

6.5.14. Prove that every nonplanar graph is homeomorphic to a graph of thickness 2.

6.5.15. Prove that $t(G) \ge \lceil m/(3n-6) \rceil$ for a connected (n, m)-graph G.

6.5.16. Prove that $t(K_n) \ge \lfloor (n+7)/6 \rfloor$.

6.5.17. Prove that $t(G) \ge \lceil m/2(n-2) \rceil$ for a connected bipartite (n, m)-graph G, $n > 2$.

6.5.18. Prove that

a) $t(K_{p,q}) \ge \lceil pq/2(p+q-2) \rceil$, $p + q \ge 2$;
b) $t(Q_n) \ge \lfloor n/4 \rfloor + 1$, $n \ge 1$.

6.5.19. What is the minimal number of edges to be deleted from Q_4 to make it planar?

6.5.20. Find the skewness for $K_{3,3}$, K_5, and the Petersen graph.

6.5.21. Prove that $sk(K_n) = C_n^2 - 3n + 6$, $n \ge 3$.

6.5.22. Find the skewness for $K_{p,q}$ and Q_n.

6.5.23. Prove that the arboricity of a planar graph is at most 3.

6.5.24. Construct a planar graph of arboricity 3.

6.5.25. Prove that every graph is homeomorphic to a graph of arboricity 1 or 2 and hence of thickness 1 or 2.

Chapter 7

Graph Traversals

7.1 Eulerian Graphs

A cycle in a graph is called *Eulerian* if it contains all edges of the graph. (Recall that by definition any edge may enter a cycle at most once.) A connected graph with an Eulerian cycle is called *Eulerian.*

Theorem 7.1.1 *A connected graph is Eulerian if and only if the degrees of all its vertices are even.*

We shall say that a set of edge-disjoint paths (not necessarily simple) *covers* a graph if every edge of the graph belongs to some of these paths.

Theorem 7.1.2 *If a connected graph contains exactly k vertices of odd degree then the minimal number of edge-disjoint paths that cover the graph is $k/2$.*

In particular, if $k = 2$ then the graph has a path connecting the odd-degree vertices and containing all edges of the graph. Such path is called *Eulerian.*

* * *

7.1.1. Prove that a connected graph is Eulerian if and only if every its block is Eulerian.

7.1.2. Prove that every graph G may be represented as the union of two edge-disjoint graphs G_1, G_2 such that G_1 is a forest and the degrees of the vertices of G_2 are even.

7.1.3. Prove that an Eulerian graph is bridgeless.

7.1.4. Prove that a connected graph is the union of edge-disjoint simple cycles if and only if it is Eulerian.

7.1.5. Let every edge e of an Eulerian graph G be provided with a real number $w(e)$. Prove that if $w(G) = \sum_{e \in EG} w(e) > 0$ then G has a simple cycle C such that $w(C) = \sum_{e \in EC} w(e) > 0$.

7.1.6. A cycle $(a_0, \ldots, a_{p-1}, a_0)$ is called a *subcycle* of cycle $(b_0, \ldots, b_{q-1}, b_0)$ if $a_0 = b_k$, $a_1 = b_{k+1}$, $\ldots$ $a_{p-1} = b_{k+p-1}$, $a_0 = b_{k+p}$ for some k (here the addition in the indices of b_i is modulo q). Is it true that every simple cycle of an Eulerian graph is a subcycle of an Eulerian cycle?

7.1.7. Let G be a tree. Under which conditions G^2 is Eulerian?

7.1.8. Let G be a connected graph distinct from K_2. Under which conditions $L(G)$ is Eulerian?

7.1.9. Find the minimal number of edge-disjoint paths covering the graph shown at
Fig. 7.1.1.

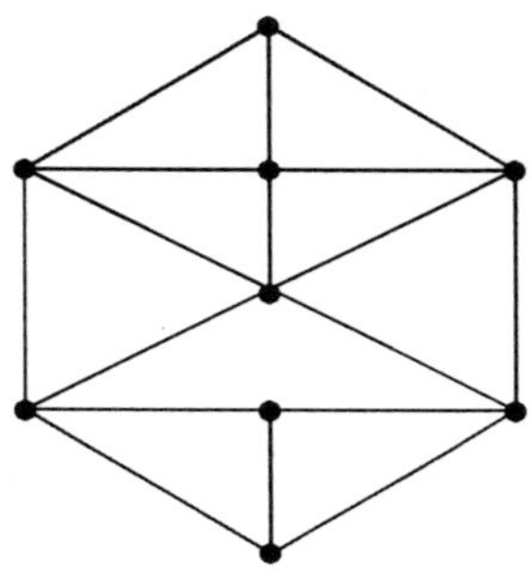

Figure 7.1.1: To Exercise 7.1.9

7.1.10. Prove that every connected graph contains a walk in which every edge of the graph is repeated at most twice.

7.1.11. Prove that an Eulerian graph contains a single Eulerian cycle if and only if the graph is a cycle graph.

7.1.12. Design an algorithm for the construction of an Eulerian cycle in an Eulerian graph.

7.2 Hamiltonian Graphs

A graph is called *Hamiltonian* if it has a simple cycle containing all vertices of the graph. Such cycle is called *Hamiltonian.* A simple path is called *Hamiltonian* if it contains all vertices of the graph. There are quite a few theorems stating sufficient conditions for the existence of a Hamiltonian cycle in a graph.

Theorem 7.2.1 (V.Chvátal) *A graph with the degree sequence $d_1 \leq d_2 \leq \ldots \leq d_n$, $d_1 \geq 1$, is Hamiltonian if either $d_k > k$ or $d_{n-k} \geq n-k$ for every k, $1 \leq k < n/2$.*

A graph is called *locally-connected* if the neighborhood of every its vertex induces a connected graph.

Theorem 7.2.2 *Let G be a connected locally-connected graph of order $n \geq 3$. If G has no induced subgraphs isomorphic to $K_{1,3}$ then G is Hamiltonian.*

Theorem 7.2.3 *Every 4-connected planar graph is Hamiltonian.*

Theorem 7.2.4 *If G is biconnected then G^2 is Hamiltonian.*

The following theorem gives a necessary condition for the Hamiltonicity.

Theorem 7.2.5 *Let f_k be the number of faces of a plane graph G with exactly k edges at the boundary. If G is Hamiltonian then for every $k \geq 3$ there exist nonnegative integers f'_k, f''_k such that*

$$f'_k + f''_k = f_k,$$

$$\sum_{k \geq 3} f'_k(k-2) = \sum_{k \geq 3} f''_k(k-2).$$

* * *

7.2.1. Find a Hamiltonian cycle a) in the graph shown at Fig. 7.2.1; b) in the dodecahedron graph.

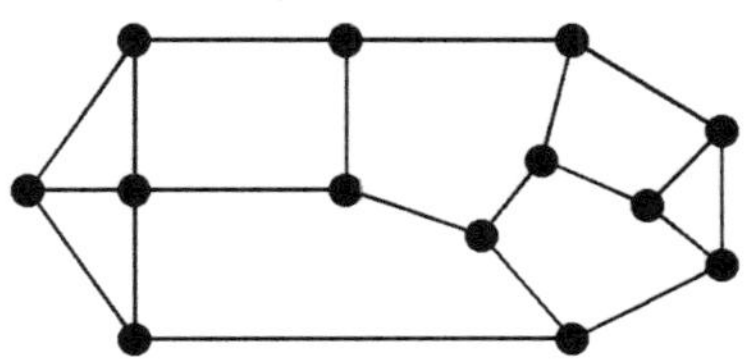

Figure 7.2.1: To Exercise 7.2.1

7.2.2. Is the Petersen graph Hamiltonian?

7.2.3. A graph is called *Hamiltonian-connected* if every two its vertices are connected by a Hamiltonian path. Is Q_3 Hamiltonian-connected?

7.2.4. A non-Hamiltonian graph G is called *hypohamiltonian* if $G - x$ is Hamiltonian for every vertex x. Prove that the Petersen graph is hypohamiltonian.

7.2.5. Find a connected graph G of order $n \geq 3$ such that G^2 is non-Hamiltonian. Find a graph of minimal order with these properties.

7.2.6. Prove that if a cubic graph is Hamiltonian then it contains three edge-disjoint perfect matchings. Is the converse statement true?

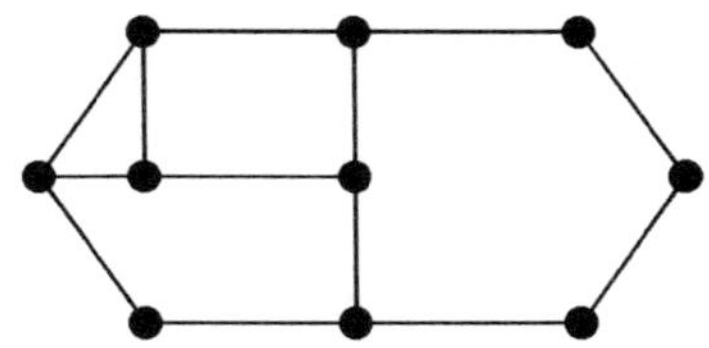

Figure 7.2.2: To Exercise 7.2.7

7.2.7. An n-vertex graph is called *pancyclic* if it contains a simple cycle of length k for every $k = 3, \dots n$. Is the graph shown at Fig. 7.2.2 pancyclic?

7.2.8. Prove that if for a graph G there exists $X \subseteq VG$ such that $k(G - X) > |X|$ then G is non-Hamiltonian. (Here $k(H)$ denotes the number of components of a graph H.)

7.2.9. Prove that if a graph G has an independent vertex subset $X \subseteq VG$ such that $|X| > |N(X)|$ then G is non-Hamiltonian.

7.2.10. Prove that $\alpha_0(G) \leq |G|/2$ for a Hamiltonian graph G.

7.2.11. Let $H_{n,m}$ be an n-vertex graph with the degree sequence

$$P_{n,m} = (\underbrace{m, \dots, m}_{m \text{ times}}, \underbrace{n - m - 1, \dots, n - m - 1}_{n-2m \text{ times}}, \underbrace{n - 1, \dots, n - 1}_{m \text{ times}}).$$

Prove that $H_{n,m}$ is non-Hamiltonian for all m, $0 < m < n/2$.

7.2.12. Let $d_1 \leq d_2 \leq \dots \leq d_n$ be the degree sequence of a graph that does not satisfy the hypotheses of Theorem 7.2.1. Prove the existence of a number m, $0 < m < n/2$, such that $d_i \leq p_{i,m}$, $i = 1, \dots, n$, where $(p_{1,m}, p_{2,m}, \dots, p_{n,m}) = P_{n,m}$ is the sequence defined in the previous exercise.

7.2.13. Prove that an n-vertex graph G is Hamiltonian if $|EG| > \dfrac{(n-2)^2 + n}{2}$.

7.2.14. Prove that every connected locally-connected line graph of order $n > 2$ is Hamiltonian.

7.2.15. Prove that if a graph G is connected and every its edge belongs to a triangle then $L(G)$ is Hamiltonian.

7.2.16. Prove that if a graph G, $|G| \geq 3$, is connected and locally-connected then $L(G)$ is Hamiltonian.

7.2.17. Prove that if a graph G, $|G| \geq 3$, is connected then $L(G^2)$ is Hamiltonian.

7.2.18. For positive integers m, n a graph G is called a *grid graph* if $VG = \{(x, y) : x = 1, \dots, m,\ y = 1, \dots, n\}$ and $ab \in EG$ if and only if $a = (x, y)$, $b = (z, t)$ and $|x - z| + |y - t| = 1$. For which m, n G is Hamiltonian?

7.2.19. Prove that a connected locally connected regular graph of degree 4 is Hamiltonian.

7.2.20. Let an n-vertex graph G satisfy the hypotheses of Theorem 7.2.1. Prove that $|EG| \geq (n-1)^2/8$.

7.2.21. Prove that if $\deg v \geq |G|/2$ for any vertex of a graph G then G is Hamiltonian.

7.2.22. For each $n \geq 4$ construct a non-Hamiltonian $(\lfloor n/2 \rfloor - 1)$-connected n-vertex graph.

7.2.23. For each theorem 7.2.1,7.2.2,7.2.3 find a graph satisfying the hypothesis of the theorem and not satisfying the hypotheses of the remaining ones.

7.2.24. Find a cubic non-Hamiltonian graph in which every pair of nonadjacent vertices is connected by a Hamiltonian path.

7.2.25. Prove that a Hamiltonian-connected graph of order $n \geq 4$ is 3-connected.

7.2.26. Prove that $\Delta(G) \leq (n-1)/2$ for a hypohamiltonian graph G.

7.2.27. Let $t(X)$ denote the minimal number of vertex-disjoint paths covering the subset $X \subseteq VG$ of a graph G. Prove that if $t(X) > |VG \backslash X|$ for some $X \in VG$, then G is non-Hamiltonian.

7.2.28. Construct a biconnected cubic bipartite planar non-Hamiltonian graph.

7.2.29. Construct a triconnected non-Hamiltonian graph of radius 1.

7.2.30. A planar triconnected k-regular graph is called *k-polytope.* Let G be a non-Hamiltonian bipartite 3-polytope with minimal number of vertices. Prove that G has no vertex that belongs only to faces bounded by 4-cycles.

7.2.31. Prove that a graph shown at Fig. 7.2.3 has no Hamiltonian cycle.

7.2.32. A graph is called a *caterpillar* if the deletion of all its pendant vertices produces a path graph. Prove than the square of a caterpillar is Hamiltonian.

7.2.33. Construct a connected locally connected non-Hamiltonian graph with more than two vertices.

7.2.34. Let $P = (p_1, \ldots, p_n)$, $Q = (q_1, \ldots, q_n)$ be degree sequences such that $q_i \geq p_i$, $i = 1, \ldots, n$. Is it true that if every graph with degree sequence P is Hamiltonian then every graph with degree sequence Q is also Hamiltonian?

7.2.35. Prove that if $\deg u + \deg v \geq |G|$ for any two nonadjacent vertices u, v of a graph G then G is Hamiltonian.

7.2.36. Prove that if $\deg u + \deg v \geq |G| - 1$ for any two nonadjacent vertices u, v of a graph G then G contains a Hamiltonian path.

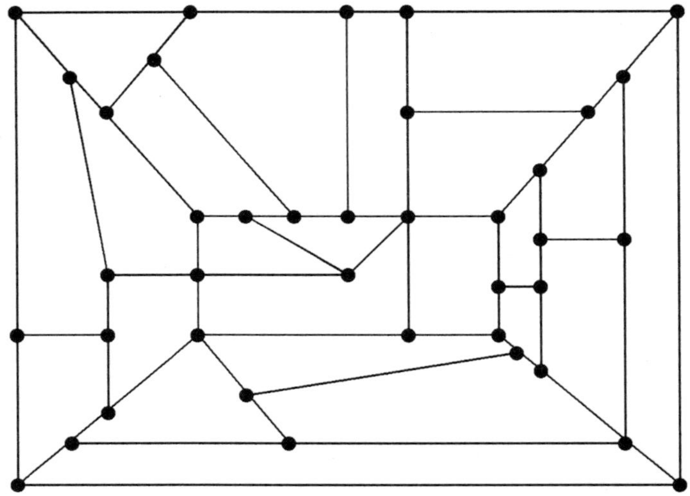

Figure 7.2.3: To Exercise 7.2.31

7.2.37. Suppose that $\deg u + \deg v \geq |G|$ for some two vertices u, v of a graph G. Prove that if G contains a Hamiltonian (u, v)-path then G is Hamiltonian.

7.2.38. For every $n \geq 4$ find an n-vertex non-Hamiltonian graph such that $\deg u + \deg v \geq n - 1$ for any two vertices u, v of a graph G

7.2.39. A graph is called *locally Hamiltonian* if the induced graph $G(N(v))$ is Hamiltonian for every vertex v. Construct a non-Hamiltonian locally Hamiltonian graph.

7.2.40. Let G be a complete n-partite graph, $|G| \geq 3$. Prove that:

a) G is Hamiltonian if and only if $|G| \geq 2\alpha_0(G)$.

b) If G is non-Hamiltonian then it has a cycle of length $2\beta_0(G)$.

7.2.41. Let a graph G have a cycle that contains a vertex covering of the graph. Prove that $L(G)$ is Hamiltonian.

7.2.42. Let I be the incidence matrix of a graph G, considered over the field $\mathbf{Z}_2$. Prove that G is Hamiltonian if and only if I has $|G| - 1$ linearly independent columns such that their sum is also a column of I.

7.2.43. Let ab be an edge of a graph G such that $G - a - b$ is disconnected. Prove then G has no Hamiltonian cycles passing through ab.

7.2.44. Prove that every biconnected outerplanar graph has a single Hamiltonian cycle.

7.2.45. Is it true that for a graph G, $|G| \geq 4$, one of $L(G)$, $L(\overline{G})$ is Hamiltonian?

Chapter 8

Degree Sequences

8.1 Graphical Sequences

A sequence $(d_1, d_2, \ldots d_n)$ of nonnegative integers will be called an *n-sequence* and denoted by the single letter d. The sequence d is called *graphical* if there exists a graph whose degree sequence is d. Such graph is called a *realization* of the sequence d.

A nonincreasing n-sequence d is called *proper* if its sum is even and $d_1 \leq n-1$. Without loss of generality any graphical sequence may be ordered and considered to be proper.

There is a convenient "shorthand" for graphical sequences: we write down only different elements and their multiplicities, for example, $(2^3, 1^2) = (2, 2, 2, 1, 1)$.

As a rule a graphical sequence may have several realizations. For example, $(2^3, 1^2)$ has two nonisomorphic realizations shown at Fig. 8.1.1.

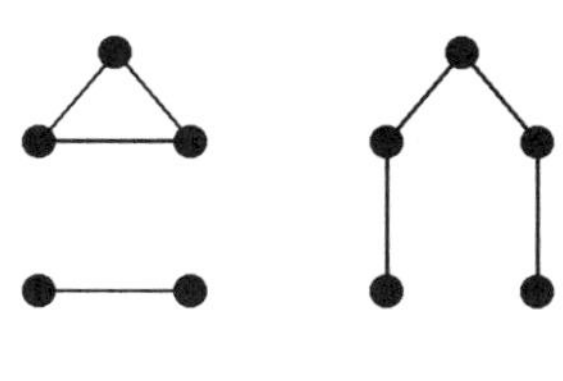

Figure 8.1.1: Two realizations of the same sequence

Let a, b, c, d be four different vertices of a graph G such that $ab, cd \in EG$ and $ac, bd \notin EG$. In such case it is said that G *admits the switching* $s = (ab, cd)$, which is the deletion of the edges ab, cd followed by the addition of the edges ac, bd. The resulting graph $G - ab - cd + ac + bd$ is denoted by sG.

Theorem 8.1.1 (Switching Theorem) *Any realization of a graphical sequence may be obtained from any other one by a finite sequence of switchings.*

If a sequence has a unique (up to isomorphism) realization G then such sequence is called *unigraphical* and G is called a *unigraph.* The structure of unigraphs is known, but its description is cumbersome. We give here only the theorem about unigraphicity for multigraphs.

Theorem 8.1.2 (K.Senior, S.Hakimi) *Let a proper n-sequence d be realized by a multigraph. The realization is unique if and only if one of the following conditions hold:*

a) $n \leq 3$;

b) $d_1 = d_2 + \ldots + d_n$;

c) $d_1 + 2 = d_2 + \ldots + d_{n-1}$, $d_2 = d_3 = \ldots = d_n$;

d) $n = 4$, $d_3 > d_4 = 1$;

e) $d_2 = d_3 = \ldots = d_n = 1$.

For a proper n-sequence, $n > 1$, the *derived sequence* d^i, $1 \leq i \leq n$ is defined as follows. The element d_i is deleted from d and the first d_i remaining elements are decreased by 1.

Theorem 8.1.3 (V.Havel, S.Hakimi) *A proper n-sequence $d \neq (0^n)$ is graphical if and only if every derived sequence d^i, $1 \leq i \leq n$, is graphical.*

Theorem 8.1.4 (P.Erdös, T.Gallai) *A proper n-sequence, $n > 1$, is graphical if and only if*

$$\sum_{i=1}^{k} d_i \leq k(k-1) + \sum_{i=k+1}^{n} \min\{k, d_i\}$$

for all $k = 1, 2, \ldots, n-1$.

Clearly, to verify the graphicity using any of the two latter theorems, n inequalities must be verified in the worst case. However it is often possible to test the graphicity of a sequence much faster, as the following sufficient condition suggests.

Theorem 8.1.5 *A proper n-sequence d without zeroes such that*

$$n \geq \frac{(d_1 + d_n + 1)^2}{4d_n}$$

is graphical.

A *pair* $d = (d_1, \ldots, d_n)$, $c = (c_1, \ldots, c_m)$ of proper sequences is called *graphical* if there exists a bipartite graph (called a *realization of the pair*) for which d, c are the lists of degrees of the vertices of its parts.

Theorem 8.1.6 (D.Gale) *A pair of proper sequences d and c with equal sums of elements is graphical if and only if*

$$\sum_{i=1}^{k} d_i \leq \sum_{i=1}^{m} \min\{k, c_i\}.$$

for all $k = 1, 2, \ldots, n$.

Theorem 8.1.7 *A pair of proper sequences d and c with equal sums of elements is graphical if and only if the sequence*

$$c_1 + m - 1, c_2 + m - 1, \ldots, c_m + m - 1, d_1, d_2, \ldots, d_n$$

is graphical.

The following algorithm called the *l-procedure* may be used for testing a proper n-sequence for graphicity and for constructing a possible realization.

Let V be an n-element set of vertices (of a graph to be constructed) such that every element $v \in V$ is provided with a nonnegative integer label $d(v) < n$. Let further $S(v) \subseteq V$ be a subset of $d(v)$ vertices distinct from v with maximal possible labels ($S(v)$ is not unique).

An *iteration* of the l-procedure consists of the following steps.

1. pick a *leading* vertex with nonzero label; denote it by v;

2. pick a subset $S(v)$;

3. connect the leading vertex by edges with all vertices from $S(v)$;

4. set d(v)=0, set $d(u) = d(u) - 1$ for all $u \in S(v)$.

If after an iteration some label becomes negative, we shall say that the iteration *fails.*

The l-procedure is applied to a proper n-sequence d. Initially it is assumed that $d(v_i) = d_i$ and then the above iterations are successively applied. Two outcomes are possible: either some iteration fails, or we obtain the zero sequence. In the first case d is not graphical. In the second case we obtain a realization of d.

* * *

8.1.1. Find all pairwise nonisomorphic realizations of the following sequences: a) $(4^2, 3^4)$; b) (2^{10}); c) $(5, 4^3, 3^4)$.

8.1.2. Find a sequence with minimal sum of elements that has more than one realization.

8.1.3. a) Prove that if a graph of order n has exactly two vertices of degree k and the degrees of the remaining vertices are different, then $k = \lfloor n/2 \rfloor$.

b) Describe the structure of a graph in which only two vertices are of the same degree.

8.1.4. Show that a proper n-sequence with some coinciding elements is not necessarily graphical.

8.1.5. Prove that the three sequences

$$(d_1, \ldots, d_n),\ (n-d_1-1, n-d_2-1, \ldots, n-d_n-1),\ (n, d_1+1, d_2+1, \ldots, d_n+1)$$

are graphical or not simultaneously and have the same number of realizations.

8.1.6. Let $d = (d_1, \ldots, d_n)$, $c = (c_1, \ldots, c_m)$ be two graphical sequences. Prove that the following sequences are also graphical.

a) the *union* $d \cup c = (d_1, \ldots, d_n, c_1, \ldots, c_m)$;

b) the *join* $d + c = (d_1 + m, \ldots, d_n + m, c_1 + n, \ldots, c_m + n)$.

8.1.7. Let graphical sequences d and c have N and M realizations respectively. Is it true that the number of realizations for the union $d \cup c$ (see the previous exercise) is NM?

8.1.8. A *molecular graph* is a multigraph whose vertices correspond to the atoms of the molecule and the edges correspond to the bonds between the atoms. In general, a molecular formula may correspond to several molecular graphs, which are called *structural isomeres.*

a) Find all structural isomeres of the compound N_2O_3 (the nitrogen N has valency 3 and the oxygen O is of valency 2.)

b) A *saturated hydrocarbonate* is a compound of the hydrogen H (valency 1) and the carbon C (valency 4) whose molecular graph is a tree. Derive the molecular formula C_nH_{2n+2} for saturated hydrocarbonates.

c) Draw the graphs of saturated hydrocarbonates with 1,2,3 and 4 atoms of carbon.

8.1.9. Let

$$d = (d_1, d_2, \ldots, d_n),\ d' = (d'_1, d'_2, \ldots, d'_n)$$

be two graphical sequences with $d_i \geq d'_i$, $i = 1, 2, \ldots, n$. Is it true that there exists a graph G realizing d such that a spanning subgraph of G realizes d'?

8.1.10. Find a sequence of switchings that transforms the graph G shown at Fig. 8.1.2 into $2K_4$.

8.1.11. Suppose that a switching s is applied to a graph G and to its subgraph H. Is it true that if $H \not\cong sH$ then $G \not\cong sG$?

8.1.12. Let the vertices $v_i, \ldots, v_n$ of a graph G of order n have degrees $d_1 \geq \ldots \geq d_n$ respectively. Prove that for every vertex v_i there exists a sequence $s_1, s_2, \ldots, s_t$ of switchings such that the vertex v_i in the graph $s_t \ldots s_2 s_1 G$ is adjacent to the first d_i vertices distinct from v_i.

8.1.13. What are the possible values of the diameter of a graph that does not admit switchings?

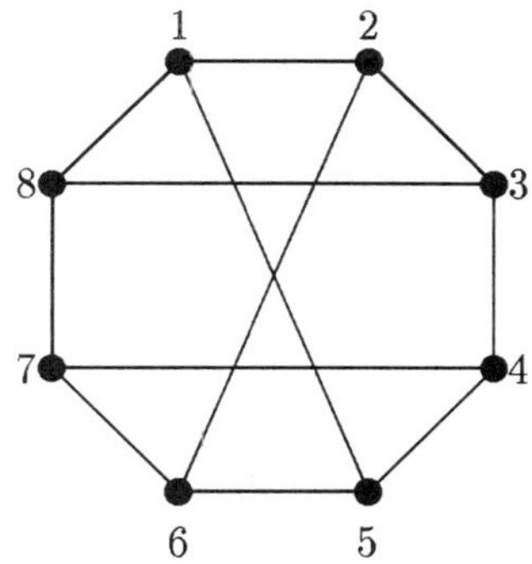

Figure 8.1.2: To Exercise 8.1.10

8.1.14. State and prove the Switching Theorem for a) multigraphs and b) pseudographs, modifying the definition of switching in a proper way.

8.1.15. How can one switching change the number of components of a graph?

8.1.16. Let d be a graphical sequence. The ***graph of realizations*** G_d is defined as follows. Its vertices bijectively correspond to pairwise nonisomorphic realizations of d. Two vertices of G_d are defined to be adjacent if the corresponding realizations may be obtained from each other by a switching.

a) Construct the graphs of realizations for the sequences $d = (2^4, 1^2)$ and $d' = (4, 2^4, 1^2)$.

b) Prove that the graphs of realizations of sequences from Exercise 8.1.5 are isomorphic.

c) Prove that the graph of realizations of any graphical sequence is connected.

d) What is the degree of the vertex corresponding to the cycle C_n in the graph G_d with $d = (2^n)$, $n \geq 3$?

8.1.17. An *alternating cycle* in a graph G is a sequence $C = (v_1, v_2, \ldots, v_{2k}, v_1)$ of vertices that constitutes a cycle in the complete graph with the vertex set VG such that every other edge of C is not in EG *(antiedge)* and the remaining edges are in EG. A *long switching* $s(C)$ is the deletion of the edges of an alternating cycle C from G and the addition of the antiedges of C to G.

a) Prove that any realization of a graphical sequence may be obtained from any other one by a finite sequence of long switchings.

b) Let $s(G, G')$ denote the minimal number of long switchings necessary to produce a realization G' of a graphical n-sequence d from another realization G of the same sequence. Let us denote $s(d) = \max s(G, G')$, where the maximum is over all pairs of realizations of d. Prove that $s(d) \leq \lfloor n/4 \rfloor$.

8.1.18. A *labelled realization* of a graphical n-sequence is a graph G with the vertex set $V = \{v_1, v_2, \ldots, v_n\}$ such that $\deg v_i = d_i$, $i = 1, 2, \ldots, n$. Labelled realizations are considered to be different if $EG \neq EH$.

a) Prove that the Switching Theorem remains valid for labelled realizations.

b) Construct all labelled realizations of the sequence $(2^3, 1^2)$.

c) Give the definition of the *graph of labelled realizations* G^l_d, similar to the graph G_d from Exercise 8.1.16. and construct it for the sequence $(2^2, 1^2)$.

d) In the graph of labelled realizations G^l_d, two vertices are said to be *equivalent* if the corresponding labelled realizations are isomorphic. Let $V_1, \ldots, V_k$ be the resulting equivalence classes. Construct the graph G_d from the graph G^l_d and the partition $V_1 \cup \ldots \cup V_k = V$.

e) Prove that $\mathrm{d}(D^l_d) \geq \mathrm{d}(D_d)$, where $\mathrm{d}(G)$ denotes the diameter of the graph G.

f) Similarly to the Exercise 8.1.17, one may define the "labelled" parameter $s_l(d)$. Let us define further $s_l(n) = \max s_l(d)$, where the maximum is over all graphical n-sequences d. Prove that $s_l(n) = O(n^2)$.

8.1.19. A graph-theoretical property $\mathcal{P}$ is said to be *s-complete* if for any graphical sequence and any two its realizations $G, G' \in \mathcal{P}$ there exists a sequence of switchings transforming G into G' such that all intermediate graphs have the property $\mathcal{P}$.

a) Prove that if a property $\mathcal{P}$ is specified in terms of the graphical sequence of a graph then it is s-complete. Give examples of such properties.

b) Restate the definition of the s-complete property in terms of graphs of realizations.

c) Prove that the properties "to be bipartite graph" and "to be cactus" are not s-complete. (A *cactus* is a connected graph in which no pair of simple cycles have common edges.)

8.1.20. Let nonincreasing n-sequences d, c, b be such that $d_i \leq c_i \leq b_i$, $i = 1, 2, \ldots, n$, the sum of the elements of c is even, and d, b are graphical. Is it true that c is also graphical?

8.1.21. Let $d = (d_1, \ldots, d_n)$, $c = (c_1, \ldots, c_m)$ be two graphical sequences. Is it true that the $(n \times m)$-sequence of the sums $d_i + c_j$, $i, j = 1, 2, \ldots, n$, is also graphical?

8.1.22. Prove the validity of the l-procedure.

8.1.23. Use the l-procedure to test the graphicity of the following sequences. a) $(9^4, 6^2, 8^3, 5^2)$, b) $(9^3, 8^3, 6, 5, 4, 2, 1, 0^{1000})$, c) $(8^7, 5^6, 3^{11})$.

8.1.24. Prove that if a proper graphical n-sequence d is realized by a graph G that has a subgraph H isomorphic to K_p then there exists a realization G' of d with a subgraph H' isomorphic to K_p such that VH' is the set of vertices with first largest degrees $d_1, \ldots, d_p$.

8.1.25. Let a sequence d satisfy the Erdös-Gallai inequalities for all $k = 1, \ldots, k(d)$, where $k(d) = \max\{i : d_i \geq i\}$. Prove that d is graphical.

8.1.26. Use the Erdös-Gallai theorem to test the graphicity of the following sequences.

a) $(7^6, 6^4, 5^6)$;

b) $(12^3, 11, 9^2, 8^2, 6^2, 4^2, 3)$;

8.1.27. Find all nonnegative integers m, n for which the following sequences are graphical.

a) (m^n); b) $(m, 1^n)$; c) (m^n, n^m); d) $((2m)^{2n}, (2m-1)^{2n-1})$.

8.1.28. For which values of n the sequence $n^n, (n-1)^{n-1}, \ldots, 2^2, 1^1$ is graphical?

8.1.29. Let $n_1, n_2, \ldots, n_k$ be natural numbers with the sum equal to n. Prove that the sequence $(d_1^{n_1}, d_2^{n_2}, \ldots, d_k^{n_k})$, $d_i = n - n_i$, $i = 1, 2, \ldots, k$, is graphical.

8.1.30. Prove that a sequence of nonnegative integers is realizable by a pseudograph if and only if the sum of its elements is even.

8.1.31. Prove that a proper n-sequence is realizable by a multigraph if and only if $d_2 + d_3 + \ldots + d_n \geq d_1$.

8.1.32. Let d be a proper n-sequence with $d_j > 0$ for some j. Prove that the sequences d and

$$d' = (d_1 - 1, d_2, \ldots, d_{j-1}, d_j - 1, d_{j+1}, \ldots, d_n)$$

are realizable or not realizable by a multigraph simultaneously.

8.1.33. The *Ferrer diagram* of a proper n-sequence d is the $(0,1)$-matrix of size $n \times n$ constructed as follows. For all i, we set the first d_i elements (excluding the i-th element) of the i-th row to one. The remaining elements are set to zero.

a) Construct the Ferrer diagram for the sequence $(4, 4, 3, 2, 2, 1)$.

b) Let $\overline{d}_i$, $i = 1, 2, \ldots, n$, denote the sum of elements of the i-th column of the Ferrer diagram. The sequence $\overline{d} = \overline{d}_1, \overline{d}_2, \ldots, \overline{d}_n$ is said to be *conjugate* to d. Prove that $\overline{\overline{d}} = d$.

c) Prove the *Fulkerson theorem: a proper n-sequence d is graphical if and only if*

$$\sum_{i=1}^{k} d_i \leq \sum_{i=1}^{k} \overline{d}_i$$

for all $k, = 1, 2, \ldots, n$.

8.1.34. Let d be a proper graphical n-sequence and let d_{n+1} be an even number such that $d_n \geq d_{n+1} \geq 0$. Prove that the sequence $d \cup (d_{n+1})$ (see Exercise 8.1.6) is graphical.

8.1.35. Prove that there exist infinitely many graphical n-sequences for which the inequality from Theorem 8.1.5 turns into the equality.

8.1.36. Prove that if a proper n-sequence d contains n_0 zero elements and $d_1 \leq 2\sqrt{n - n_0} - 2$ then it is graphical.

8.1.37. Consider the class K of sequences whose elements are the numbers between 1000 and 1500 and whose sums of elements are even. Find the minimal number N such that a sequence $d \in K$ is graphical of and only if it length is at least N.

8.1.38. Which of the following pairs are graphical?

a) $(4^4, 2)$, $(5, 4^2, 3, 1^2)$; b) $(5^4, 2^4)$, $(7^2, 6, 5, 3)$.

8.1.39. Propose an algorithm for testing the graphicity of pairs of sequences and constructing a realization (if it exists). Is it possible to apply the l-procedure?

8.1.40. Let $d = (a^n)$ and $c = (b^m)$ be two sequences. For which positive integers a, b and $m, n \geq 1$ the pair (d, c) is graphical?

8.1.41. Prove that an n-sequence d is realizable by a multigraph if and only if the pair (d, c) is graphical, where $c = (2^m)$, $m = (d_1 + \ldots + d_n)/2$.

8.1.42. Let G be a bipartite graph. A switching (ab, cd) is called *bipartite* if a, d are in one part and b, c are in the other part. Using this notion, state and prove the counterpart of the Switching Theorem for graphical pairs of sequences.

8.1.43. For which n, m the graphs a) K_n, b) P_n, c) C_n, d) $K_{m,n}$ are unigraphs?

8.1.44. Is the sequence $(16^3, 14, 12, 11, 8^5, 6^2, 4^2, 3^2)$ unigraphical?

8.1.45. Is it true that a graph is a unigraph if and only if it admits no switchings?

8.1.46. Prove that for $n \equiv 2 \pmod 4$ or $n \equiv 3 \pmod 4$ the number of unigraphs of order n is even.

8.1.47. Find all regular unigraphs of degree two.

8.1.48. Find all selfcomplementary unigraphs of orders $n \leq 7$.

8.1.49. Is the property "to be a unigraph" preserved under the transition to a subgraph?

8.1.50. A set X of vertices of a graph G is called *d-complete* if $\deg u \neq \deg v$ for any $u \in X$, $v \in VG \backslash X$. Prove that for every d-complete subset X of vertices of a unigraph G the subgraph $G(X)$ is a unigraph.

8.1.51. Let $k_1, \ldots, k_l$ be positive integers with sum s. Prove that the sequence

$$(s^s, k_1, \ldots, k_l)$$

is unigraphical.

8.1.52. Prove that a graph is a unigraph if and only if its complement is a unigraph.

8.1.53. Is it true that the join $G + H$ of unigraphs is a unigraph?

8.1.54. Is the conjunction of the properties "to be a bipartite graph" and "to be a unigraph" a hereditary property?

8.1.55. Prove that every disconnected unigraph without isolated vertices has no cycles. Describe the components of such graph.

8.1.56. Which of the following statements are valid?

a) Every component of a unigraph is a unigraph.

b) The disjoint union of unigraphs is a unigraph.

8.1.57. An n-sequence is called *l-unigraphical* if it has exactly one labelled realization (called an *l-unigraph*).

a) Give examples of l-unigraphical n-sequences for all $n \geq 1$.

b) Prove that the property "to be l-unigraphical" is hereditary and find a minimal FIS set.

8.1.58. Prove that if an n-sequence d satisfies the conditions

$$d_1 + 2 = d_2 + \ldots + d_{n-1}, \; d_2 = d_3 = \ldots = d_n$$

then it is realizable by a single multigraph.

8.1.59. The *degree of an edge* is the unordered pair of degrees of its endvertices. The *edge sequence* of a graph is the list of degrees of its edges.

a) Does there exist a graph with edge sequence

$$((2,3)^2, (2,4)^2, (3,3), (3,4)^2)?$$

b) Prove that if a graph has no isolated vertices then it is possible to reconstruct the degree sequence of the graph from its edge sequence.

c) Prove that a graph is not specified up to isomorphism by its edge sequence.

d) Construct an infinite series of graphs which are specified up to isomorphism by their edge sequences but which are not unigraphs.

8.1.60. For a vertex v let $\deg^1(v)$ denote the list of degrees of the vertices of its neighborhood. If $\deg v = 0$ then we set $\deg^1(v) = \emptyset$.

Prove that the sequence $(\deg^1(v) \, : \, v \in VG)$ uniquely specifies the edge sequence.

8.2 $\mathcal{P}$-graphical Sequences

Let $\mathcal{P}$ be a graph-theoretical property and let d be a graphical sequence. Let $\mathcal{I}(d, \mathcal{P})$ denote the set of realizations of d with the property $\mathcal{P}$ and let $\mathcal{I}(d)$ denote the set of all realizations of d. If $\mathcal{I}(d, \mathcal{P}) \neq \emptyset$ then d is called *potentially $\mathcal{P}$-graphical.* If $\mathcal{I}(d, \mathcal{P}) = \mathcal{I}(d)$ then d is called *forcibly $\mathcal{P}$-graphical.*

Characterizations of forcibly and potentially $\mathcal{P}$-graphical sequences are known for many properties $\mathcal{P}$. We give here several theorems.

Theorem 8.2.1 (G.Edmonds) *A proper graphical n-sequence d is realizable by a k-edge-connected graph, $k \geq 2$, if and only if $d_n \geq k$.*

Theorem 8.2.2 (D.Wang, D.Kleitman) *A proper graphical n-sequence d is realizable by a k-connected graph, $k \geq 2$, if and only if $d_n \geq k$ and*

$$\sum_{i=1}^{n} d_i - 2\sum_{i=1}^{k-1} d_i + (k-1)(k-2) \geq 2(n-k).$$

Theorem 8.2.3 (V.Chungphaisan) *If a proper n-sequence d has a realization with a Hamiltonian path starting at a vertex of degree d_i then this realization is produced by the l-procedure in which at the first iteration the leading vertex is a vertex of degree d_i and at all remaining iterations the leading vertex is the one with the minimal label in the set $S(v)$, where v is the vertex leading at the previous iteration.*

Theorem 8.2.4 (V.Chungphaisan) *A proper n-sequence d without zeroes is realizable by a Hamiltonian graph if and only if there exists a realization of d with a Hamiltonian path starting at a vertex of degree d_1.*

A graphical sequence is called *$\mathcal{P}$-unigraphical* if $|\mathcal{I}(d, \mathcal{P})| = 1$. Notice that the set of $\mathcal{P}$-unigraphical sequences contains all unigraphical sequences whose realizations have the property $\mathcal{P}$.

* * *

8.2.1. Prove that a graphical sequence is forcibly $\mathcal{P}$-graphical if and only if it is not potentially $\overline{\mathcal{P}}$-graphical, where $\overline{\mathcal{P}}$ is the negation of $\mathcal{P}$.

8.2.2. Let $\mathcal{P}_1$, $\mathcal{P}_2$ be two graph-theoretical properties. Interpret the following statements in terms of potentially and forcibly $\mathcal{P}_1$- and $\mathcal{P}_2$-graphical sequences.

a) The sequence is forcibly $(\mathcal{P}_1 \& \mathcal{P}_2)$-graphical.

b) The sequence is potentially $(\mathcal{P}_1 \& \mathcal{P}_2)$-graphical.

c) The sequence is forcibly $(\mathcal{P}_1 \vee \mathcal{P}_2)$-graphical.

d) The sequence is potentially $(\mathcal{P}_1 \vee \mathcal{P}_2)$-graphical.

8.2.3. Let a class of graphs with property $\mathcal{P}$ be contained in class of graphs with property $\mathcal{P}'$. How are the following pairs of classes related to each other?

a) Classes of potentially $\mathcal{P}$-graphical and potentially $\mathcal{P}'$-graphical sequences.

b) Classes of forcibly $\mathcal{P}$-graphical and forcibly $\mathcal{P}'$-graphical sequences.

8.2.4. Let A, C, and T be the properties of "being acyclic", "being connected", and "being tree" respectively. Let P_X and F_X, $X = A, C, T$, denote respectively the sets of potentially and forcibly X-graphical sequences. Prove that

a) $F_T = F_A \cap F_C$; a) $P_T = P_A \cap P_C$; a) $F_A \cap P_C = P_A \cap F_C = F_T$.

8.2.5. For which properties $\mathcal{P}$ the sets of potentially and forcibly $\mathcal{P}$-graphical sequences coincide?

8.2.6. Prove that for every graph-theoretical property $\mathcal{P}$ the set of all graphical sequences is partitioned into three subsets:

1) forcibly $\mathcal{P}$-graphical sequences;

2) forcibly $\overline{\mathcal{P}}$-graphical sequences;

3) the sequences that are simultaneously $\mathcal{P}$- and $\overline{\mathcal{P}}$-graphical.

8.2.7. Design algorithms for recognition of graphical sequences which are realizable by the complete k-partite graph in the following cases:

a) the number k is known;

b) the sizes $n_1, \ldots, n_k$ of parts are known;

c) the number k is unknown.

8.2.8. Prove that a proper graphical n-sequence without zeroes is potentially connected if and only if $\sum_{i=1}^{n} d_i \geq 2(n-1)$.

8.2.9. a) Prove that a proper n-sequence without zeroes is realizable by a tree if and only if $\sum_{i=1}^{n} d_i = 2(n-1)$.

b) Let a sequence d be realizable by a tree. Design an algorithm for the construction of a tree realizing d.

c) Find necessary and sufficient conditions under which the sequence is realizable by a unique tree.

8.2.10. Find necessary and sufficient conditions under which the sequence is realizable by a graph with at least one cycle.

8.2.11. Test whether the sequence $(7, 6, 5^2, 4, 3, 2^3)$ is realizable by a Hamiltonian graph.

8.2.12. A graph is called *unicyclic* if it is connected and has exactly one cycle. Prove that a proper n-sequence d without zeroes is potentially unicyclic if and only if $n \geq 3$, $d_3 > 1$, and $\sum_{i=1}^{n} d_i = 2n$.

8.2.13. Find all graphical n-sequence d with $d_1 = d_2 = n-1 \geq 3$ that are realizable by line graphs.

8.2.14. Let d be a potentially acyclic n-sequence.

a) Prove that $d_1^2 + \ldots + d_n^2 \leq n^2 - n$.

b) Give an infinite series of examples for which this inequality turns into the equality.

8.2.15. Let d be a proper potentially planar n-sequence. Prove that $d_1 + d_2 + d_3 \leq 2n - 2$.

8.2.16. Characterize the forcibly bipartite sequences.

8.2.17. Prove that a proper n-sequence d without zeroes is forcibly realizable by a tree only in the following two cases:

a) $n = 2$ and $d_1 = d_2 = 1$;

b) $n \geq 3$, $d_3 = 1$, and $d_1 + d_2 = n$.

8.2.18. Characterize the forcibly unicyclic sequences.

8.2.19. Is the sequence $(6^2, 3^6)$ a) forcibly planar; a) potentially planar?

8.2.20. Characterize proper forcibly outerplanar n-sequences d such that $d_1 = n - 1$.

8.2.21. Let d, c be two graphical sequences. We shall say that c is a *hereditary subsequence* of d (and we shall denote this relation by $c \leq d$) if there exist realizations G and H of d and c respectively such that H is isomorphic to an induced subgraph of G.

a) Which of the three sequences $d = (3^6)$, $c = (3^4)$, $b = (2, 1^2)$ are related by " $\leq$ "?

b) Prove that the relation " $\leq$ " defines a partial order on the set of graphical sequences.

8.2.22. Let $\mathcal{P}$ be a hereditary graph-theoretical property and let $I_m(\mathcal{P})$ be the corresponding minimal FIS set. Let $d_m(\mathcal{P})$ denote the set of degree sequences of graphs from $I_m(\mathcal{P})$ and let $d_m^0(\mathcal{P})$ denote the set of minimal elements of $d_m(\mathcal{P})$ with respect to the partial order " $\leq$ " (see Exercise 8.2.21).

a) Find a property $\mathcal{P}$ such that $d_m^0(\mathcal{P}) \neq d_m(\mathcal{P})$.

b) Prove that a graphical sequence d is forcibly $\mathcal{P}$-graphical if and only if $c \leq d$ (see Exercise 8.2.21) for no sequence $c \in d_m^0(\mathcal{P})$.

8.2.23. The following *Rao Conjecture* is known: for every hereditary property $\mathcal{P}$ the set $d_m^0(\mathcal{P})$ is finite. Verify its validity for a) acyclic graphs; b) bipartite graphs.

8.2.24. Let $U_{\mathcal{P}}$ denote the set of unigraphical sequences whose realizations are with property $\mathcal{P}$ and let $\mathcal{P}_U$ denote the set of $\mathcal{P}$-unigraphical sequences.

a) prove that $U_{\mathcal{P}} \subseteq \mathcal{P}_U$.

b) Find examples showing that these two sets may coincide, as well as be different.

8.2.25. Let T denote the property "being tree". Characterize the T-unigraphical sequences.

8.2.26. Suppose that a graph-theoretical property $\mathcal{P}$ is specified by the degree sequence of a graph. Prove that $\mathcal{P}_U = U_{\mathcal{P}}$ (see Exercise 8.2.24). Is the converse true?

8.2.27. Suppose that every graph that has a property $\mathcal{P}$ also has a property $\mathcal{P}'$. Is it true that $\mathcal{P}_U \subseteq \mathcal{P}'_U$ (see Exercise 8.2.24)?

8.3 Split and Threshold Graphs

This section considers two properties that may be determined from the degree sequence of a graph: splitness and thresholdness.

A graph G is called *split* if there exists a partition $VG = A \cup B$, where A is a clique and B is an independent set. A is called the *upper part* of G, and B is called the *lower part* of G. One of these parts may be empty.

Theorem 8.3.1 (P.Hammer, B.Simeone) *Let d be a proper graphical n-sequence. A realization G of d is a split graph if and only if*

$$\sum_{i=1}^{m} d_i = \sum_{i=m+1}^{n} d_i + m(m-1),$$

where $m = m(d) = \max\{i : d_i \geq i-1\}$.

An interesting subclass of split graphs is the threshold graphs. Let the vertices of a graph G be indexed by the numbers $1, 2, \ldots, n$, i.e., $VG = \{1, 2, \ldots, n\}$. Let IG denote the set of all independent sets of vertices of G plus the empty set. If there exist nonnegative numbers $\alpha_1, \alpha_2, \ldots, \alpha_n, \beta$ such that the set of $(0,1)$-solutions of the inequality

$$\alpha_1 x_1 + \alpha_2 x_2 + \ldots + \alpha_n x_n \leq \beta \tag{8.3.1}$$

is the set of characteristic vectors of the elements of IG then G is called a *threshold graph* and (8.3.1) is called the *separating inequality* (see Exercise 3.1.37).

Let $K \circ G$ and $O \circ G$ denote the graphs obtained from G by the addition of a dominating and an isolated vertex respectively.

Theorem 8.3.2 *A graph G is a threshold graph if and only if it is of the form $G = G_1 \circ (G_2 \circ (\ldots \circ G_n) \ldots)$, where G_n is K_1 and for $i = 1, 2, \ldots, n-1$ the terms G_i are either K or O.*

* * *

8.3.1. Prove that for $n \leq 3$ all graphs are split.

8.3.2. Is the graph shown at Fig. 8.3.1 split?

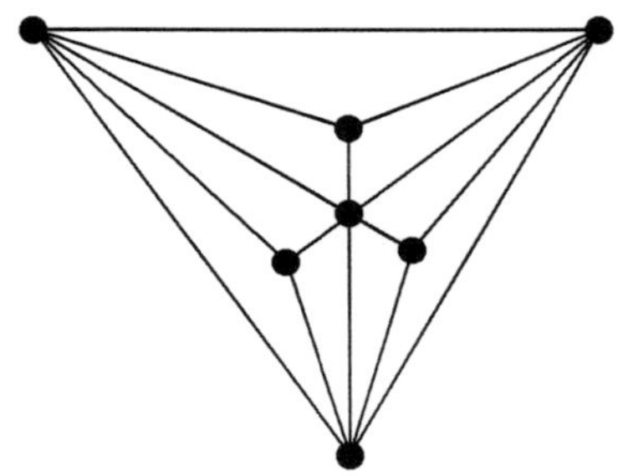

Figure 8.3.1: To Exercise 8.3.2

8.3.3. Give examples of pairs of nonisomorphic split graphs with the same degree sequences.

8.3.4. Which of the following sequences are realizable by split graphs?

a) $(8, 7, 6^2, 5, 4, 3, 2, 1^3)$; b)$(7, 6, 5^2, 4, 3^2, 1^2)$.

8.3.5. Prove that the graph G is split if and only if its complement $\overline{G}$ is split.

8.3.6. Let A, B be the upper and lower parts of a split graph G. Prove that $\deg u \geq \deg v$ for any $u \in A,\ v \in B$.

8.3.7. How many partitions into parts may a split graph of order n have?

8.3.8. Prove that if one realization of a graphical sequence is split then all its realizations are split.

8.3.9. Prove that the splitness property is hereditary and find a corresponding minimal FIS set.

8.3.10. Construct an algorithm for

a) recognition of split graphs by the degree sequence;

b) construction of all partitions of a split graph into parts.

8.3.11. Reduce the problem of recognition of isomorphism of arbitrary graphs to the isomorphism problem for split graphs.

8.3.12. Is it true that every split selfcomplementary graph has parts of equal size?

8.3.13. Find all split trees.

8.3.14. Which regular graphs are split?

8.3.15. Prove that every threshold graph is split, but not conversely.

8.3.16. Which graphs of order four are threshold? Find their separating inequalities.

8.3.17. Give a geometric interpretation of the separating inequality and of the thresholdness of a graph.

8.3.18. Prove that if a graph G is threshold then $G+K_1$ and $G \cup K_1$ are threshold.

8.3.19. Prove that the vertices of a threshold graph may be numbered in such a way that the closed neighborhood $N(v) \cup \{v\}$ of any vertex v contains neighborhoods of all other vertices with smaller numbers.

8.3.20. Design algorithms for recognition of

a) threshold graphs;

b) graphical sequences that have a threshold realization.

8.3.21. Prove that the property "to be threshold graph" is hereditary and find a corresponding minimal FIS set.

8.3.22. Design an algorithm for the construction of a separating inequality for a threshold graph.

8.3.23. Prove that the number of threshold graphs of order n is 2^{n-1}.

8.3.24. Which graphs without triangles are threshold?

8.3.25. Prove that

a) threshold graphs are unigraphs;

b) the l-unigraphs are the threshold graphs.

8.3.26. Find all selfcomplementary threshold graphs.

8.3.27. Prove that a graph G is threshold if and only if for every subset $S \subseteq VG$ there exists a vertex $u \in S$ such that it is either adjacent to all vertices of $S - \{u\}$, or it is adjacent to none of them.

8.3.28. Prove that a graph G is threshold if and only if there exists a labelling of the vertices $c: VG \to \mathbf{N}$ and an integer t such that

$$(uv \in EG) \iff (u \neq v,\ c(u) + c(v) > t).$$

8.3.29. Prove that a split graph is threshold if and only if it has no induced subgraphs isomorphic to P_4.

8.3.30. Prove that a split graph G is threshold if and only if the vertices of its parts may be ordered $A = \{u_1, \ldots, u_l\}$, $B = \{v_1, \ldots, v_m\}$ in such a way that

$$(u_i v_j \in EG) \iff (u_{i'} v_{j'} \in EG,\ i' = i, \ldots, l,\ j' = j, \ldots, m).$$

8.3.31. Let a graph G be represented as the union $G = \bigcup_{i=1}^{t} G_i$ of threshold graphs with the same vertex set. This representation is called *threshold decomposition.* Prove that

a) every graph admits a threshold decomposition;

b) there exists a threshold decomposition of G in which $t \leq |VG|$.

8.3.32. The minimal number of components over the threshold decompositions of a graph G is called the *threshold number* of G and is denoted by $\mathrm{th}(G)$. Prove that

a) $\mathrm{th}(G)$ is the minimal value of t such that there exists a system of t inequalities

$$\begin{cases} a_{11}x_1 + a_{12}x_2 + \ldots + a_{1n}x_n \le b_1, \\ \quad \ldots \\ a_{t1}x_1 + a_{t2}x_2 + \ldots + a_{tn}x_n \le b_t, \end{cases}$$

such that $I \in IG$ if and only if the characteristic vector of the set I is a $(0,1)$-solution of the system;

b) if H is an induced subgraph of a graph G then $\mathrm{th}(H) \le \mathrm{th}(G)$.

8.3.33. Determine a) $\mathrm{th}(C_n)$, $n > 3$; b) $\mathrm{th}(K_{m,n})$; c) $\mathrm{th}(P_n)$.

8.3.34. Prove the following statements. (α_0, α_1, β_0 are introduced in Sect. 3.1.)

a) $\mathrm{th}(G) + \alpha_0(G) \le |G|$ for every nonempty graph G.

b) If a graph G is triangle-free then $\mathrm{th}(G) = \beta_0(G)$.

c) $\mathrm{th}(G) = \alpha_1(G)$ for bipartite graphs.

8.3.35. Using the graphs shown at Fig. 8.3.2 prove that the equality $\mathrm{th}(G) = \mathrm{th}(\overline{G})$ is not always true.

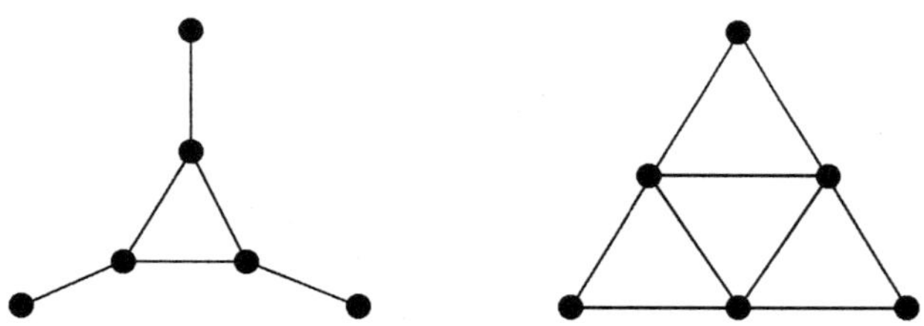

Figure 8.3.2: To Exercise 8.3.35

8.4 Degree Sets and Arity Partitions

Let us define two more objects basing on the degree sequence, namely the degree set (the set of degrees of vertices) and the arity partition (the list of the multiplicities of the degrees).

Suppose that a graphical sequence d realizable by a graph G is written in the form $d = (k_1^{p_1}, k_2^{p_2}, \ldots, k_n^{p_n})$, where $k_i \ne k_j$ for $i \ne j$ and p_i is the multiplicity of k_i. Then the set $S = \{k_1, \ldots, k_n\}$ is called the *degree set* of the graph G and the list $F = (p_1, \ldots, p_n)$ is called the *arity partition* of G. Accordingly, the graph G is called a *realization* of the set S and the list F. Notice that the arity partition, as well as the degree sequence, is the *multiset,* i.e., it is considered up to permutations of the elements, and the elements may repeat.

* * *

8.4.1. Construct realizations of the sets $\{4\}$, $\{4,2\}$, $\{4,1,0\}$ with minimal numbers of vertices.

8.4.2. Prove that

a) any finite set S of positive integers is the degree set of some graph;

b) the minimal order of a realization of a set S is $k_1 + 1$, where k_1 is the maximal element of S.

8.4.3. Prove that the claim of the previous exercise remains valid

a) in the class of connected graphs;

b) in the class of threshold graphs.

8.4.4. Let $S = \{k_1, \ldots, k_n\}$ be a set of positive integers such that $k_1 > \ldots > k_n$.

a) Prove that if $k_n = 1$ then S is realizable by a tree.

b) Find the minimal order of such tree.

c) Design an algorithm for the construction of a tree realization of minimal order.

8.4.5. Prove that two threshold graphs without isolated vertices are isomorphic if and only if their degree sets coincide.

8.4.6. Does there exist a graph which is uniquely specified by its degree set?

8.4.7. Let $S = \{k_1, \ldots, k_n\}$ be a set of integers such that $k_1 > \ldots > k_n \geq 3$. Prove that there exists a Hamiltonian graph of order $k_1 + 1$ realizing S.

8.4.8. Find necessary and sufficient conditions under which a degree set S is realized by a planar graph.

8.4.9. Suppose that $k_1 > k_2$ and let $\mu_p(k_1, k_2)$ denote the minimal number of vertices among planar realizations of the degree set $\{k_1, k_2\}$. Prove that

a)$\mu_p(k_1, k_2) = k_1 + 1$ if $1 \leq k_2 \leq 3$,

b) $\mu_p(k_1, k_2) = k_1 + 2$ if $k_2 = 4$,

c) $\mu_p(k_1, k_2) \leq 2k_1 + 2$ if $k_2 = 5$.

8.4.10. For which positive integers k, n there exists a graph of order n realizing the degree set $\{1, 2k\}$?

8.4.11. Is it true that every set of nonnegative integers is realizable by a bipartite graph?

8.4.12. Let S_1, S_2 be two sets of positive integers. Prove that

a) there exists a bipartite graph such that S_1, S_2 are the degree sets of its parts;

b) if $S_1 = \{l\}$, $S_2 = \{k_1, \ldots, k_n\}$ then the minimal order of graphs from item a is at most $nl + \sum_{i=1}^{n} k_i$.

8.4.13. A graph with the degree set $S = \{k_1, \dots, k_n\}$ and girth g is called a $(k_1, \dots, k_n; g)$-*cell.*

Construct a) a $(3, 4; 5)$-cell of order 13; b) a $(3, 4; 6)$-cell of order 18.

8.4.14. A graph G is called *homotraceable* if for any its vertex v there exists a Hamiltonian path starting at v. Construct homotraceable non-Hamiltonian graphs with degree sets a) $\{2, 3\}$; b) $\{2\}$; c) $\{3\}$.

8.4.15. For $n \geq 3$ find arity partitions of the graphs a) P_n; b) C_n; c) $K_{m,n}$.

8.4.16. Let d be a graphical sequence and let S, F be the corresponding degree set and arity partition respectively. Is it true that the set of realizations of d is the intersection of the sets of realizations of S and F?

8.4.17. Prove that

a) a graph and its complement have equal arity partitions;

b) there are no graphs with arity partition $(1, 1, \dots, 1)$.

8.4.18. Find necessary and sufficient conditions under which a list $(p_1, \dots, p_n)$ of positive integers is the arity partition of a connected graph.

8.4.19. Is it true that for any $n \geq 2$ there exists a graph of order n that is uniquely specified by its arity partition?

8.4.20. Is it true that every list $(p_1, \dots, p_n) \neq (1^n)$ of positive integers is the arity partition of a threshold graph?

Chapter 9

Graph Colorings

9.1 Vertex Coloring

A *proper coloring* of a graph G in k colors, or *k-coloring*, is a function $f : VG \to \{1, 2, \ldots, k\}$ such that $f(u) \neq f(v)$ for any adjacent vertices u, v of G. The number $f(u)$ is called the *color* of the vertex u. A graph that has a proper k-coloring is called *k-colorable.*

Notice that a k-coloring does not necessarily use all k colors. The set of vertices of the same color is called *color class.*

The minimal k such that a graph G is k-colorable is called the *chromatic number* of G and is denoted by $\chi(G)$. If $\chi(G) = k$ then G is called *k-chromatic.* A proper $\chi(G)$-coloring of a graph G is called its *minimal coloring.*

Clearly, a graph is 1-colorable if and only if it is empty, and it is 2-colorable (bicolorable) if and only if it is bipartite.

Theorem 9.1.1 (A.Korshunov) *The chromatic number of almost all graphs G of order n satisfies the relation $\chi(G) \sim \frac{n}{2\log_2 n}$.*

Theorem 9.1.2 *For every graph G, $\chi(G) \leq 1 + \max \delta(H)$, where the maximum is over all induced subgraphs H of G.*

Corollary 9.1.3 *For every graph G, $\chi(G) \leq 1 + \Delta(G)$.*

Theorem 9.1.4 (R.Brooks) *If G is a non-complete connected graph with $\Delta(G) \geq 3$ then $\chi(G) \leq \Delta(G)$.*

Theorem 9.1.5 (E.Nordhouse, J.Gaddum) *For an n-vertex graph G and its complement the following inequalities hold:*

$$2\sqrt{n} \leq \chi + \overline{\chi} \leq n + 1, \quad n \leq \chi\overline{\chi} \leq (n+1)^2/4,$$

where $\chi = \chi(G)$, $\overline{\chi} = \chi(\overline{G})$.

A. Zykov proved that there exist triangle-free graphs with arbitrarily large chromatic number. P.Erdös generalized this result as follows.

Theorem 9.1.6 *For any positive integers g, χ there exists a χ-chromatic graph with girth at least g.*

Recall that the girth is the length of the shortest cycle in a graph.

A labelled graph G is called *uniquely k-colorable* if $\chi(G) = k$ and all minimal colorings induce the same partition of the vertex set into color classes up to the numbering of colors.

Theorem 9.1.7 *Every uniquely k-colorable graph is $(k-1)$-connected.*

Theorem 9.1.8 *For every $k \geq 3$ there exists a uniquely k-colorable graph without complete subgraphs of order k.*

A graph G is called *critical* if $\chi(G - v) < \chi(G)$ for any its vertex v. If in addition $\chi(G) = k$ then G is called *k-critical.*

Characterizations of k-critical graphs for $k \geq 4$ are unknown. Only some their properties are studied.

The *circumference* of a graph is the length of the longest cycle of the graph.

Theorem 9.1.9 (H.Dirac) *If G is a k-critical graph, $k \geq 3$, then either G is Hamiltonian or its circumference at least $2k - 2$.*

A graph G is called *edge-critical* if $\chi(G - e) < \chi(G)$ for any its edge e. If in addition $\chi(G) = k$ then G is called *k-edge-critical.*

Theorem 9.1.10 *If a k-edge-critical (n, m)-graph G is not complete, and $k \geq 4$, then $2m \geq (k-1)n + k - 3$.*

* * *

9.1.1. Consider the following algorithm of *sequential coloring* of the vertices of a graph.

1. An arbitrary vertex v_1 obtains color 1.
2. If the vertices $v_1, v_2, \ldots, v_i$ are already colored then an arbitrary next vertex v_{i+1} obtains the minimal color not used for the vertices of its neighborhood.

a) Prove that the resulting coloring is proper.

b) Prove that a sequential coloring of a complete k-partite graph is minimal.

c) Give an example showing that the difference between the number of colors in a sequential coloring and the chromatic number may be arbitrarily large.

9.1.2. Consider one more algorithm of coloring of the vertices of a graph.

1. Set $G_1 = G$.
2. If the graph G_k is already constructed, select a maximal cardinality independent vertex set I_k in it and color the vertices of I_k in color k.

3. Set $G_{k+1} = G_k - I_k$.
4. Repeat the two last steps until all vertices become colored.

a) Prove that this algorithm does not necessarily produce a minimal coloring.

b) Construct a sequence of bichromatic graphs $H_1, H_2, \dots$ such that the described algorithm applied to H_i produces a coloring of the graph with at least i colors.

9.1.3. Prove that in some minimal coloring of a graph G the neighborhood $N(v)$ of every vertex v is colored in the same color if and only if G is bipartite.

9.1.4. Find the chromatic numbers of graphs a) K_n; b) $K_{n,m}$; c) P_n; d) C_n; e) Petersen graph.

9.1.5. Prove that if every block of a graph is k-colorable then the graph is k-colorable.

9.1.6. Prove that the difference $\Delta(G) - \chi(G)$ may be arbitrarily large.

9.1.7. Find all counterexamples to the proposition: every connected graph G contains a vertex v such that $\deg v \geq \chi(G)$.

9.1.8. Prove that for every minimal coloring of the vertices of a connected graph which is neither a complete graph nor C_{3k} the graph has two vertices of the same color at distance 2.

9.1.9. Prove that for any two different colors i, j of any minimal coloring of the vertices of a graph G there exists an edge $uv \in EG$ such that u has color i and v has color j.

9.1.10. A *chess piece graph* is a graph whose vertices are the squares of the chessboard and the edges are the pairs of squares linked by a move of the given piece.

a) Find chromatic numbers for the castle, the knight, the bishop, and the king graphs.

b) Prove that $\chi(G_Q) \geq 8$ for the queen graph G_Q.

c) Extend the problem a) to the $(n \times n)$-chessboard case.

9.1.11. Prove that in every proper coloring of a line graph every vertex is adjacent to at most two vertices of the same color.

9.1.12. Prove that the octahedron graph $L(K_4)$ cannot be properly colored in four colors in such a way that every two-colored subgraph has no cycles.

9.1.13. Let a graph G have a proper coloring in which every color is used at least twice. Prove that G has a coloring in $\chi(G)$ colors with the same property.

9.1.14. Prove that $\chi(G) \geq \dfrac{n^2}{n^2 - 2m}$ for an (n, m)-graph G.

9.1.15. Prove that $\chi(G) \geq \frac{n}{n-r}$ for an r-regular graph G of order n.

9.1.16. Prove that the number of edges of a graph G is at least $\chi(G)(\chi(G)-1)/2$.

9.1.17. Let $(d_1, \ldots, d_n)$ be the degree sequence of a graph G, $d_1 \geq \ldots \geq d_n$. Prove that

$$\chi(G) \leq \max_{1 \leq i \leq n} \min\{i, d_i + 1\}.$$

9.1.18. a) Prove that $n/\alpha_0 \leq \chi(G) \leq n - \alpha_0 + 1$ for every graph G, where $n = |G|$, α_0 is the independence number of G.

b) Characterize the graphs with $\chi(G) = n - \alpha_0 + 1$.

9.1.19. Is it true that for any positive integers n, α_0, χ that satisfy the inequalities from the Exercise 9.1.18a there exists an n-vertex graph with independence number α_0 and chromatic number χ?

9.1.20. Prove that the chromatic number $\chi(G)$ is equal to the minimal number m such that $\alpha_0(G \times K_m) = |G|$, where α_0 denotes the independence number and $G_1 \times G_2$ denotes the Cartesian product of graphs defined in Sect. 1.4.

9.1.21. Is it true that if $\deg v < \chi(G) - 1$ for some vertex v of a graph G then $\chi(G) = \chi(G - v)$?

9.1.22. Prove that $\chi(G_1 + G_2) = \chi(G_1) + \chi(G_2)$ holds for all graphs G_1, G_2, where $G_1 + G_2$ denotes the join of graphs defined in Sect. 1.4.

9.1.23. a) Prove that $\chi(G_1 \cup G_2) \leq \chi(G_1)\chi(G_2)$ even if G_1, G_2 have common vertices.

b) Give an example when $\chi(G_1 \cup G_2) = \chi(G_1)\chi(G_2)$.

c) Express the chromatic number $\chi(G_1 \cup G_2)$ for the disjoint union of G_1 and G_2 in terms of $\chi(G_1)$ and $\chi(G_2)$.

9.1.24. Prove the Nordhouse-Gaddum Theorem 9.1.5.

9.1.25. Construct infinite series of graphs such that

a) $\chi + \overline{\chi} = n + 1$ and $\chi\overline{\chi} = n$;

b) $\chi + \overline{\chi} = 2\sqrt{n}$ and $\chi\overline{\chi} = n$;

c) $\chi + \overline{\chi} = n + 1$ and $\chi\overline{\chi} = (n+1)^2/4$,

where $\chi, \overline{\chi}$ are the chromatic numbers of a graph and its complement.

9.1.26. Prove that there are no nontrivial graphs such that $\chi + \overline{\chi} = 2\sqrt{n}$ and $\chi\overline{\chi} = (n+1)^2/4$.

9.1.27. The merging of two nonadjacent vertices of a graph into one is called an *elementary homomorphism* of the graph. Prove that $\chi(G) \leq \chi(\epsilon G) \leq \chi(G) + 1$ for any graph G and for any elementary homomorphism ϵ of G.

9.1.28. The merging of two adjacent vertices of a graph into one is called an *elementary contraction* of the graph.

a) Prove that $|\chi(G) - \chi(\nu G)| \leq 1$ for any graph G and for any elementary contraction ν of G.

b) Give examples of graphs for all three possible values of $\chi(G) - \chi(\nu G)$ for an elementary contraction ν.

9.1.29. Prove the following statements:

a) For every nonempty graph G there exists a partition $VG = V_1 \cup V_2$ such that $\chi(G(V_1)) + \chi(G(V_2)) = \chi(G)$.

b) For every non-complete graph G there exists a partition $VG = V_1 \cup V_2$ such that $\chi(G(V_1)) + \chi(G(V_2)) > \chi(G)$.

9.1.30. Let a graph G admit a partition $VG = V_1 \cup \ldots \cup V_k$ such that for any i, j, $1 \leq i < j \leq k$, there exist nonadjacent vertices $x \in V_i$, $y \in V_j$. Prove that $\chi(G) \leq |G| - k + 1$.

9.1.31. Prove that $\chi(G) \leq l + 1$, where l is the length of the longest simple path of G.

9.1.32. Construct 4-regular 4-chromatic graphs with girth a) 4; b) 5.

9.1.33. Which graphs shown at Fig. 9.1.1 are uniquely 3-colorable?

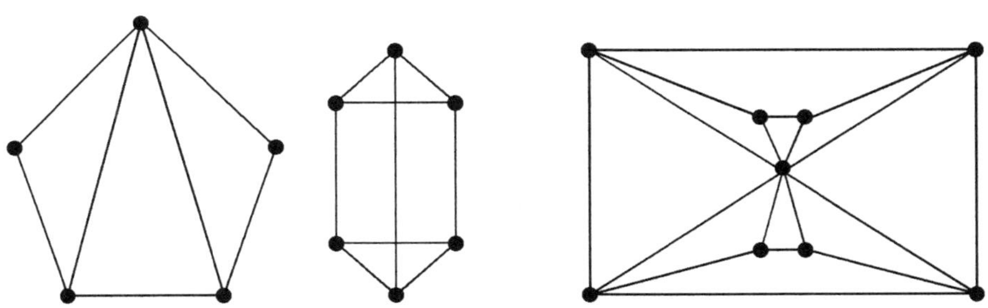

G_1 G_2 G_3

Figure 9.1.1: To Exercise 9.1.33

9.1.34. Is it true that the union of any l color classes of a uniquely k-colorable graph, $1 \leq l \leq k$, induces a uniquely l-colorable graph?

9.1.35. Construct a uniquely 3-colorable graph without triangles.

9.1.36. Prove that if G is a uniquely k-colorable graph then $\delta(G) \geq k - 1$.

9.1.37. Prove that for k-coloring of a uniquely k-colorable graph the subgraph induced by the union of any two color classes is connected.

9.1.38. Prove that a uniquely k-colorable graph of order n has at least $(k-1)n - C_k^2$ edges.

9.1.39. Prove that if a graph G is connected and $\chi(G) > k$ then $G * K_k$ is a uniquely k-colorable graph, where "$*$" denotes here the *weak product of graphs* defined as follows: $V(G * H) = VG \times VH$, $E(G * H) = \{((u_1, u_2), (v_1, v_2)) : u_1 v_1 \in EG,\ u_2 v_2 \in EH\}$.

9.1.40. The *merging* of G_1 and G_2 *by a common subgraph* K is the operation producing a graph $H = G_1' \cup G_2'$ such that $G_1' \cong G_1$, $G_2' \cong G_2$, and $H(VG_1' \cap VG_2') \cong K$. The set of *k-trees* is defined inductively.

1. K_{k+1} is a k-tree.
2. If G is a k-tree then G' obtained by merging of G and K_{k+1} by a complete subgraph of order k is also a k-tree.

Is it true that the k-trees are uniquely $(k+1)$-colorable?

9.1.41. Prove that if $\chi(G - u - v) = \chi(G) - 2$ for every pair u, v of vertices of a graph G, then G is a complete graph.

9.1.42. Prove that $\chi(G - v) = \chi(G) - 1$ for every vertex v of a critical graph G.

9.1.43. Find all 2-critical and 3-critical graphs.

9.1.44. Prove that critical graphs other than K_2 are biconnected.

9.1.45. Prove that $\delta(G) \geq k - 1$ for a k-critical graph G.

9.1.46. Prove that $\chi(G - I) = \chi(G) - 1$ for any independent set I of vertices of a critical graph G.

9.1.47. Is it true that every k-chromatic graph, $k \geq 2$, contains a k-critical subgraph?

9.1.48. Prove that the join $G_1 + G_2$ is a critical graph if and only if both G_1 and G_2 are critical graphs.

9.1.49. Let u, v be two nonadjacent vertices of a non-complete k-critical graph G. Prove that there exists a k-coloring of G such that the vertices u, v belong to a) the same color class; b) different color classes.

9.1.50. Which of the following statements are valid?

a) Every edge-critical graph is critical.

b) Every critical graph is edge-critical.

9.1.51. Find all 3-edge-critical graphs.

9.1.52. Is it true that the join (defined in Sect. 1.4) of a k_1- and a k_2-edge-critical graph is a $(k_1 + k_2)$-edge-critical graph?

9.1.53. Construct 6-edge-critical graphs of order $4n-2$, $n \geq 3$ with $\delta(G) > 2n$.

9.1.54. Let G be an edge-critical graph. Let us construct a graph G' as follows. For every vertex $v \in VG$ we introduce a new vertex v' and connect it with all vertices from $N(v)$. Then we introduce one more vertex w and connect it with all vertices of type v'. Prove that G' is edge-critical and $\chi(G') = \chi(G) + 1$.

9.2 Chromatic Polynomial

Two proper t-colorings $\varphi : VG \to \{1, 2, \ldots, t\}$ and $\xi : VG \to \{1, 2, \ldots, t\}$ of a labelled graph G are considered to be *different* if there exists a vertex $v \in VG$ such that $\varphi(v) \neq \xi(v)$. The number of different proper t-colorings of G is denoted by $f(G, t)$ and is called the *chromatic function of graph G*. Clearly, if $t < \chi(G)$ then $f(G, t) = 0$.

The chromatic function of the complete graph K_n is easily computable:

$$f(K_n, t) = \begin{cases} t(t-1)\ldots(t-n+1), & \text{if } t \geq n; \\ 0, & \text{otherwise.} \end{cases}$$

Theorem 9.2.1 *Let u, v be nonadjacent vertices of a graph G. If $G_1 = G + uv$ and G_2 is produced by merging u and v then $f(G, t) = f(G_1, t) + f(G_2, t)$.*

Since $|EG_1| > |EG|$ and $|VG_2| < |VG|$, multiple application of Theorem 9.2.1 eventually reduces the computation of $f(G, t)$ to the computation of chromatic polynomials of complete graphs (the technique due to Zykov). Noticing that $f(K_n, t)$ is a polynomial in t, we have a corollary.

Corollary 9.2.2 *The chromatic function $f(G, t)$ of a graph G is a polynomial in t.*

Therefore $f(G, t)$ is usually called the *chromatic polynomial* of G.

Theorem 9.2.3 *If G is an (n, m)-graph with k connected components $G_1, \ldots, G_k$ then*

(i) $f(G, t)$ is of degree n;

(ii) the coefficient at t^n in $f(G, t)$ is 1;

(iii) the coefficient at t^{n-1} in $f(G, t)$ is $-m$;

(iv) $f(G, t) = \prod_{i=1}^{k} f(G_i, t)$;

(v) $a_k t^k$ is the lowest term of $f(G, t)$ with nonzero coefficient; in particular, the constant term of $f(G, t)$ is zero.

A graph is called *chromatically unique* if it is specified by its chromatic polynomial up to isomorphism.

* * *

9.2.1. Find the chromatic polynomials of the graphs shown at Fig. 9.2.1.

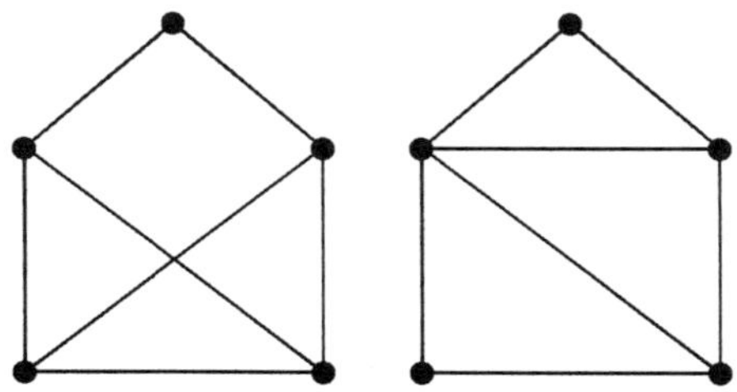

Figure 9.2.1: To Exercise 9.2.1

9.2.2. Prove that

a) $f(K_n, t) = t(t-1)\dots(t-n+1)$, if $t \geq n$;

b) $f(O_n, t) = t^n$.

9.2.3. Find the chromatic polynomials of a) C_n, $n \geq 3$; b) the wheel $W_n = K_1 + C_n$, $n \geq 3$.

9.2.4. Does there exist a graph with the chromatic polynomial a) $t^4 - 3t^3 + 3t^2$? b) $t^4 - 3t^3 + 2t^2$?

9.2.5. Prove that if $G_1, \dots, G_k$ are the connected components of a graph G then $f(G, t) = \prod_{i=1}^{k} f(G_i, t)$.

9.2.6. Let a graph G be obtained from graphs G_1, G_2 by merging by a common subgraph K_n (as defined in the Exercise 9.1.40). Express $f(G, t)$ in terms of $f(G_1, t)$, $f(G_2, t)$.

9.2.7. Prove the Theorem 9.2.1

9.2.8. Prove that a graph G of order n is a tree if and only if $f(G, t) = t(t-1)^{n-1}$.

9.2.9. Prove that $f(G, t) \leq t(t-1)^{n-1}$ for any n-vertex graph G and any positive integer t.

9.2.10. Prove that the maximal real root of a chromatic polynomial of order n is at most $n - 1$.

9.2.11. Prove that if for a vertex v of a graph G the neighborhood $N(v)$ induces the complete subgraph then $f(G, t) = (t - \deg v) f(G - v, t)$.

9.3 Edge Coloring

An *edge coloring* of a graph G in k colors is a function $\varphi : EG \to \{1, 2, \dots, k\}$ such that adjacent edges are colored differently. The number $\varphi(e)$ is called the *color* of the edge e. An *edge k-coloring* of a graph uses at most k colors. The *chromatic index* $\chi' = \chi'(G)$ of a graph G is the minimal k such that G has an edge k-coloring. Other notions concerning edge colorings (*edge color class, minimal edge coloring,* etc.) are defined similarly to the vertex-related case.

Theorem 9.3.1 (V.Vizing) *For every graph G, $\Delta(G) \le \chi'(G) \le \Delta(G)+1$.*

* * *

9.3.1. Find all minimal edge colorings of graphs shown at Fig. 9.3.1. (Colorings differ up to renaming the colors.)

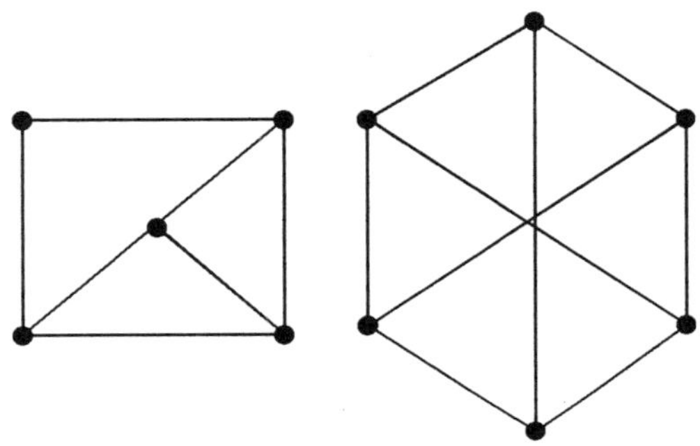

Figure 9.3.1: To Exercise 9.3.1

9.3.2. Evaluate a) $\chi'(C_n)$, $n \ge 3$; b) $\chi'(K_n)$, $n \ge 2$.

9.3.3. Design an algorithm constructing a minimal edge coloring of K_n.

9.3.4. Fig. 9.3.2 shows two edge colorings of a graph G with $\chi'(G) = 4$ colors. Is it possible to convert one coloring into another by means of recoloring of maximal under inclusion bicolored paths?

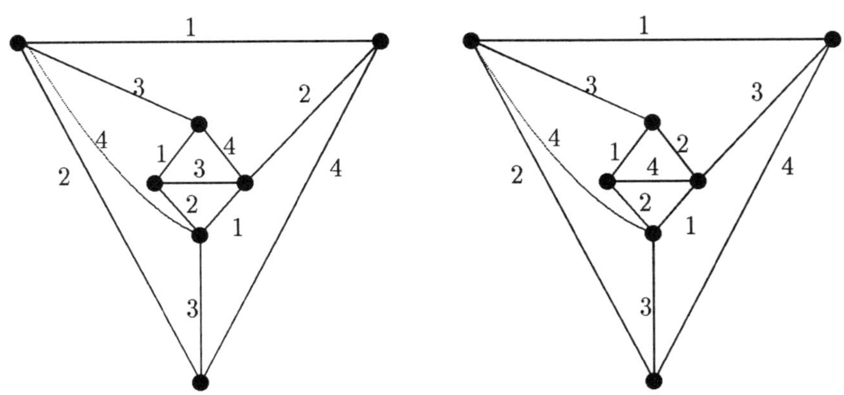

Figure 9.3.2: To Exercise 9.3.4

9.3.5. Describe the graphs with $\chi' \le 2$.

9.3.6. Are the following equalities valid?

a) $\chi(G) = \chi'(L(G))$. b) $\chi'(G) = \chi(L(G))$.

9.3.7. Design an algorithm for finding a minimal edge coloring of a tree.

9.3.8. Prove that $\chi'(G) = \Delta(G)$ for a bipartite graph G.

9.3.9. Let G be a connected graph which is not a simple cycle of odd length. Prove that $\chi'(G) = \Delta(G)$ if the lengths of all simple cycles of G are of the same parity.

9.3.10. Let a cubic graph G have exactly one edge coloring with $\chi'(G)$ colors. Prove that G has three Hamiltonian cycles and $\chi'(G) = 3$.

9.3.11. Prove that $\chi'(G) = \Delta(G) + 1$ for nonempty regular graphs of odd order.

9.3.12. A graph G is called *isochromatic* if $\chi(G) = \chi'(G)$. Find all isochromatic graphs among a) complete graphs; b) bipartite graphs.

9.3.13. Find a cubic graph G of minimal order with $\chi'(G) = 4$.

9.3.14. A graph is called ***uniquely edge-colorable*** if all its minimal edge colorings induce the same partition of the edge set. Construct a uniquely edge-colorable graph with $\chi'(G) = \Delta(G) + 1$.

9.3.15. We shall say that a graph G is of *class* 1 if $\chi'(G) = \Delta(G)$ and of *class* 2 if $\chi'(G) = \Delta(G) + 1$. A graph is called χ'-critical if it is connected, belongs to class 2 and the deletion of any edge decreases the chromatic index.

a) Construct a χ'-critical graph G of fifth order with $\Delta(G) = 3$.

b) For what values of n K_n is χ'-critical?

c) Prove that there are no χ'-critical graphs of orders 4 or 6 with $\Delta(G) = 3$.

9.3.16. If in the definition of the χ'-critical graph we replace edges by vertices, we obtain the definition of the χ'-vertex-critical graph.

a) Is it true that every χ'-vertex-critical graph is critical?

b) For which n the graph K_n is χ'--vertex-critical?

c) Is it true that if a graph G is χ'-vertex-critical then $\chi'(G - v) = \chi'(G) - 1$ for any $v \in VG$?

9.4 Colorings of Planar Graphs

A connected plane graph without bridges is called a ***planar map.*** Two faces of a map are said to be *adjacent* if they have at least one edge in common. A *k-coloring* of a map is a function f from the set of faces of the map into the set $\{1, 2, \ldots, k\}$ of colors such that the colors of adjacent faces are different. A map is called *k-colorable* if it has a k-coloring.

The following conjecture has been known for more than 100 years.

Four-Color Conjecture: *Every planar map is four-colorable.*

The following theorem allows to cast it in terms of graphs.

Theorem 9.4.1 *A map G is k-colorable if and only if its geometric dual graph is vertex-k-colorable.*

In 1890 P.Heawood proved the "Five-Color Conjecture":

Theorem 9.4.2 *Every planar graph is 5-colorable.*

In 1976 K.Appel and W.Haken proved the Four-Color Conjecture with the help of the computer.

Below is a generalization of the Four-Color Conjecture.

Hadwiger Conjecture, 1943: *Every connected n-chromatic graph is contractible to K_n.*

The notions of a planar map and its coloring are naturally extendable to other surfaces. The *chromatic number* $\chi(S)$ of a surface S is the number $\max\chi(G)$, where the maximum is over all graphs embeddable onto S.

Theorem 9.4.3 (Ringel, Youngs) *If S_p is the sphere with $p > 0$ handles then*

$$\chi(S_p) = \left\lfloor (7 + \sqrt{1 + 48p})/2 \right\rfloor .$$

For $p = 0$ this formula amounts to the Four-Color Conjecture.

* * *

9.4.1. What is the minimal number of colors to color the map shown at Fig. 9.4.1?

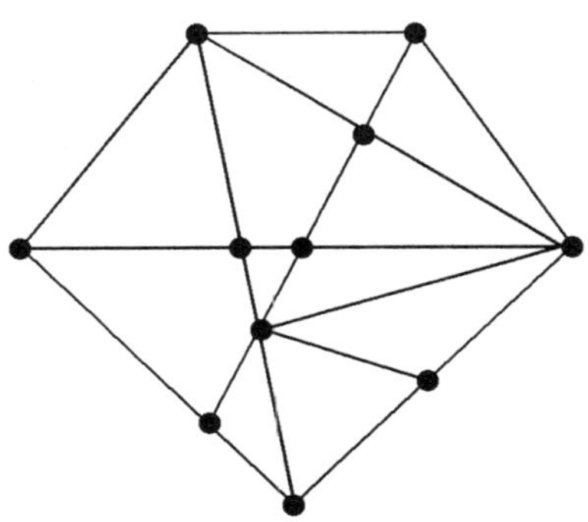

Figure 9.4.1: To Exercise 9.4.1

9.4.2. Construct a map with minimal number of faces that cannot be colored in less than four colors.

9.4.3. Let A be an arrangement of nonoverlapping but possibly tangent disks of equal radius in the plane. Let $\chi(A)$ denote the minimal number of colors to color the disks in such a way that tangent disks are colored differently. Prove that $\chi(A) \leq 4$ for all A.

9.4.4. Prove that a map produced by the subdivision of the plane by a set of arbitrary circles is colorable in two colors.

9.4.5. Does the statement of the previous exercise remain valid if we replace circles by straight lines?

9.4.6. Prove that a planar map admits a coloring in two colors if and only if the degree of every vertex is even.

9.4.7. Prove that every outerplanar graph is tricolorable.

9.4.8. Consider an arrangement of straight lines in the plane such that no three lines pass through the same line. Assume that the intersection points are the vertices of a graph whose edges are the line segments between the neighboring intersection points. Prove that the resulting plane graph is tricolorable.

9.4.9. Prove that for a planar graph G with $\delta(G) < \Delta(G) = 5$ the chromatic number is at most four.

9.4.10. Prove that a Hamiltonian planar map is 4-colorable.

9.4.11. Prove that the Four-Color Conjecture is true for planar cubic graphs.

9.4.12. Prove that it is sufficient to prove the Four-Color Conjecture for planar bridgeless cubic maps.

9.4.13. Prove that a planar 3-regular 2-connected map is 4-colorable if and only if its chromatic index is 3.

9.4.14. Prove that every uniquely four-colorable planar graph is a maximal planar graph.

9.4.15. Verify the validity of the Hadwiger conjecture for $n = 1, 2, 3$.

9.4.16. Prove that the Hadwiger conjecture for $n = 5$ implies the Four-Color Conjecture.

Note: The converse is also true (the Wagner Theorem).

9.4.17. Does Theorem 9.4.1 remain valid for an arbitrary surface?

9.4.18. Prove that the chromatic number of the sphere is equal to the chromatic number of the plane.

9.4.19. Prove that $\chi(T) \geq 7$ for the torus T.

9.5 Perfect Graphs

A graph G is called *χ-perfect* if $\chi(H) = \varphi(H)$ for every induced subgraph H of G, where φ denotes the toughness.

Similarly, G is called *α-perfect* if $\alpha_0(G) = c(H)$ for every induced subgraph H of G, where c denotes the clique covering number.

Theorem 9.5.1 (L.Lovasz, D.Fulkerson) *The following statements are equivalent for a graph G :*

1. *G is χ-perfect;*
2. *G is α-perfect;*
3. *Every nonempty induced subgraph H of G has an independent set I such that $\varphi(H - I) < \varphi(H)$.*

By virtue of the above theorem we may simply speak about *perfect graphs.*

Corollary 9.5.2 *If a graph is perfect then its complement is also perfect.*

The most famous conjecture about perfect graphs is the following one.

Strong Berge Conjecture: *A graph is perfect if and only if neither it nor its complement contain induced subgraphs C_{2k+1}, $k \geq 2$.*

A graph G is called *triangulated* if it has no induced subgraphs C_k, $k \geq 4$.

Theorem 9.5.3 *Every triangulated graph is perfect.*

Theorem 9.5.4 *A connected graph is triangulated if and only if every its minimal under inclusion separating set of vertices is a clique.*

The following theorem is a tool for the analysis of triangulated graphs. A vertex of a graph is called *simplicial* if its neighborhood induces a complete graph.

Theorem 9.5.5 *Every triangulated graph has at least two simplicial vertices.*

Let the vertices of a graph G be ordered as follows: $v_0, v_1, \ldots, v_{n-1}$. Let us construct the sequence of sets

$$Y_0 = VG,\ Y_i = Y_{i-1} \backslash \{v_{i-1}\},\ \ i = 1, 2, \ldots, n-1.$$

A *dismantling* of a graph G is a sequence of graphs obtained form G by a successive deletion of the vertices $v_0, v_1, \ldots, v_{n-1}$. A dismantling is called *perfect* if every vertex v_i is simplicial in the subgraph $G(Y_i)$. A dismantling is called *normal* if every vertex v_i is simplicial either in $G(Y_i)$ or in its complement.

It is known that a triangulated graph admits a perfect dismantling. A graph admitting a normal dismantling is called *quasitriangulated.*

* * *

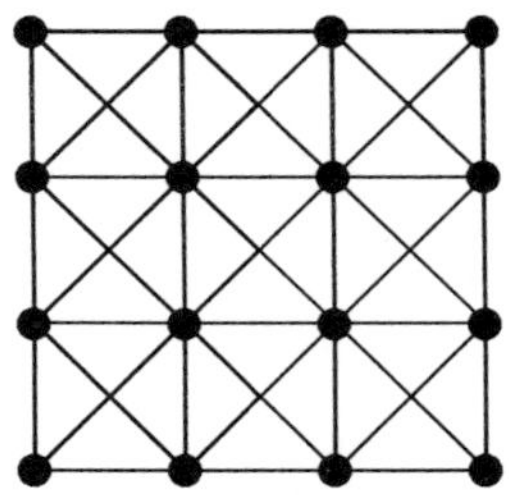

Figure 9.5.1: To Exercise 9.5.1

9.5.1. Is the graph shown at Fig. 9.5.1 perfect?

9.5.2. Verify directly that the graphs C_{2k+1} and $\overline{C}_{2k+1}$, $n \geq 2$, are not perfect.

9.5.3. Prove that the join (see Sect. 1.4) of perfect graphs is perfect.

9.5.4. Let G, H be two perfect graphs whose intersection is a complete graph. Prove that $G \cup H$ is perfect.

9.5.5. Prove that the following graphs are perfect: a) bipartite graphs; b) split graphs.

9.5.6. Prove that $\chi(\overline{G}) = \varphi(\overline{G})$ for any bipartite graph G.

9.5.7. Characterize the set of triangle-free perfect graphs.

9.5.8. Let $(P, \leq)$ be a finite partially ordered set. A *comparability graph* G is the graph with $VG = P$ such that $x, y \in VG$ are adjacent if and only if either $x \geq y$ or $y \geq x$.

a) Prove that $\chi(G) = \varphi(G)$.

b) Is a comparability graph perfect?

9.5.9. The *strong product* $G_1 \cdot G_2$ of graphs G_1, G_2 is the graph with $V(G_1 \cdot G_2) = VG_1 \times VG_2$ and $((u_1, u_2), (v_1, v_2)) \in E(G_1 \cdot G_2)$ if and only if

- either $u_1v_1 \in EG_1$ and $u_2v_2 \in EG_2$, or
- $u_1 = v_1$ and $u_2v_2 \in EG_2$, or
- $u_1v_1 \in EG_1$ and $u_2 = v_2$.

a) Prove that if $\chi(G_i) = \varphi(G_i)$, $i = 1, 2$, then $\chi(G_1 \cdot G_2) = \varphi(G_1 \cdot G_2)$.

b) Is it true that the strong product of perfect graphs is perfect?

9.5.10. Prove that a graph G is perfect if and only if every its induced subgraph H has an independent vertex set I which intersects all maximum cardinality cliques of H.

9.5.11. Let $I = \{I_1, \ldots, I_n\}$ be a collection of closed segments of the real line. The *interval graph* of the collection I is the intersection graph of I, i.e., its vertices bijectively correspond to the segments from I and two vertices are adjacent if and only if the corresponding segments intersect.

a) Prove that the property "to be an interval graph" is hereditary.

b) Prove that interval graphs are perfect.

9.5.12. Which of graphs shown at Fig. 9.5.2 are triangulated?

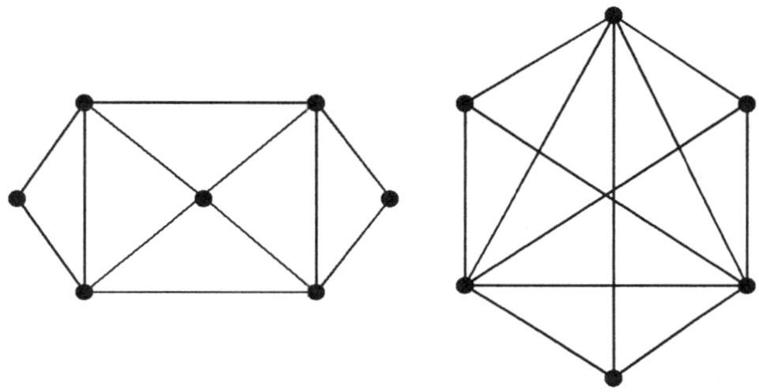

G H

Figure 9.5.2: To Exercise 9.5.12

9.5.13. Is it true that a graph G is triangulated if and only if every its induced subgraph has a simplicial vertex?

9.5.14. a) Is it true that every plane triangulation is a triangulated graph?

b) Is it true that every plane triangulated graph is a plane triangulation?

9.5.15. Under what conditions a bipartite graph is triangulated?

9.5.16. Prove that every split graph is triangulated.

9.5.17. Is it true that a complement of a triangulated graph is triangulated?

9.5.18. Prove that the join $G_1 + G_2$ is a triangulated graph if both G_1 and G_2 are triangulated graphs.

9.5.19. Prove the Theorem 9.5.5.

9.5.20. Prove that the interval graphs are triangulated, but not conversely.

9.5.21. The *Vizing-Wilf number* $w(G)$ of a graph G is the value $w(G) = \max \delta(H)$, where the maximum is over all induced subgraphs H of G. Prove that a graph G is triangulated if and only if $w(H) = \varphi(H) - 1$ for every induced subgraph H of G.

9.5.22. Let $T_1, \ldots, T_k$, $k \geq 2$, be a family of subtrees of a tree T. Prove that

a) if every two subtrees of the family have a common vertex then all subtrees of the family have a common vertex;

b) the intersection graph of the family of subtrees is a triangulated graph.

9.5.23. Prove that the comparability graph (see Exercise 9.5.8) has no induced cycles C_{2n+1}, $n \geq 2$.

9.5.24. Prove that a graph is triangulated if and only if it admits a perfect dismantling.

9.5.25. Prove that every simplicial vertex of a triangulated graph may be the first vertex of a perfect dismantling.

9.5.26. Prove that every subgraph of a quasitriangulated graph is quasitriangulated.

9.5.27. Find a graph which is not quasitriangulated but every its component is quasitriangulated.

9.5.28. Prove that every quasitriangulated graph is perfect.

Chapter 10

Directed Graphs

10.1 Directed Graphs: Basic Notions

Let V be a finite nonempty set and let V^2 be its Cartesian square. A *directed graph,* or *digraph* is a pair (V, A), where $A \subseteq V^2$. The elements of V are called the *vertices* of the digraph, and the elements of A are called its *arcs.* The sets of vertices and arcs of a digraph G are denoted by VG and AG respectively. The number $|VG|$ is called the *order of the digraph* G and is also denoted by $|G|$. If $|G| = n$ and $|AG| = m$ then G is said to be an (n, m)-digraph.

If $x = (u, v)$ is an arc then the vertices u, v are called the *endvertices,* or *ends* of the arc; u is the *tail* of the arc, v is the *head* of the arc. It is said that an arc is *incident* to its ends. It is also said that an arc *outgoes* from its tail and *ingoes* into its head. An arc with coinciding head and tail, i.e., of kind (u, u), is called a *loop.* One may consider digraphs (*multidigraphs*) that have several arcs with common head and common tail. Such arcs are called *multiple,* or *parallel.*

In drawings, an arc is depicted by an arrowed line pointing from the tail and to the head. An example digraph is shown at Fig. 10.1.1. At this figure, x_7, x_8 are parallel arcs and x_9 is a loop.

Two vertices of a digraph are called *adjacent* if they are the ends of the same arc. Arcs are called *adjacent* if they have a common endvertex.

The digraph $(V, \emptyset)$ is called *empty.* The digraph (V, V^2) is called *complete.*

The digraph $\overleftarrow{G}$ obtained from a digraph G by reversing the directions of all arcs is called the *digraph opposite to* G.

The set of heads of the arcs outgoing from a vertex v is denoted by Γv. The set of tails of the arcs ingoing into a vertex v is denoted by $\Gamma^{-1} v$. For any subset $W \subseteq VG$ we denote

$$\Gamma W = \bigcup_{w \in W} \Gamma w, \quad \Gamma^{-1} W = \bigcup_{w \in W} \Gamma^{-1} w.$$

The *outdegree* $d^+(v)$ of a vertex v is the number of arcs outgoing from it. If there are no parallel arcs, $d^+(v) = |\Gamma v|$.

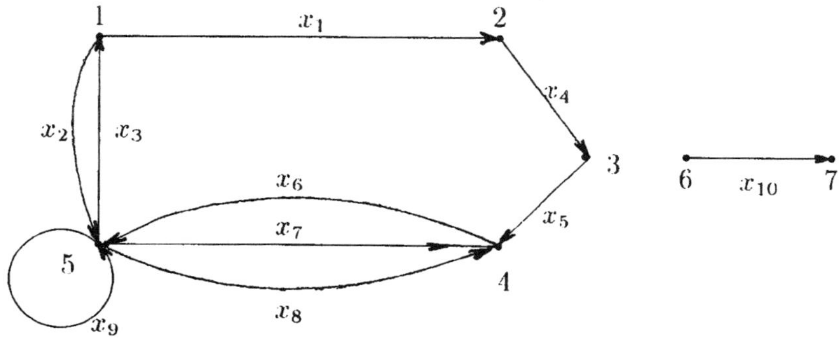

Figure 10.1.1: Example digraph

The *indegree* $d^-(v)$ of a vertex v is the number of arcs ingoing into it. If there are no parallel arcs, $d^-(v) = |\Gamma^{-1}v|$. The *degree* $\deg v$ of a vertex v of a digraph is the number of arcs incident to it, i.e., $\deg v = d^+(v) + d^-(v)$.

Theorem 10.1.1 *The sum of indegrees of the vertices of a digraph G is equal to the sum of outdegrees and is equal to the number of arcs of G :*

$$\sum_{v \in VG} d^+(v) = \sum_{v \in VG} d^-(v) = |AG|.$$

Many graph-theoretical notions, such as isomorphism, connectivity, may be naturally extended to digraphs.

* * *

10.1.1. Find the maximal number of arcs in a digraph of order n without parallel arcs.

10.1.2. Prove that for any integer numbers n, m, $0 \le m \le n^2$ there exists an (n, m)-digraph without parallel arcs.

10.1.3. Find the number of different labelled digraphs of order n without parallel arcs.

10.1.4. For every positive integer n find a connected digraph of order n isomorphic to the opposite digraph.

10.1.5. Prove that the automorphism groups of a digraph and of the opposite digraph coincide.

10.1.6. Find the sets Γv_i and $\Gamma^{-1} v_i$ for the vertices v_1, v_2, v_3 for the digraphs shown at Fig. 10.1.2. Find also the sets $\Gamma(v_1, v_2)$, $\Gamma^{-1}(v_1, v_2)$, $\Gamma(v_1, v_2, v_3)$, $\Gamma^{-1}(v_1, v_2, v_3)$ for them.

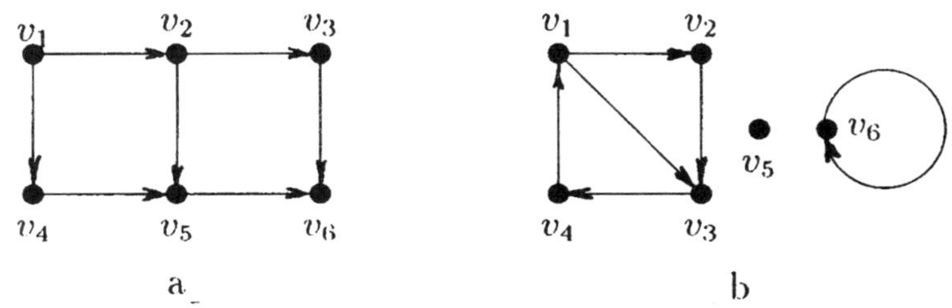

Figure 10.1.2: To Exercise 10.1.6

10.1.7. Describe the structure of a digraph of order n without parallel arcs such that every its vertex satisfies one of the following conditions: a) $d^+(v) = 0$; b) $d^-(v) = 0$; c) $d^+(v) = n$; d) $d^-(v) = n$.

10.1.8. Prove that for isomorphic digraphs

a) the orders are equal;

b) the numbers of arcs are equal;

c) the collections of indegrees and outdegrees coincide.

Are these conditions together sufficient for the isomorphism of digraphs?

10.1.9. Which of digraphs shown at Fig. 10.1.3 are isomorphic?

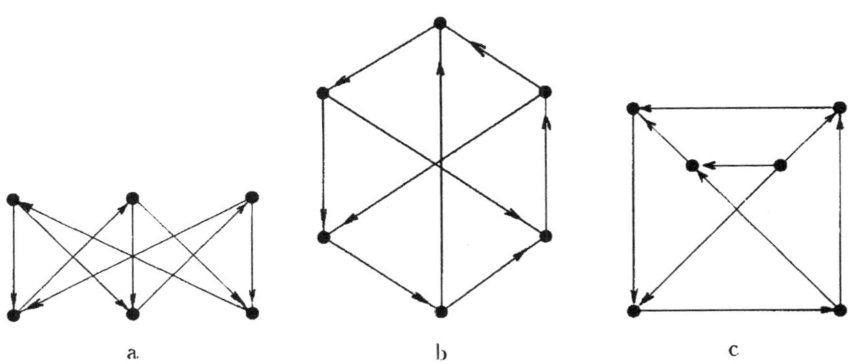

Figure 10.1.3: To Exercise 10.1.9

10.1.10. Find all pairwise nonisomorphic digraphs of third order without loops and parallel arcs.

10.1.11. The *line digraph* for a digraph G is the digraph $L(G)$ such that its vertices bijectively correspond to the arcs of G and $(u, v) \in AL(G)$ if and only if the head of u is the tail of v in G.

a) Express the numbers of vertices and arcs of the line digraph $L(G)$ in terms of the parameters of G.

b) Prove that any isomorphism of G and H induces an isomorphism of their line digraphs $L(G)$ and $L(H)$.

10.1.12. Prove that the conditions $\sum_{i=1}^{n} c_i = \sum_{i=1}^{n} d_i$, $0 \leq c_i \leq n$, $0 \leq d_i \leq n$, $i = 1, 2, \ldots, n$, are insufficient for the existence of a digraph without parallel arcs with the indegrees $(c_1, \ldots, c_n)$ and outdegrees $(d_1, \ldots, d_n)$.

10.1.13. Construct all digraphs without loops and parallel arcs for which the list $(3, 2, 1, 1)$ is the list both of the indegrees and the outdegrees.

10.2 Reachability and Components

A *walk*, or a (u, v)-*walk* in a digraph G is an alternating sequence

$$S = (v_0, x_1, v_1, x_2, \ldots, x_n, v_n), \; v_0 = u, v_n = v, \tag{10.2.1}$$

of vertices v_i and arcs x_i of G such that $x_i = (v_{i-1}, v_i)$, $i = 1, 2, \ldots, n$. The vertices v_0, v_n are called the *terminal vertices* and the remaining vertices of the walk (10.2.1) are called *inner*, or *intermediate*. The *length of the walk* is the number of arcs contained in it. The length of the shortest (u, v)-walk is called the *distance from u to v* and is denoted by $\rho(u, v)$. A walk is called a *chain* if all its arcs are different. A walk is called a *path* if all its vertices are different, except that the terminal ones may coincide.

A walk may be specified by the sequence of its arcs; in a digraph with no parallel arcs a walk may be also specified by the sequence of its vertices.

A walk (chain, path) is called *cyclic* if its terminal vertices coincide, otherwise it is called *noncyclic*. A cyclic path is called a *contour*. A digraph without contours is called *contourless*, or *acyclic*.

An alternating sequence (10.2.1) of vertices and arcs of a digraph G such that either $x_i = (v_{i-1}, v_i)$, or $x_i = (v_i, v_{i-1})$, $i = 1, 2, \ldots, n$, is its arc is called a *semiwalk*. A *semichain, semipath,* and *semicontour* are defined in a similar way.

If a digraph has a (u, v)-walk then it is said that v is *reachable* from u. Every vertex is assumed to be reachable from itself.

A digraph is said to be *strongly connected*, or *strong* if any two vertices are mutually reachable. A digraph is said to be *unilaterally connected*, or *unilateral* if for any two vertices at least one of them is reachable from the other one. A digraph is said to be *weakly connected*, or simply *weak* if any two its vertices are connected by a semiwalk.

Subgraphs, induced subgraphs, and spanning subgraphs are defined in the same way as for the ordinary graphs.

A *strong component* of a digraph is a maximal under inclusion strong subgraph. A *weak component*, or *connected component* is a maximal under inclusion weak subgraph.

A *condensation* of a digraph G is the digraph G^* with vertex set $\{S_1, \ldots, S_m\}$ of all strong components of G, and $(S_i, S_j) \in EG^*$ if and only if G has an arc with the tail in S_i and the head in S_j. Fig. 10.2.1 shows a digraph and its condensation.

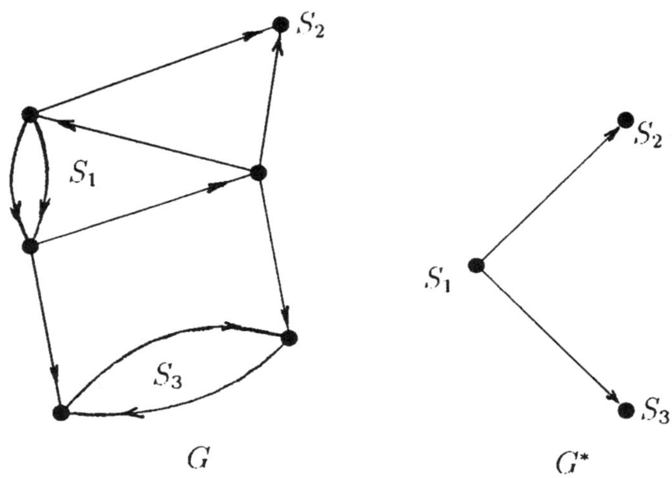

Figure 10.2.1: A digraph and its condensation

If we ignore the orientations of the arcs of a digraph G, we obtain an ordinary (non-oriented) pseudograph which is called the *substrate* of G and is denoted by G_s. Accordingly, the digraph G is called an *orientation* of the pseudograph G_s. Clearly a digraph is connected if and only if its substrate is a connected pseudograph.

A digraph is called *disconnected* if its substrate is a disconnected pseudograph.

* * *

10.2.1. Prove that every (u, v)-walk contains a (u, v)-path, which is a contour if $u = v$.

10.2.2. Prove that every digraph containing both a (u, v)-walk and a (v, u)-walk contains a contour. Give an example showing that this digraph does not necessarily contain a contour passing through both u and v.

10.2.3. Prove that every arc of a cyclic walk belongs to at least one contour contained in this walk.

10.2.4. Prove that if $d^+(v) \geq 1$ for every $v \in VG$ of a digraph G then G contains at least one contour.

10.2.5. Prove that every contourless digraph has a vertex with zero indegree and a vertex with zero outdegree.

10.2.6. Is it true that if $d^+(v) \geq 1$ and $d^-(v) \geq 1$ for all vertices v of a digraph G then every vertex of G belongs to at least one contour?

10.2.7. Prove that if a vertex v is reachable from u and w is reachable from v then w is reachable from u.

10.2.8. Prove the equivalence of the following statements about digraph:

a) The vertices u, v are mutually reachable.

b) The sets of vertices reachable from u and from v coincide.

c) If u, v are different vertices then they are connected by a path whose every arc belongs to a contour.

d) There exists a sequence of contours $C_1, \ldots, C_k$ such that $u \in C_1$, $v \in C_k$ and every two consecutive contours have a common vertex.

10.2.9. Prove that a digraph

a) is strong if and only if it has a spanning cyclic walk;

b) is unilateral if and only if it has a spanning walk;

b) is weak if and only if it has a spanning semiwalk.

10.2.10. A digraph is called *irreducible* if its vertex set cannot be partitioned into disjoint sets V_1, V_2 in such a way that every arc of the digraph connecting the vertices from different subsets is directed from V_1 to V_2.

Prove that a digraph is strong if and only if it is irreducible.

10.2.11. Prove that a digraph G is strong if and only if one of the following conditions hold:

a) there exists a cyclic walk that contains all arcs of G;

b) there is no nonempty set $V_1 \subset VG$ such that $\Gamma V_1 \subseteq V_1$;

c) G may be represented as the union of contours $C_1, \ldots, C_k$ such that every two consecutive contours of this list have a common vertex.

10.2.12. An arc of a digraph is called *acyclic* if it does not belong to any contour of the digraph. Prove that a digraph is strong if and only if it is connected and has no acyclic arcs.

10.2.13. Prove that

a) a non-strong digraph has arcs that belong to no strong component;

b) all such arcs are acyclic.

10.2.14. Prove that a connected non-unilateral digraph has no vertices whose deletion produces a strong digraph.

10.2.15. Let G be a (n, m)-digraph without loops and parallel arcs. Prove that

a) if G is a connected non-strong digraph then $n - 1 \leq m \leq (n-1)^2$;

b) if G is a strong digraph then $n \leq m \leq n(n-1)$;

c) if G is a connected contourless digraph then $n - 1 \leq m \leq n(n-1)/2$.

10.2.16. A vertex of a digraph is called *source* if every vertex of the digraph is reachable from it. Prove that every unilateral digraph has a source.

10.2.17. Prove that every unilateral digraph may be converted into a strong one by the addition of a single arc.

10.2.18. A *topological sorting* of the vertices of a digraph G is a numbering of its vertices by numbers $\{1, 2, \ldots, |G|\}$ in such a way that for every arc the number of the tail is less than the number of the head. Prove that a topological sorting exists if and only if the digraph is contourless.

10.2.19. Construct the condensations of digraphs shown at Fig. 10.2.2.

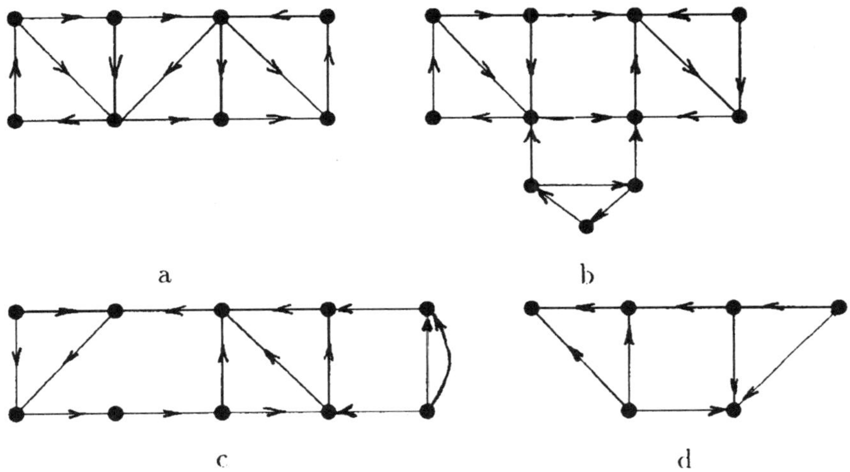

Figure 10.2.2: To Exercise 10.2.19

10.2.20. Prove that the following properties of a digraph G are equivalent:

a) G is a contourless digraph;

b) the digraph G and its condensation G^* are isomorphic;

c) every walk in G is a path;

d) the vertices of G may be numbered in such a way that its adjacency matrix (see Sect. 10.3) will be triangular.

10.2.21. Let $V \subseteq VG$ be a subset of vertices of a connected digraph G such that no arcs are ingoing into V. Then the set of arcs (v, u), $v \in V$, $u \in VG \backslash V$, is called a *directed cut* of G. Prove that every arc of a digraph belongs either to a directed cut or to a contour, but not to both simultaneously.

10.2.22. Prove that a digraph is contourless if and only if every its arc belongs to a directed cut.

10.2.23. Find maximal possible number of arcs in an n-vertex digraph without parallel arcs that has: a) k connected components; b) k strong components.

10.2.24. A graph G is called *directable* if al its edges may be directed in such a way that the resulting digraph will be strong.

a) Prove that the Petersen graph and the graphs of Platonic solids are directable.

b) Prove that a connected graph is directable if and only if it is bridgeless.

10.2.25. Prove that every graph has an acyclic orientation.

10.2.26. Prove that the minimal number of strong components achievable when orienting an ordinary graph is equal to the number of bridges of the graph plus one.

10.2.27. A digraph is called *minimally connected* if it is strong and the deletion of any its edge destroys this property. Construct a minimally connected digraph of order n for every $n \geq 2$.

10.2.28. A digraph is called *quasistrong* if any two its vertices v_1, v_2 are reachable from some vertex v_3 (not necessarily distinct from v_1, v_2). For every n construct a quasistrong digraph of order n that is not strong.

10.2.29. Prove that a digraph has a source (see Exercise 10.2.16) if and only if it is quasistrong.

10.2.30. A *rooted tree* is an acyclic digraph G that has a vertex v with $d^-(v) = 0$, and $d^-(u) = 1$ for every vertex $u \in VG \backslash v$. The vertex v is called the *root* of the tree. A vertex u with $d^+(u) = 0$ is called a *leaf* of the tree.

Prove that the following statements are equivalent for a digraph G:

a) G is a rooted tree;

b) G has a vertex that reaches every remaining vertex by a single walk;

c) there exists a vertex $v \in VG$ such that

1. $d^-(v) = 0$,
2. for every other vertex $u \in VG$ there exists a (v, u)-walk and $d^-(u) = 1$.

d) G is a quasistrong digraph which loses this property after the deletion of any edge.

10.2.31. A *binary tree* is a rooted tree such that $d^+(v) \leq 2$ for every vertex v. The *depth* of a vertex is the length of the longest path starting from it. The *height* of a tree is the depth of its root. Prove that a binary tree of height n has at most 2^n leaves and at most $2^{n+1} - 1$ vertices.

10.2.32. A $2-3$-*tree* is a rooted tree in which $d^+(v) = 2$ or 3 for every non-root vertex v and the lengths of all paths from the root to the leaves are equal.

Prove the following bounds for the number n of vertices and the number l of leaves of a 2-3-tree of height h :

$$2^{h+1} - 1 \le n \le (3^{h+1} - 1)/2;\ 2^h \le l \le 3^h.$$

10.2.33. A subgraph of a digraph G is called a *rooted spanning tree* if it is a rooted tree and contains all vertices of G. Prove that a digraph has a rooted spanning tree if and only if it is quasistrong.

10.2.34. A (di)graph with m edges (arcs) is called *edge-labelled* if its edges (arcs) are provided with labels, e.g., numbers $1, 2, \ldots, m$. Two edge-labelled (di)graphs G, H are said to be isomorphic if there exists a bijection $f : VG \to VH$ such that $uv \in EG$ and $f(u)f(v) \in EH$ have the same label (respectively, $(u, v) \in AG$ and $(f(u), f(v)) \in AH$ have the same label) for all edges (arcs).

Find:

a) the number of edge-labelled rooted trees of order n.

b) edge-labelled trees of order n.

10.2.35. The *transitive closure* of a digraph G is a digraph $\tilde{G}$ such that $VG = V\tilde{G}$ and $(u, v) \in \tilde{G}$ if and only if the vertex v is reachable from u in G. The *transitive reduction* of a digraph G is a subgraph of G minimal under inclusion whose transitive closure coincides with that of G.

a) Find the transitive closure and transitive reduction of a digraph shown at Fig. 10.2.3.

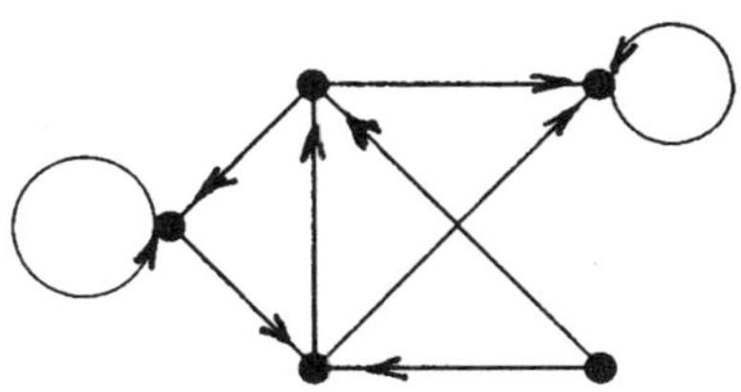

Figure 10.2.3: To Exercise 10.2.35

b) Find a digraph with minimal number of vertices that has nonisomorphic transitive reductions.

c) Prove that a contourless digraph has a single transitive reduction.

10.2.36. A digraph G is called *transitive* if $((x, y) \in AG,\ (y, z) \in AG)$ implies $(x, z) \in AG$. Prove that a transitive digraph coincides with its transitive closure.

10.2.37. Prove that the distance $\rho(u, v)$ between vertices in a digraph satisfies the first, second, and fourth axioms of metric from Exercise 1.2.6, but not necessarily the third one.

10.2.38. Let every arc x of a digraph G without loops be provided with a positive number $q(x)$. Let us define a function over VG^2 :

$$\rho_q(u,v) = \min_P \sum_{x \in L(P)} q(x),$$

where the minimum is over all (u,v)-paths P of G and $L(P)$ is the set of arcs of the path P. Prove that $\rho_q(u,v)$ satisfies the first, second, and fourth metric axioms from Exercise 1.2.6, but not necessarily the third one.

10.2.39. Let f be a permutation over the set $V = \{1,2,\ldots,n\}$. The *permutation digraph* G_f is the one with $VG_f = V$ and $(i,j) \in AG_f$ if and only if $f(i) = j$.

a) Construct permutation digraphs for the permutations

$$f_1 = (1,2,3,4,5,6)\,;\ f_2 = (2,3,4,5,6,1)\,;\ f_3 = (4,5,2,1,3,6)\,.$$

b) Describe the structure of the permutation digraph.

c) State a criterion of isomorphism of permutation digraphs in terms of the underlying permutations.

10.3 Matrices Associated with Digraph

In this section it is convenient to consider digraphs with the vertex set (numbered by) $\{1,2,\ldots,n\}$.

The $(n \times n)$-matrix $A = A(G)$ such that $A_{ij} = 1$ if $(i,j) \in AG$ and $A_{ij} = 0$ otherwise is called the *adjacency matrix* of the digraph G.

Let G be a digraph without loops and with $AG = \{a_1,\ldots,a_m\}$. The $(n \times m)$-matrix $I = I(G)$ such that

$$I_{ij} = \begin{cases} 1 & \text{if the vertex } i \text{ is the tail of the arc } a_j; \\ -1 & \text{if the vertex } i \text{ is the head of the arc } a_j; \\ 0 & \text{if the vertex } i \text{ and the arc } a_j \text{ are not incident.} \end{cases}$$

is called the *incidence matrix* of the digraph G

The $(n \times n)$-matrix $R = R(G)$ such that $R_{ij} = 1$ if the vertex j is reachable from i and $R_{ij} = 0$ otherwise is called the *reachability matrix* of the digraph G.

* * *

10.3.1. Find the matrices $A(G)$, $I(G)$, $R(G)$ for the digraph shown at Fig. 10.3.1.

10.3.2. Prove that

a) the number of ones in the i-th row (i-the column) of $A(G)$ is equal to the indegree (outdegree) of the digraph G;

b) the number of elements of the i-th row of $I(G)$ equal to 1 (-1) is equal to the indegree (outdegree) of the digraph G.

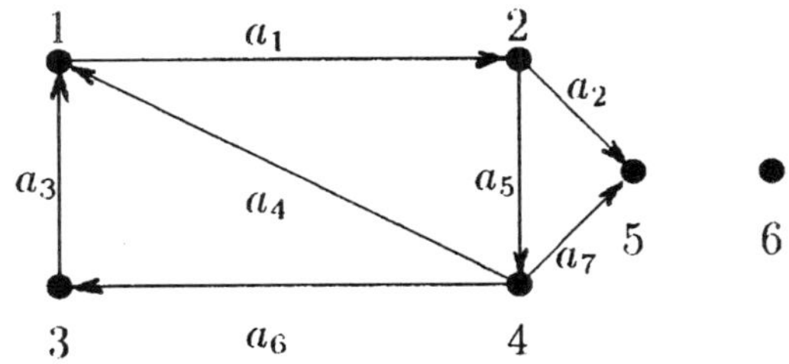

Figure 10.3.1: To Exercise 10.3.1

10.3.3. How are the adjacency, incidence and reachability matrices of the opposite digraph $\overleftarrow{G}$ related with those of G?

10.3.4. Prove the following statements:

a) If the digraphs G and H are isomorphic then the following conditions are valid:

(1) $A(G)$ may be obtained from $A(H)$ by equal permutations of rows and columns;

(2) $I(G)$ may be obtained from $I(H)$ by permutations of rows and columns;

b) if one of the above conditions (1), (2) is valid then $G \cong H$.

10.3.5. Demonstrate that $R(G) = R(H)$ does not imply $G \cong H$.

10.3.6. Prove that the matrix $A(G)$ can be reduced to the block-diagonal form

$$\mathrm{diag}[A_1, \ldots, A_k],\ k > 1,$$

where A_i, $i = 1, \ldots, k$ are square matrices, by equal permutations of rows and columns if and only if G is not connected.

10.3.7. Let $G_1, \ldots, G_k$ be the strong components of a digraph G. Prove that

a) $A(G)$ may be reduced to the following form

$$\begin{bmatrix} A_1 & * & \ldots & * \\ O & A_2 & \ldots & * \\ \ldots & \ldots & \ldots & \ldots \\ O & O & \ldots & A_k \end{bmatrix}$$

by equal permutations of rows and columns, where A_i is the adjacency matrix of the digraph G_i, $i = 1, \ldots, k$, and O is the zero matrix;

a) $R(G)$ may be reduced to the following form

$$\begin{bmatrix} I_1 & * & \ldots & * \\ O & I_2 & \ldots & * \\ \ldots & \ldots & \ldots & \ldots \\ O & O & \ldots & I_k \end{bmatrix}$$

by equal permutations of rows and columns, where I_i is the square matrix of ones of order $|G_i|$, $i = 1, \ldots, k$.

10.3.8. Prove that for an acyclic digraph G the matrix $A(G)$ may be reduced to the triangular form by equal permutations of rows and columns.

10.3.9. Prove that the rank of the incidence matrix $I(G)$ of a digraph G of order n is

a) $n - 1$, if G is connected;

b) $n - p$, if G has p connected components.

10.3.10. A matrix is called *totally unimodular* if every its minor is equal to 0, 1, or -1. Prove that the incidence matrix of a digraph is totally unimodular.

10.3.11. Prove that the (i, j)-th element of the matrix $(A(G))^k$ is equal to the number of (i, j)-walks of length k in the digraph G.

10.3.12. Let G be a digraph of order n. Let $A^{(k)}(G)$ denote the binary $(n \times n)$-matrix such that $A_{ij}^{(k)}$ is equal to 1 if and only if G has an (i, j)-path of length at most k.

a) Find the matrices $A^{(1)}(G)$, $A^{(2)}(G)$, $A^{(3)}(G)$, $A^{(4)}(G)$ for a digraph G shown at Fig. 10.3.2.

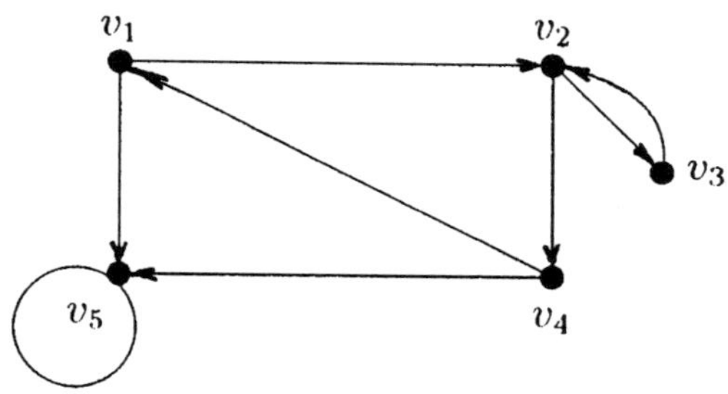

Figure 10.3.2: To Exercise 10.3.12

b) prove that $A^{(k)}(G) = A(G)$, $k = 1, 2, \ldots, n$, for a transitive digraph G (see Exercise 10.2.36) of order n.

c) Prove that a digraph G of order n is transitive if and only if $A^{(k)}(G) = A(G)$ for some $k \geq 2$.

d) Prove that $A(\tilde{G}) = A^{(n)}(G)$, where $\tilde{G}$ is the transitive closure of a digraph G of order n.

10.3.13. Prove that for a digraph, its transitive closure, and its transitive reduction (see Exercise 10.2.35) the strong components partition the vertex set into subsets in the same way.

10.3.14. Prove that the (i,i)-th element of the matrix $(R(G))^2$ is equal to the number of vertices in the strong component of G containing the vertex i.

10.3.15. For a loopless digraph G we define an $(n \times n)$-matrix $M(G)$ with $M_{ii} = d^-(i)$, $M_{ij} = -k$, $i \neq j$, where k is the number of parallel arcs (i,j). Prove that G is a rooted tree with the root r (see Exercise 10.2.30) if and only if $M_{rr} = 0$, $M_{ii} = 1$ for $i \neq r$ and the complementary minor of the (r,r)-th element of $M(G)$ is equal to 1;

10.4 Tours and Paths

The definitions of Eulerness and Hamiltonicity for digraphs are similar to those for ordinary graphs.

An *Eulerian chain* is a chain containing all arcs of a digraph. A connected digraph is called *Eulerian* if it has a cyclic Eulerian chain.

Theorem 10.4.1 *For a connected digraph G the following statements are equivalent:*

1. *G is Eulerian;*
2. *$d^+(v) = d^-(v)$ for every $v \in VG$;*
3. *G is the union of contours such that no pair of them have common arcs.*

A contour or path of a digraph is said to be *Hamiltonian* if it contains all vertices of the digraph. A digraph is called *Hamiltonian* if it has a Hamiltonian contour.

Theorem 10.4.2 *Let G be a strong digraph of order $n > 1$ without loops and parallel arcs. If*

$$\deg u + \deg v \geq 2n - 1 \tag{10.4.1}$$

for any pair of different nonadjacent vertices of G then G is Hamiltonian.

The independence number $\alpha_0(G)$, proper coloring, and the chromatic number $\chi(G)$ of a digraph G are defined in terms of its substrate graph G_s : $\alpha_0(G) = \alpha_0(G_s)$, $\chi(G) = \chi(G_s)$.

Let $M = \{P_1, P_2, \ldots, P_l\}$ be a set of vertex-disjoint paths in a digraph G. If $VG = \bigcup_{i=1}^{l} VP_i$ then M is called a *path partition of digraph G*. The minimal number of paths that constitute a path partition of G is denoted by $l(G)$.

Theorem 10.4.3 $l(G) \leq \alpha_0(G)$ *for any digraph G.*

Theorem 10.4.4 *If k is the maximal length of a path in a digraph G then $\chi(G) \leq k + 1$.*

* * *

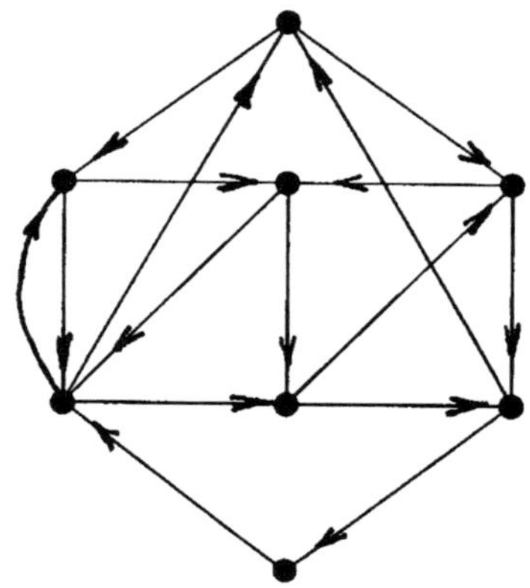

Figure 10.4.1: To Exercise 10.4.1

10.4.1. For a digraph shown at Fig. 10.4.1 find contours of lengths 2, 3, 4, 5, 6; a cyclic Eulerian chain; a Hamiltonian contour.

10.4.2. Find a digraph that is not Eulerian but has a cyclic Eulerian chain.

10.4.3. Find a digraph that

a) is Eulerian but is not Hamiltonian;

b) is Hamiltonian but is not Eulerian;

c) is neither Eulerian, nor Hamiltonian.

10.4.4. Prove the Theorem 10.4.1.

10.4.5. Prove that a connected digraph contains a noncyclic Eulerian chain if and only if it has two vertices v_1, v_2 such that $d^-(v_1) = d^+(v_1) + 1$, $d^-(v_2) = d^+(v_2) - 1$, and $d^-(v) = d^+(v)$ for all vertices v other than v_1, v_2.

10.4.6. Let a digraph G have two vertices u, v such that $d^+(v) - d^-(v) = k = d^-(u) - d^+(u)$, and $d^-(w) = d^+(w)$ for all vertices w other than u, v. Prove that G may be partitioned into k edge-disjoint (v, u)-chains.

10.4.7. Prove that the edges of any graph may be directed in such a way that $|d^+(v) - d^-(v)| \leq 1$ for any vertex v in the resulting digraph.

10.4.8. Prove that the inequality (10.4.1) of Theorem 10.4.2 cannot be relaxed.

10.4.9. Prove that a strong digraph of order n without loops and parallel arcs has a Hamiltonian contour if one of the following conditions holds for every its vertex v :

(a) $\deg v \geq n$;

(b) $d^+(v) \geq n/2,\ d^-(v) \geq n/2$.

10.4.10. Prove that if $\deg u + \deg v \geq 2n - 3$ for any two nonadjacent vertices u, v of a digraph G of order n then G has a Hamiltonian path.

10.4.11. A set S of vertices of a digraph G is called *incomparable* if no vertex from S is reachable from some other vertex from S. Prove that the maximal cardinality of an incomparable set of a digraph G is equal to the independence number of the transitive closure of G.

10.4.12. Prove that for any orientation of edges of the graph shown at Fig. 10.4.2

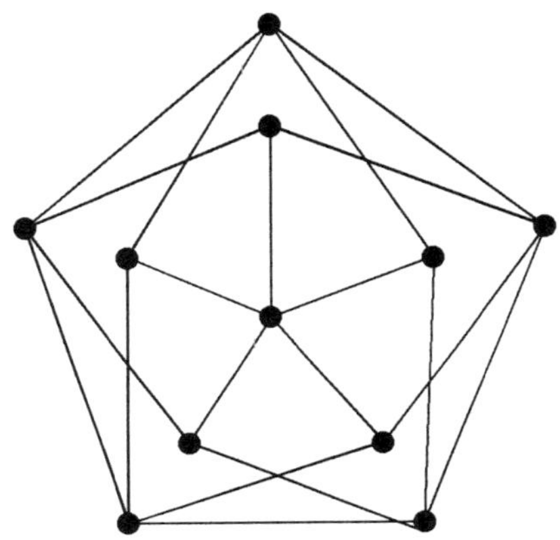

Figure 10.4.2: To Exercise 10.4.12

a) it may be partitioned into at most 5 paths;

b) there exists a path of length at least 3.

10.4.13. Prove that the minimal number of paths in a path partition of a transitive digraph G is equal to $\alpha_0(G)$.

10.4.14. Let $\tau(G)$ denote the minimal integer such that every orientation of the graph G has a path of length $\tau(G)$ and there exists an orientation of G that has no path of length $\tau(G) + 1$. Prove that $\chi(G) = \tau(G) + 1$.

10.4.15. Prove that $\chi(L(G)) \geq \log_2 \chi(G)$, where $L(G)$ is the line digraph of a digraph G, see Exercise 10.1.11.

10.4.16. For a digraph G without loops and parallel arcs we denote

$$k_1 = \min_{v \in VG} d^+(v), \; k_2 = \min_{v \in VG} d^-(v), \; k = \max\{k_1, k_2\}.$$

Prove that

a) for any vertex v of a digraph there exists at least $k_1!$ different paths of length at least k_1 starting at v;

b) for any vertex v of a digraph there exists at least $k_2!$ different paths of length at least k_2 ending at v;

c) if $k > 0$ then the digraph has a contour of length at least $k + 1$.

10.4.17. Let G be a digraph. For $A \subseteq AG$ define:

$$\delta(A) = \begin{cases} |A| - |\Gamma A|, & \text{if } A \neq \emptyset, \\ 0, & \text{otherwise} \end{cases} \qquad \delta_0(G) = \max_{\emptyset \subseteq A \subseteq VG} \delta(A).$$

Prove that G has a spanning subgraph whose all connected components are contours if and only if $\delta_0(G) = 0$.

10.5 Tournaments

A digraph whose substrate is a complete graph is called a *tournament.*

Theorem 10.5.1 *For every vertex of a strong tournament of order n and for every integer k, $3 \leq k \leq n$, there exists a contour of length k passing through this vertex.*

Corollary 10.5.2 *Every strong tournament is Hamiltonian.*

* * *

10.5.1. Draw all pairwise nonisomorphic tournaments of orders 2, 3, 4.

10.5.2. Prove that every tournament has a Hamiltonian path.

10.5.3. Prove that every tournament has a vertex such that all other vertices are reachable from it.

10.5.4. Prove that every tournament either is strong or may be converted into strong by reversing a single arc.

10.5.5. Prove that a transitive tournament has a single Hamiltonian path.

10.5.6. Prove that a tournament is strong if and only if it is Hamiltonian.

10.5.7. Prove that a Hamiltonian tournament T has a vertex v such that the tournament $T - v$ is Hamiltonian.

10.5.8. Prove that the automorphism group of a tournament is of odd order.

10.5.9. Prove that for every tournament

a) the distance from a vertex with maximal outdegree to any other vertex is at most two;

b) the distance from any vertex to a vertex with minimal indegree is at most two.

10.5.10. Let $(s_1 \le s_2 \le \dots \le s_n)$ be the sequence of outdegrees of vertices of a tournament T. Prove that

a) $\sum_{i=1}^{n} s_i = n(n-1)/2$;

b) $\sum_{i=1}^{k} s_i \ge k(k-1)/2,\ k = 1, 2, \dots, n-1$, and these inequalities are strict if and only if T is strong;

c) $(k-1)/2 \le s_k \le (n+k-2)/2,\ k = 1, 2, \dots, n-1$;

d) T is transitive if and only if $s_k = k-1,\ k = 1, 2, \dots, n$.

10.5.11. Prove that

$$\sum_{v \in VT} (d^+(v))^2 = \sum_{v \in VT} (d^-(v))^2$$

for a tournament T.

10.5.12. Let $\tilde{T}$ be a maximal contourless subgraph of a tournament T. Prove that $|A\tilde{T}| \ge \lfloor |VT|/2 \rfloor \lfloor (|VT|+1)/2 \rfloor$.

10.5.13. A digraph is called a *bipartite tournament* if its substrate is a complete bipartite graph. Is it true that a strong bipartite tournament is Hamiltonian?

10.6 Base and Kernel

Let G be a digraph and let B be a subset of vertices such that every vertex from $VG \backslash B$ is reachable from some vertex from B. If B is minimal under inclusion among the subsets with this property then B is called a *base of the digraph* G.

$D(v)$ will denote the set of vertices of G reachable from $v \in VG$ and $D(v, A)$ will denote the set of vertices of G reachable from $v \in VG$ via the arcs from $A \subseteq AG$.

A minimal under inclusion set A of arcs of G is called an *arc base of the digraph* G if $D(v) = D(v, A)$ for every $v \in VG$. The minimality property for A implies that for every $A' \subseteq A$ there exists a vertex $u \in VG$ such that $D(u, A) \backslash D(u, A') \ne \emptyset$.

A set S of vertices of a digraph G is called *dominating* if for every $w \in VG \backslash S$ there exists $v \in S$ such that $(w, v) \in AG$. The number of vertices in a dominating set of minimal cardinality will be denoted by $\beta(G)$.

A set S of vertices of a digraph G is called *independent* if no two vertices from S are adjacent.

A set of vertices of a digraph G which is both dominating and independent is called a *kernel of the digraph* G.

Theorem 10.6.1 *A digraph without contours of odd length has a kernel.*

* * *

10.6.1. Prove that no two vertices of a base are connected by a walk.

10.6.2. Prove that every vertex with zero indegree belongs to a base.

10.6.3. A *base component* of a digraph is a strong component such that the corresponding vertex in the condensation of the digraph is of zero indegree. Prove that a subset B of vertices of a digraph is a base if and only if it consists of vertices belonging to base components and every base component contains exactly one vertex from B.

10.6.4. Prove that all bases of a digraph are of the same cardinality.

10.6.5. Which vertices constitute a base of a contourless digraph? Prove that such digraph has a unique base.

10.6.6. Prove that a digraph has a unique base if and only if every base component has exactly one vertex.

10.6.7. Prove that every base of a digraph has a vertex such that maximal number of vertices of the digraph is reachable from it.

10.6.8. An *antibase* $\overline{B}$ *of a digraph* G is a minimal under inclusion set of vertices of G such that for every vertex $v \in VG \backslash \overline{B}$ some vertex from $\overline{B}$ is reachable from v.

Prove that an antibase of a digraph G coincides with some base of the opposite digraph $\overleftarrow{G}$.

10.6.9. Prove that for a digraph

a) all vertices of zero outdegree belong to every antibase;

b) no two vertices of an antibase are connected by a walk;

c) all antibases are of the same cardinality.

10.6.10. Prove that a vertex v of a digraph G may belong both to a base B and to an antibase $\overline{B}$ if and only if the strong component of G containing v corresponds to an isolated vertex in the condensation G^*.

10.6.11. Prove that a set $A \subseteq AG$ is an arc base of a digraph G if and only if it satisfies the following two conditions:

(1) for every arc $(x, y) \in AG \backslash A$ there exists a path from x to y with all arcs in A;

(2) for no arc $(x, y) \in AG \backslash A$ there exists a path from x to y with all arcs in $A \backslash \{(x, y)\}$.

10.6.12. Prove that every digraph has at least one arc base.

10.6.13. Find a digraph with two arc bases of different cardinalities.

10.6.14. Prove that a contourless digraph without parallel arcs has a single arc base.

10.6.15. Prove that in a connected digraph G a subgraph induced by the set of arcs from an arc base is connected; it is strong if G is strong.

10.6.16. Prove that a connected (strong) digraph of order n has an arc base with at least $n-1$ (respectively, n) arcs.

10.6.17. Prove that an arc base of a strong digraph of order n has at most $2n-2$ arcs.

10.6.18. For every positive integer n construct a strong digraph of order n with an arc base of size $2n-2$.

10.6.19. Prove that a digraph of order n has a Hamiltonian contour if and only if it is strong and has an arc base of size n.

10.6.20. Prove that $\beta(G) \leq |N| \leq \alpha_0(G)$ for any kernel N of a digraph G.

10.6.21. Find a loopless digraph of minimal order without kernel.

10.6.22. Prove that a set K of vertices of a digraph is a kernel if and only if K is both maximal independent set and minimal dominating set.

10.6.23. Find a digraph in which

(a) some maximal independent vertex set is not a kernel;

(b) some minimal dominating vertex set is not a kernel.

10.6.24. A set $N \subseteq VG$ for a digraph G is called an *outkernel* if $\Gamma N = VG \backslash N$ and an *inkernel* if $\Gamma^{-1} N = VG \backslash N$.

Prove that an outkernel of a digraph is an inkernel of the opposite digraph.

10.6.25. Prove that a set N of vertices of a digraph is a kernel if and only if N is an inkernel.

10.6.26. Find a digraph that has

(a) an outkernel but no inkernel;

(b) an inkernel but no outkernel;

(c) different outkernel and inkernel;

(d) coinciding outkernel and inkernel;

(e) two outkernels of different cardinalities.

10.6.27. For any integers k, n, $1 \leq k \leq n$, find a digraph of order n with k-vertex kernel.

10.6.28. Describe a digraph in which every vertex constitutes a kernel.

10.6.29. Prove that a subset N of vertices of a digraph G is an outkernel if and only if the following three conditions are satisfied (where V^+ is the set of vertices with zero indegree):

(a) $V^+ \subseteq N$;

(b) $N \cap \Gamma V^+ = \emptyset$;

(c) if $V' = VG \backslash (V^+ \cup \Gamma V^+) \neq \emptyset$ then the set $N' = N \backslash V^+$ is an outkernel of a subgraph $G' \subseteq G$ induced by V'.

10.6.30. State and prove a criterion for the inkernel similar to the previous exercise.

10.6.31. Prove that a digraph with no semicontours of odd length and

(a) with no vertices of zero indegree has at least two outkernels;

(b) with no vertices of zero outdegree has at least two inkernels.

Show that these statements will fail if we discard the requirements about vertices.

10.6.32. Prove that a contourless digraph cannot have

(a) more than one inkernel;

(b) more than one outkernel.

10.6.33. Prove that the set V^+ of vertices of a digraph G of zero indegree is the only outkernel of G if $V^+ \cap \Gamma^{-1}v \neq \emptyset$ for every $v \in VG \backslash V^+$.

State and prove a similar statement about inkernels.

10.6.34. Prove that a digraph whose substrate is a bipartite graph has both an inkernel and outkernel.

10.6.35. Prove that every graph has an orientation with outkernel.

10.6.36. A function g with nonnegative integer values defined over the vertex set of a (di)graph G such that for every vertex v $g(v)$ is the minimal positive integer outside the set $\{g(u) : u \in \Gamma v\}$ is called a *Grundy function* for the (di)graph G.

(a) Find Grundy functions for the digraphs shown at Fig. 10.6.1.

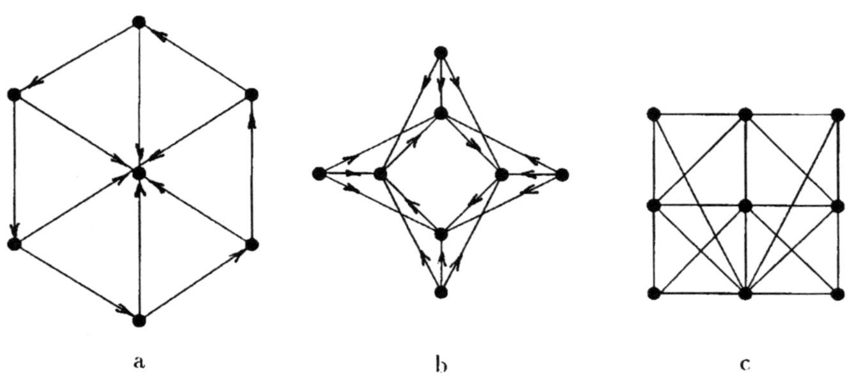

Figure 10.6.1: To Exercise 10.6.36a

(b) Prove that every graph admits a Grundy function.

(c) Find a loopless digraph of minimal order that has no Grundy functions.

(d) Find a digraph of minimal order that has several Grundy functions.

(e) Find a digraph that has a Grundy function but the opposite digraph has no Grundy functions.

(f) Prove that a digraph with a Grundy function has a kernel.

(g) Prove that a digraph has a Grundy function if every its subgraph has a kernel.

(h) Prove that a contourless digraph has a Grundy function.

(i) Find a digraph that has a Grundy function whose opposite digraph also has a Grundy function.

(j) Prove that is a (di)graph G has a Grundy function g then $\chi(G) \leq \max_{v \in VG} g(v) + 1$.

(k) Prove that if a graph G is p-chromatic then it has a Grundy function g such that $\max_{v \in VG} g(v) \leq p - 1$.

Chapter 11

Hypergraphs

11.1 Hypergraphs: Basic Notions

A hypergraph is a generalization of a graph such that any nonempty subset of vertices may be an edge.

Let V be a finite nonempty set and let $\mathcal{E}$ be a collection of any (not necessarily different) subsets of V.

The pair $(V, \mathcal{E})$ is called the *hypergraph* with the *vertex set* V and *edge (or hyperedge) collection* $\mathcal{E}$. Equal subsets in $\mathcal{E}$ are called *multiple edges* of the hypergraph. The set of vertices and the collection of edges of a hypergraph H are denoted by VH and $\mathcal{E}H$ respectively. The number $|VH|$ of vertices of H is called the *order* of the hypergraph and is denoted by $|H|$. If $|H| = n$ and $|\mathcal{E}H| = m$ then H is called an *(n, m)-hypergraph.*

If a vertex $v \in VH$ belongs to an edge $e \in \mathcal{E}H$ then we shall say that v and e are *incident (to each other).* The collection of edges incident to a vertex v will be denoted by $\mathcal{E}(v)$. The number $|\mathcal{E}(v)|$ is called the *degree* of the vertex v and is denoted by $\deg v$. The number $|e|$ is called the *degree of the edge e*. Since only nonempty sets of vertices may be edges, the degrees of edges are at least 1. A vertex of a hypergraph incident to no edge is called *isolated.*

Two vertices of a hypergraph are called *adjacent* if some edge contains both of them. Two edges e_1 and e_2 of a hypergraph are called *adjacent* if $e_1 \cap e_2 \neq \emptyset$.

In drawings it is convenient to depict an edge e of a hypergraph by an area bounded by a closed curve encompassing the vertices of e if $|e| \neq 2$. If $|e| = 2$, then e is depicted by a curve segment connecting its vertices, as in the case of an ordinary graph.

Similarly to ordinary graphs, we may consider both *labelled hypergraphs* (the vertices have different labels) and *unlabelled* ones.

If a hypergraph has no multiple edges and the degrees of all edges are equal to k then the hypergraph is called *k-uniform.* Clearly, the pseudograph is a special case of the hypergraph and the ordinary graphs are the 2-uniform hypergraphs. A *complete k-uniform hypergraph* is a hypergraph with the edge set being the set of all k-element subsets of vertices.

Let H_1, H_2 be hypergraphs and let $\varphi : VH_1 \to VH_2$ and $\xi : \mathcal{E}H_1 \to \mathcal{E}H_2$ be bijections. If for any $v \in VH_1$, $e \in \mathcal{E}H_1$ the images $\varphi(v)$ and $\xi(e)$ are incident in H_2 if and only if v and e are incident in H_1 then H_1 and H_2 are said to be *isomorphic hypergraphs,* and this relation is denoted by $H_1 \cong H_2$.

Every (n, m)-hypergraph $H = (V, \mathcal{E})$ without isolated vertices may be related with an (m, n)-hypergraph $H^* = (V^*, \mathcal{E}^*)$ as follows.

If $V = \{v_1, \ldots, v_n\}$ and $\mathcal{E} = \{e_1, \ldots, e_m\}$ then $V^* = \{v_1^*, \ldots, v_m^*\}$ and $\mathcal{E}^* = \{e_1^*, \ldots, e_n^*\}$ and v_i^* is incident to e_j^* if and only if e_i and v_j are incident in H. The hypergraph H^* is called *dual* to H.

For a hypergraph $H = (V, \mathcal{E})$, we define the *incidence graph* to be the bipartite graph $K(H)$ with the vertex set $V \cup \mathcal{E}$ and the edge set

$$\{(v, e) : (v, e) \in V \times \mathcal{E}, v \in e\}.$$

The graph $K(H)$ is also called the *König representation of hypergraph H.*

A *path* of *length* $l > 0$ in a hypergraph H, or a (v_1, v_{l+1})*-path* is an alternating sequence

$$v_1, e_1, v_2, e_2, \ldots, e_l, v_{l+1}$$

of vertices and edges of a hypergraph H that satisfies the following conditions:

(1) $e_1, \ldots, e_l$ are different edges of H;

(2) $v_i, v_{i+1} \in e_i$, $i = 1, \ldots, l$.

Notice *different* edges of a hypergraph H are considered to be different elements of the collection $\mathcal{E}H$ (they may coincide as subsets of VH).

A path is called *simple* if all its vertices are different, with the possible exception that the first and the last vertices may coincide.

If $l > 1$ and $v_1 = v_{l+1}$ then the path is called *cycle* and the simple path is called *simple cycle.*

If a hypergraph H has a (u, v)-path for every two vertices u, v, then H is called a *connected hypergraph.*

A hypergraph P is called a *part* of a hypergraph G if $VP \subseteq VH$ and $\mathcal{E}P \subseteq \mathcal{E}H$. A maximal under inclusion connected part of a hypergraph is called a *component* of the hypergraph.

A *deletion of edge e* of a hypergraph $H = (V, \mathcal{E})$ is an operation producing the hypergraph $H' = H - e = (V, \mathcal{E}')$ with $\mathcal{E}' = \mathcal{E} \backslash \{e\}$.

A *deletion of vertex v* of a hypergraph $H = (V, \mathcal{E})$ is an operation producing the hypergraph $H' = H - v = (V', \mathcal{E}')$ with $V' = V \backslash \{v\}$, $\mathcal{E}' = \{e' : e' = e \backslash \{v\} \neq \emptyset, e \in \mathcal{E}\}$.

A vertex v of a hypergraph H is called a *cutpoint* if the hypergraph $H - v$ has more components than H.

Let H be a connected hypergraph with $|H| > 1$. A maximal under inclusion connected part of H without cutpoints is called a *block.* A connected hypergraph without cutpoints has a single block, which is the hypergraph itself.

The *vertex connectivity number* $\kappa(H)$ of a connected hypergraph H is the minimal number of vertices whose deletion produces a disconnected or a single-vertex hypergraph. A hypergraph H is called *k-connected* if $\kappa(H) \geq k$.

The *edge connectivity number* $\lambda(H)$ of a connected hypergraph H is the minimal number of edges whose deletion produces a disconnected hypergraph.

A *matching* of a hypergraph H is a subset $\mathcal{E}' \subseteq \mathcal{E}H$ such that $e_1 \cap e_2 = \emptyset$ for any $e_1, e_2 \in \mathcal{E}'$, i.e., no two edges from $\mathcal{E}'$ are adjacent. A matching is called *maximum cardinality matching* if it has maximal possible number of edges among the matchings of the hypergraph; this number is called *matching number* of the hypergraph and denoted by $\alpha_1(H)$. A *covering* of a hypergraph H is a subset $U \subseteq VH$ such that $U \cap e \neq \emptyset$ for every edge $e \in \mathcal{E}H$. A covering is called *minimum cardinality covering* if it has minimal possible number of vertices among the coverings of the hypergraph; this number is called the *covering number* of the hypergraph and is denoted by $\beta_0(H)$.

A set of vertices of a hypergraph is called *independent* if no its subset is an edge of the hypergraph. The empty vertex set is defined to be independent. An independent vertex set of maximal size is called *maximum cardinality independent vertex set,* and its size is called the *independence number* and is denoted by $\alpha_0(H)$.

A *(vertex) coloring* of a hypergraph is a partition of its vertex set into *color* classes. A vertex coloring of a hypergraph H is called *proper* if every edge $e \in \mathcal{E}H$ contains two vertices colored differently. Clearly a proper coloring is possible only for hypergraphs in which every edge is of degree at least 2.

A hypergraph is called *k-colorable* if it admits a proper coloring in k colors. The *chromatic number* $\chi(H)$ of a hypergraph H is the minimal number of colors necessary for proper coloring H.

The *line graph of a hypergraph* $H = (V, \mathcal{E})$ is the graph $L(H) = (W, E)$ such that there is a bijection f between the set W and the collection $\mathcal{E}$, and two vertices $w_1, w_2 \in W$ are adjacent if and only if the edges $f(w_1)$ and $f(w_2)$ are adjacent. In other words, $L(H)$ is the graph of intersections of edges of H.

* * *

11.1.1. Find a 3-uniform hypergraph of order 10. How many are there labelled hypergraphs of such type?

11.1.2. How many are there labelled $(6,3)$-hypergraphs a) without multiple edges; b) with possible multiple edges?

11.1.3. Generalize the Handshake Lemma for the case of a) hypergraphs; b) k-uniform hypergraphs.

11.1.4. A matroid M with the support V may be related with the hypergraph $\mathcal{C}(M) = (V, C)$, where C is the set of cycles of M.

a) Characterize the independent sets of M in terms of the hypergraph $\mathcal{C}(M)$.

b) What is the independence number of $\mathcal{C}(M)$?

c) What can you say about the components of $\mathcal{C}(M)$ if M is the partition matroid defined in the Exercise 5.2.10?

11.1.5. Draw all hypergraphs H with a) $K(H) \cong K_{5,5}$; b) $K(H) \cong K_{2,3}$; c) $K(H) \cong K_{1,5}$; d) $K(H) \cong K_{1,3} \cup K_{2,2}$. Here $\cup$ denotes the disjoint union of graphs.

11.1.6. Which bipartite graph is the König representation for the cycle C_n?

11.1.7. Is it true that for every bipartite graph G without isolated vertices there exists a hypergraph H such that $K(H) \cong G$?

11.1.8. Find a hypergraph of tenth order that has a single cycle. Construct the König representation of this hypergraph and verify that it has a single cycle too.

11.1.9. Prove that the number of simple cycles of a hypergraph is equal to the number of simple cycles of its König representation.

11.1.10. Prove that the König representations of hypergraphs H and H^* are isomorphic to each other.

11.1.11. A hypergraph H is called *selfdual* if $H \cong H^*$. Construct a selfdual hypergraph of order n for every positive integer n.

11.1.12. Is it true that a hypergraph in which every vertex and every edge is at least of degree 2 always has a cycle?

11.1.13. A hypergraph is called *planar* if its König representation is planar. Prove that

a) a hypergraph H is planar if and only if its dual is planar;

b) for a planar k-uniform (n, m)-hypergraph $n \geq (k-3)m/3 + 2$.

11.1.14. For hypergraphs shown at Fig. 11.1.1 find a) a maximum cardinality matching; b) a maximum cardinality independent set.

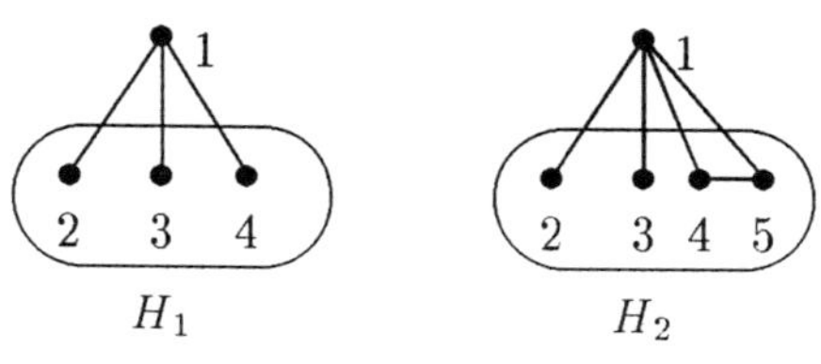

Figure 11.1.1: To Exercise 11.1.14

11.1.15. Draw line graphs for hypergraphs shown at Fig. 11.1.1.

11.1.16. The following theorem due to Whitney holds for graphs: if for connected graphs of orders at least four the line graphs are isomorphic then the graphs are isomorphic themselves. Does a similar theorem hold for hypergraphs?

11.1.17. Which graph may be a line graph of a) a hypergraph; b) a connected hypergraph; c) a disconnected hypergraph?

11.1.18. Find a line graph for the hypergraph $H_n = (V, \{E_1, \ldots, E_n\})$, where $V = \{\{i,j\} : i,j = 1,2,\ldots,n,\ i \neq j\}$, $E_i = \{\{i,j\} : j = 1,2,\ldots,n,\ i \neq j\}$.

11.1.19. Construct a hypergraph dual to the graph K_n.

11.1.20. A hypergraph without isolated vertices is called *linear* if every pair of different vertices belongs to at most one common edge. Is it true that a hypergraph dual to a linear one is also linear?

11.1.21. Let H be a linear hypergraph with all vertices of degree two. Prove that $H^* \cong L(H)$.

11.1.22. Prove that every graph is a line graph of some linear hypergraph.

11.1.23. Prove that for a graph H without isolated vertices the line graph $L(H^*)$ of the dual hypergraph H^* is isomorphic to H.

11.1.24. For a hypergraph $H = (V, \mathcal{E})$ we may define a graph $(H)_2$ with the same vertex set such that two vertices of $(H)_2$ are adjacent if and only if they are adjacent in H. Prove that if H has no isolated vertices then $(H)_2 \cong L(H^*)$.

11.1.25. A hypergraph H is called *conformal* if the set of maximal cliques of $(H)_2$ is the set of maximal edges of H. Prove the following Gilmor theorem: a hypergraph is conformal if and only if for any triple $e_1, e_2, e_3 \in \mathcal{E}H$ there exists $e_0 \in \mathcal{E}H$ such that $(e_1 \cap e_2) \cup (e_1 \cap e_3) \cup (e_2 \cap e_3) \subseteq e_0$.

11.1.26. Prove that for a hypergraph H without isolated vertices

a) a subset $U \subseteq VH$ is a covering of H if and only if $VH \backslash U$ is independent;

b) a subset $U \subseteq VH$ is a minimum cardinality covering of H if and only if $VH \backslash U$ is a maximum cardinality independent set of H;

c) if $|H| = n$ then $\alpha_0(H) + \beta_0(H) = n$.

11.1.27. A vertex v of a hypergraph $H = (V, \mathcal{E})$ is called *hypopendant* if there exists another vertex v' such that $\mathcal{E}(v) \subseteq \mathcal{E}(v')$ (recall that $\mathcal{E}(x)$ denotes the set of edges containing a vertex x). Prove that if v is a hypopendant vertex of a hypergraph $H \not\cong K_2$ without isolated vertices then

a) v belongs to some maximum cardinality independent set of H;

b) $\beta_0(H) = \beta_0(H - v)$;

c) $\alpha_0(H) = \alpha_0(H - v) + 1$;

d) $\alpha_1(H) = \alpha_1(H - v)$.

11.1.28. Prove the validity of the following statement: a graph L is a line graph of some k-uniform hypergraph if and only if there exists a covering of L by complete graphs $L_1, \ldots, L_s$ such that

a) every $k + 1$ different graphs from $L_1, \ldots L_s$ have empty intersection;

b) every k different graphs from $L_1, \ldots L_s$ have at most one common vertex.

11.1.29. Prove that there exists a graph with arbitrarily many vertices which is not the line graph of a 3-uniform linear hypergraph but every its induced subgraph is.

11.1.30. Prove that in a hypergraph

a) every two blocks have at most one common vertex;

b) every edge of size at least two belongs to exactly one block.

11.1.31. Prove that

a) if a vertex v of a hypergraph belongs to more than one block then it is a cutpoint;

b) if a block of a hypergraph contains vertices u, v then it contains every simple (u, v)-path of the hypergraph.

11.1.32. Is it true that a hypergraph is biconnected if and only if every two its vertices belong to a common simple cycle?

11.1.33. Is it true that a connected hypergraph is biconnected if and only if every two its edges of degrees at least two belong to a common simple cycle?

11.1.34. Are the hypergraphs shown at Fig. 11.1.2 3-colorable?

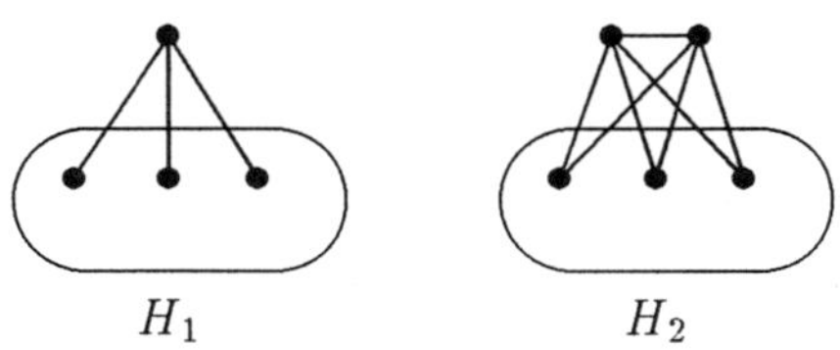

Figure 11.1.2: To Exercise 11.1.34

11.1.35. Prove that the following inequalities are valid for a connected hypergraph H :

a) $\lambda(H) \leq \delta(H)$;

b) $\delta(H) \leq |\mathcal{E}H|r(H)/|H|$.

11.1.36. Find the chromatic number for a) the complete k-uniform hypergraph H_n^k of order n, $2 \leq k \leq n$; b) hypergraphs H_1, H_2, H_3 shown at Fig. 11.1.3.

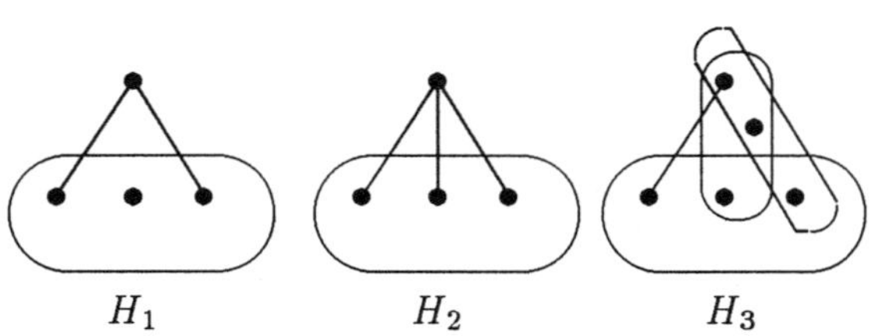

Figure 11.1.3: To Exercise 11.1.36

11.1.37. How may the chromatic number of a hypergraph change after the deletion of a) an edge b) a vertex which is not incident to an edge of degree two? Give examples.

11.1.38. Prove that a hypergraph is bicolorable if it has no cycles of odd length. Is the converse statement true?

11.2 Hypergraph Realizations

A *realization of a hypergraph* $H = (V, \mathcal{E})$ is a graph G that satisfies the following conditions:

(1) $VG = VH$;

(2) every edge of G is contained in some edge of H;

(3) for every edge $e \in \mathcal{E}$ the induced subgraph $G(e)$ is connected.

A *realization of an edge* $e \in \mathcal{E}$ of a hypergraph H is any connected graph G_e with $VG_e = e$. Thus, every realization of a hypergraph is the union of some realizations of all its edges.

If a realization of a hypergraph is a planar graph then we speak about a *planar realization.* When a realization is a tree or a cycle, etc., we call it a *realization of the hypergraph by tree, by cycle, etc.,* and the hypergraph is said to be *realizable by tree, cycle, etc.*

It is said that a hypergraph satisfies the *Helly conditions* if the intersection of any subset of pairwise adjacent edges is nonempty.

Theorem 11.2.1 *A connected hypergraph H is realizable by tree if and only if H satisfies the Helly condition and the line graph $L(H)$ is triangulated.*

A realization G of a hypergraph $H = (V, \mathcal{E})$, $\mathcal{E} = \{e_1, e_2, \ldots, e_m\}$, is called *strict* if there exists a partition of EG into subsets $E_1, \ldots, E_m$ such that $G(E_i)$ is a realization of e_i, $i = 1, 2, \ldots, m$.

* * *

11.2.1. Are the hypergraphs H_1, H_2 from Fig. 11.2.1 realizable by tree?

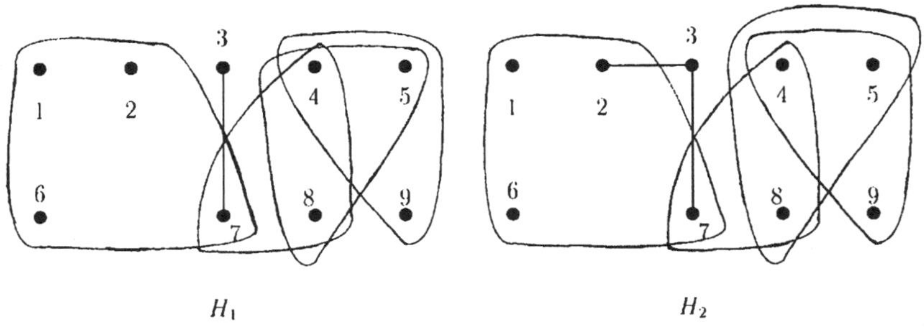

Figure 11.2.1: To Exercise 11.2.1

11.2.2. Do the hypergraphs shown at Fig. 11.2.2 and Fig. 11.2.3 have planar realizations?

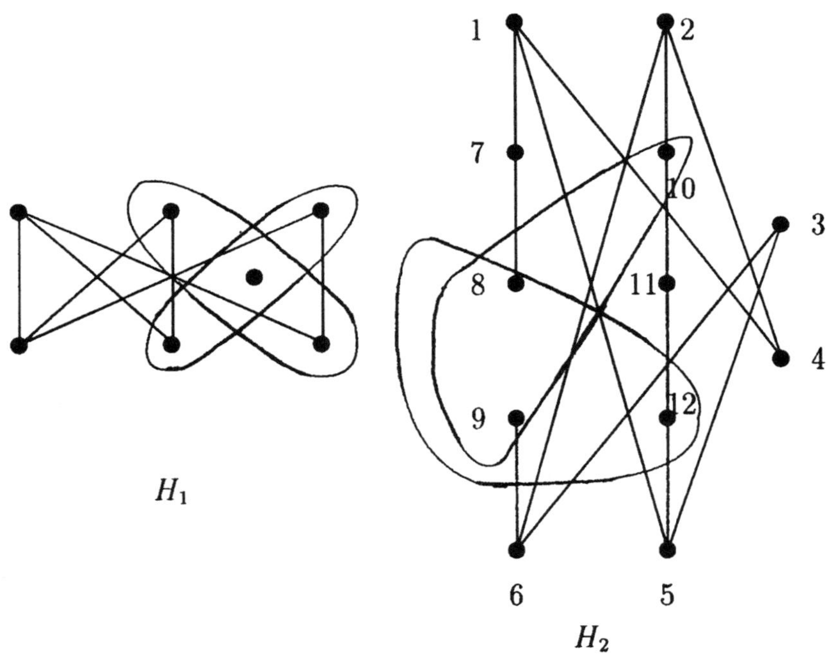

Figure 11.2.2: To Exercise 11.2.2

11.2.3. The *contraction of an edge* e_0 of a hypergraph $H = (V, \mathcal{E})$ is the operation that produces the hypergraph $H' = (V', \mathcal{E}')$ such that $V' = (V \backslash e_0) \cup \{v_0\}$, where $v_0 \notin V$, and $\mathcal{E}' = \mathcal{E}_1 \cup \mathcal{E}_2$, where

$$\mathcal{E}_1 = \{e' : e' = (e \backslash e_0) \cup \{v_0\},\ e \in \mathcal{E} \backslash \{e_0\},\ e \cap e_0 \neq \emptyset\},$$

$$\mathcal{E}_2 = \{e \in \mathcal{E} : e \cap e_0 = \emptyset\}.$$

The *insertion of a vertex* $v_0 \notin VH$ *into an edge* e_0 of a hypergraph H is an operation producing the hypergraph $H' = (V', \mathcal{E}')$ such that $V' = V \cup \{v_0\}$, $\mathcal{E}' = (\mathcal{E} \backslash \{e_0\}) \cup e^*$, $e^* = e_0 \cup \{v_0\}$.

Prove the following statements.

a) If there exists a planar realization of a hypergraph H then there exists a planar realization of a hypergraph $\tilde{H}$ obtained from H by (1) edge deletion; (2) edge contraction; (3) insertion of vertex into edge; (4) deletion of vertex of degree 1;

b) If there exists a planar realization of a hypergraph $\tilde{H}$ obtained from H by the contraction of the edges of a subset $\mathcal{E} \subseteq \mathcal{E}H$ then there exists a planar realization of a hypergraph obtained from H by the contraction of the edges of any subset $\mathcal{E}''$, $\mathcal{E}' \subseteq \mathcal{E}'' \subseteq \mathcal{E}H$.

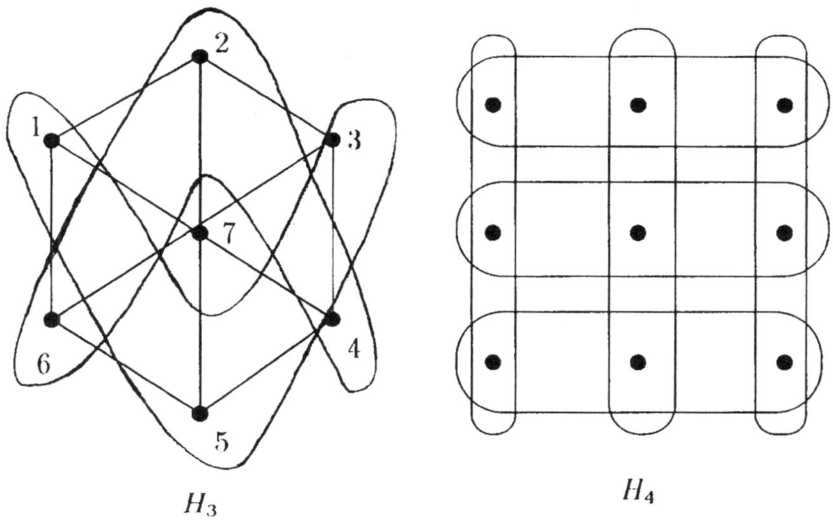

Figure 11.2.3: To Exercise 11.2.2

11.2.4. Find a hypergraph H such that it has a planar realization but no hypergraph $H - v$, $\deg v \geq 2$, has a planar realization.

11.2.5. Prove that a planar realization of a hypergraph H exists if the König representation of H is planar. Give an example when a planar realization of a hypergraph exists but its König representation is nonplanar.

11.2.6. a) Prove that a connected hypergraph with at most three edges is realizable either by tree or by simple cycle.

b) Under what conditions it is realizable by tree?

11.2.7. Prove that a hypergraph with at most four edges has a planar realization.

11.2.8. Prove that a hypergraph has a planar realization if and only if every its block has a planar realization.

11.2.9. Is it true that if the König representation of a hypergraph is planar then the dual hypergraph has a planar realization?

11.2.10. Is it true that if a hypergraph has a planar realization then the dual hypergraph has a planar realization?

11.2.11. Prove that if the hypergraph obtained from a hypergraph H by the deletion of two vertices is realizable by path graph then H has a planar realization.

11.2.12. Suppose that a hypergraph H is realizable by a path graph P. Prove that for any two adjacent edges e', e'' of H the induced subgraph $P(e' \cap e'')$ is connected.

11.2.13. Suppose that a hypergraph H' is obtained from a hypergraph H by the deletion of two vertices which do not constitute an edge of degree two. Prove that if H' is realizable by cycle graph then H has a planar realization.

11.2.14. Suppose that a hypergraph has a planar realization. Is it true that any its minimal under inclusion realization is a planar graph?

11.2.15. Suppose that a cycle graph C is a realization of a hypergraph $H = (V, \mathcal{E})$ and there exist two adjacent edges $e', e'' \in \mathcal{E}$ such that the induced subgraph $C(e' \cap e'')$ is not connected. Prove that $e' \cup e'' = V$.

11.2.16. Prove that a hypergraph is k-colorable if and only if it is realizable by k-colorable graph.

11.2.17. What is the chromatic number of a hypergraph realizable by tree?

11.2.18. Prove that the chromatic number of a hypergraph with planar realization is at most four.

11.2.19. Do the following hypergraphs have strict realizations:

a) $H = (V, \mathcal{E})$ with $V = \{1, 2, 3, 4\}$, $\mathcal{E} = \{\{1, 2, 3\}, \{1, 2, 4\}, \{1, 3, 4\}, \{2, 3, 4\}\}$;

b) the hypergraph H_1 shown at Fig. 11.2.1?

11.2.20. Find a non-strict realization of the hypergraph H_1 shown at Fig. 11.2.1.

11.2.21. Prove that a strict realization of a connected hypergraph by tree exists if and only if the hypergraph has no cycles.

11.2.22. Let v be a hyperpendant vertex of a hypergraph H (see Exercise 11.1.27) and let $t(H)$ denote the minimal number of edges in a realization of H. Prove that a) $t(H) = t(H - v) + 1$; b) if H is realizable by tree then $H - v$ is also realizable by tree.

11.2.23. Prove that if a hypergraph is realizable by tree then it satisfies the Helly condition.

11.2.24. A cycle of a hypergraph that contains an edge incident to three vertices of this cycle is called *balanced.* A hypergraph is called *balanced* if every cycle of odd length $l \geq 3$ is balanced. Prove that every balanced hypergraph satisfies the Helly condition.

11.2.25. A hypergraph is called *strongly balanced* if every cycle of length $l \geq 3$ is balanced. Prove that a hypergraph is strongly balanced if and only if every its connected part is realizable by tree.

11.2.26. Prove that a hypergraph is conformal (see Exercise 11.1.25) if and only if its dual satisfies the Helly conditions.

Answers to Chapter 1:

ABC of Graph Theory

1.1 Graphs: Basic Notions

1.1.3. c: The empty graph.

1.1.4. K_n.

1.1.5. C_n^2.

1.1.6. a: No.

1.1.7. No.

1.1.8. a: 9; b: 7; c: There are no such graphs.

1.1.9. Hint: See Exercise 1.1.7 and proceed by contradiction.

1.1.10. No.

1.1.11. a: Hint: Count the number of edges in such graph.

b: Yes. Hint: Consider an appropriate bipartite graph.

c: Hint: See Exercise 1.1.11a.

1.1.12. a: Hint: Use the fact that if a, b are real numbers and $a + b = n$ then $ab \leq n^2/4$.

b: $K_{p,p}$ for $n = 2p$.

1.1.13. Complete bipartite graphs.

1.1.14. Selfcomplementary graphs: O_1, P_4, C_5.

1.1.16. a: Bipartite r-regular graphs of order n exist only for $n = 2m$, $0 \leq r \leq m$. One of them with parts A and B may be defined as follows. Let $A = \mathbf{Z}_m = \{0, 1, \ldots, m-1\}$ be the additive group of residuals modulo m and let B be another copy of this group. If $\overline{x} \in B$ then we set $N(\overline{x}) = \{x + y \in A : y = 0, 1, \ldots, r-1\}$. Here $\overline{x} \in B$ denotes the copy of $x \in A$.

b: There are no such graphs.

c: Hint: If G is the required graph then $|E\overline{G}| = n - 1$.

1.1.17. For even n the graph is unique, for odd n there are no such graphs.

1.1.18. a: $2^{C_n^2}$; b: $C_{C_n^2}^m$ if $m \le C_n^2$, and 0 otherwise.

1.1.20. Hint: Two nonisomorphic graphs satisfying conditions a)-c) are shown at Fig. A1.1.1.

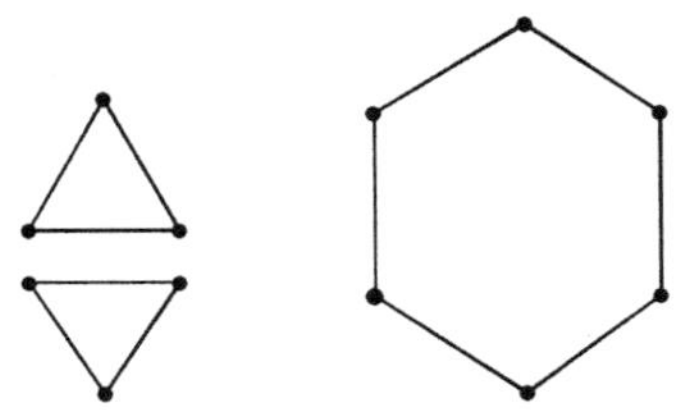

Figure A1.1.1: To Answer 1.1.20

1.1.21. Hint: If $f : VG \to VH$ is an isomorphism of graphs G and H then for any edge $e = uv \in EG$ the set $f(e) = \{f(u), f(v)\}$ is an edge of H. Prove that the mapping $e \mapsto f(e)$ specifies an isomorphism of line graphs $L(G)$ and $L(H)$.

1.1.22. b: $G_1 \cong G_2 \cong G_3$, and there are no other pairs of isomorphic graphs at Fig. 1.1.8.

1.1.24. Hint: The numbers of pairwise nonisomorphic graphs of order n are: 4 for $n = 3$; 11 for $n = 4$; 34 for $n = 5$.

1.1.25. a: The complete graph K_4 is the unique cubic graph of order 4.

b: The number of pairwise nonisomorphic cubic graphs of order 6 is 2, see Fig. A1.1.2. Hint: Consider the complementary graphs.

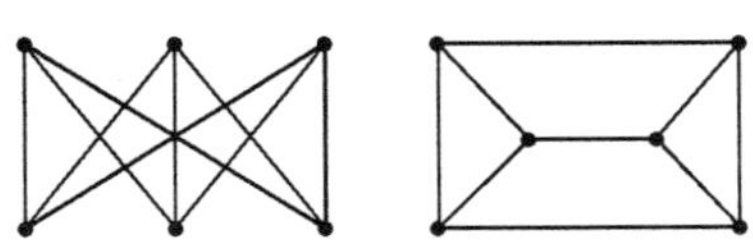

Figure A1.1.2: Cubic graphs of order 6

c: The number of pairwise nonisomorphic cubic graphs of order 8 is 6, see Fig. A1.1.3.

Hint: Let G be the required graph. First prove that G contains either the triangle C_3, i.e., three pairwise adjacent vertices, or the cycle C_4, i.e., four vertices a, b, c, d such that $\{ab, bc, cd, da\} \subset EG$. Further, consider the following five possibilities:

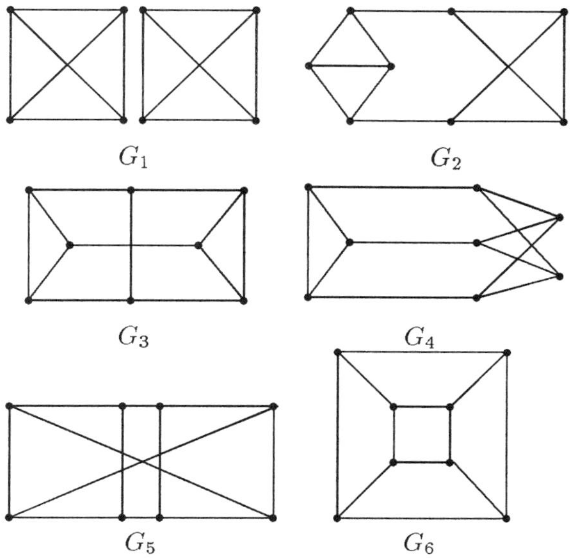

Figure A1.1.3: Cubic graphs of order 8

(1) G contains K_4, i.e., four pairwise adjacent vertices;

(2) G does not contain K_4, but it contains $K_4 - e$, i.e., four vertices a, b, c, d such that $\{ab, bc, cd, da, ac\} \subseteq EG$ and $bd \notin EG$;

(3) G does not contain neither K_4, nor $K_4 - e$, but it contains two triangles;

(4) G contains exactly one triangle;

(5) G contains no triangles.

Prove that $G \cong G_i$ (Fig. A1.1.3) in the case (i), $i = 1, \ldots, 6$. In the case (5) $G \cong G_5$ or $G \cong G_6$. G_5 contains the cycle C_5 and G_6 does not.

1.1.26. b: The number of pairwise nonisomorphic cubic circulant graphs of order 6 is 2.

c: Yes.

1.1.27. a: Hint: See Fig. 1.2.4a.

b: A regular graph of degree C_n^2. The number of edges is $C_n^2 \frac{n!}{2}$.

c: Solution: For $n = 1$ the statement is trivial. Assume that $n \geq 2$. Since the application of a single transposition changes the parity of the permutation, no two permutations of the same parity are adjacent in the n-permutation graph. Hence it is a bipartite graph, one part being all even permutations, the other one being all odd permutations.

1.1.28. Solution: Consider the set of all prime numbers in the ascending order: 2, 3, 5.... Let p_i be the i-th prime number. Let further G be an

(n, m)-graph. Let us label its vertices by numbers $p_1, \ldots p_n$ and let us label its edges by the numbers $p_{n+1}, \ldots, p_{n+m}$. Afterwards we multiply the label of each vertex by the labels of the edges incident to it. This gives a common divisor graph isomorphic to G.

1.1.30. a: The number of edges of $\overline{G}$ is $C_n^2 - m$. The number of edges of a selfcomplementary graph of order n is $n(n-1)/4$.

b: Hint: See Exercise 1.1.30a.

1.1.31. a: P_4; b: C_5 and the graph shown at Fig. A1.1.4; c: there are no such graphs.

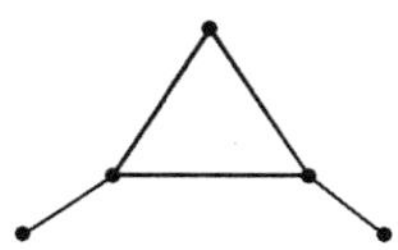

Figure A1.1.4: To Exercise 1.1.31

1.1.33. Hint: Use Exercise 1.1.32.

1.1.34. No. Hint: Such graph would contain a single vertex of degree 49. Let u be a vertex of degree 50 and let v be of degree 49. Is there an edge uv in the graph?

1.1.35. Solution: The number of vertices of the graph $L(G)$ is $m = |EG| = \frac{1}{2}\sum_{i=1}^{n} d_i$. Two edges of G are adjacent as vertices of $L(G)$ if and only if G has a vertex incident to both of them. Therefore all edges of G incident to the same vertex are pairwise adjacent in $L(G)$. Hence

$$|EL(G)| = \sum_{v \in VG} C_{\deg v}^2 = \frac{1}{2}\sum_{i=1}^{n}(d_i^2 - d_i) = \frac{1}{2}\sum_{i=1}^{n} d_i^2 - m.$$

1.1.37. a: K_3; b: The line graph of the tetrahedron is isomorphic to the octahedron.

1.1.38. a: Yes; b: No.

1.1.40. Solution: Suppose that $V = VK_5 = \{1, 2, 3, 4, 5\}$. Then

$$V\overline{L(K_5)} = VL(K_5) = \{xy : x, y \in V, x \neq y\}.$$

The vertices xy and uv are adjacent in $\overline{L(K_5)}$ if and only if $\{x, y\} \cap \{u, v\} = \emptyset$. Hence $G = \overline{L(K_5)}$ is a 3-regular graph. The induced subgraphs (see Sect. 1.3) $G(\{12, 34, 15, 23, 45\})$ and $G(\{13, 24, 35, 14, 25\})$ are isomorphic to C_5. Further see Fig. A1.1.5.

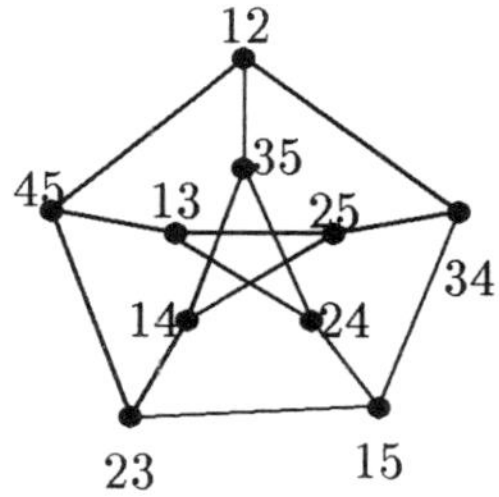

Figure A1.1.5: To Exercise 1.1.40

1.2 Walks, Paths, Components

1.2.2. a: $\frac{1}{2}\sum_{k=3}^{n}(k-1)!C_n^k$ if $n \geq 3$, and 0 for $n \leq 2$;

b: $\frac{1}{2}\sum_{k=2}^{\min(p,q)} k!(k-1)!C_p^k C_q^k$ if $p \geq 2$, $q \geq 2$, and 0 otherwise.

Hint: Let A, B be the parts of $K_{p,q}$, $p \leq q$, The length of any simple cycle of $K_{p,q}$ is an even number $2k$, $2 \leq k \leq p$. In such cycle the vertices from A and from B alternate. Finally, the number of all simple cycles with the same vertex set V, $|V| = 2k$, is $(k-1)!k!/2$.

1.2.4. a: 3 for $n \geq 3$, ∞ for $n = 1, 2$;

b: 4 for $p, q \geq 2$; ∞ for p or q equal to 1;

c: 5.

1.2.5. a: Hint: Delete the "excessive" pieces of the walk. b: False; c: False.

1.2.10. Hint: Proof is by contradiction using the Handshake Lemma.

1.2.11. Hint: Starting from any vertex v, move along an incident edge e into the adjacent vertex w. Similarly, move from w further along an edge other than e. Continue this process until some vertex will be visited twice. Select a cycle in the traversed walk.

1.2.12. a: Hint: Suppose that

$$P = (u_1, \ldots, u_k),\ Q = (v_1, \ldots, v_m)$$

are given paths, $u_1 = v_1 = u$, $u_k = v_m = v$, u_α and v_α are the first (counting from u) different vertices of these paths, u_β and v_γ are the first coinciding vertices of these paths after u_α and v_α. Consider the union of the $(u_{\alpha-1}, u_\beta)$-subpath of P and $(v_{\alpha-1}, v_\gamma)$-subpath of Q.

1.2.13. a: Hint: Use König's theorem.

b: False.

1.2.14. See Fig. A1.2.1. The graphs $G_1 = K_1$ and $G_6 = P_4$ are selfcomplementary.

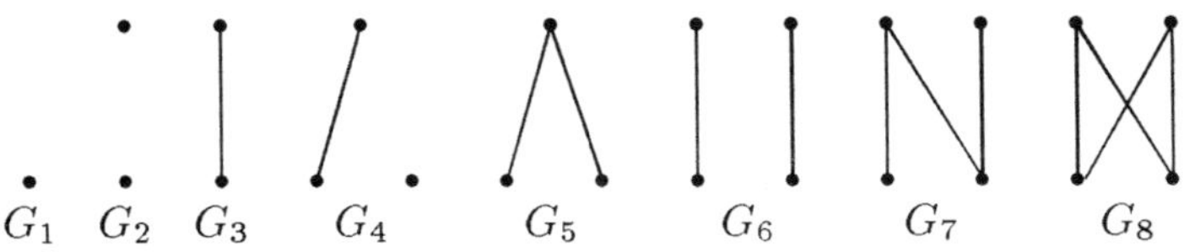

Figure A1.2.1: To Exercise 1.2.14

1.2.15. $K_{m,n}$, $m, n \geq 1$, $m + n \geq 3$. Hint: Demonstrate that after the deletion of any edge $e = uv$ from $K_{m,n}$ either $d(u, v) = 3$ in the resulting graph or it becomes disconnected.

1.2.16. Hint: Use the previous exercise.

1.2.17. Hint: The number of such graphs is 5.

1.2.18. a: the star $K_{1,n}$, $n \geq 2$;

b: K_n;

c: K_2;

d: the star $K_{1,n}$;

e: the star $K_{1,n}$ or the triangle C_3;

f: the star $K_{1,n}$, $n \geq 2$;

g: the path graph P_n, $n \geq 3$;

h: the cycle graph C_n. Hint: Using the Handshake Lemma show that G has no vertices of degrees greater than 2. Then notice that $|VG| = |EG|$.

1.2.19. a: K_1; b: K_2; c: C_n.

1.2.20. Hint: In any walk that contains the deleted edge one may replace this edge by the remaining edges of the cycle. The deletion of an edge that does not belong to any cycle of a connected graph produces a graph with two connected components.

1.2.21. $n - 1$. Hint: Use the previous exercise and then the induction over the number of edges of the graph.

1.2.22. Hint: Consider the shortest path connecting the ends of the edges e_1, e_2.

1.2.23. a: The lower bound is attained for the graphs with no cycles. The upper bound is attained only for the graph with one component isomorphic to K_{n-k+1} and the remaining components trivial.

Hint: To prove the lower bound, use the Exercise 1.2.21. For the upper bound, notice that in a graph with maximal number of edges all components are complete graphs and then use the fact that if $p \geq q \geq 1$ then

$$|EK_{p+1}| + |EK_{q-1}| > |EK_p| + |EK_q|.$$

b: If $n - k \neq 2$ then there exists a unique n-vertex bipartite graph G with maximal number of edges. It is obtained by the addition of $k - 1$ isolated vertices to $K_{p,q}$, where $p = (n - l)/2$, $q = p + 1$ or $p = q = (n - l + 1)/2$, depending on parity. For $n - l = 2$ there is one more such graph, i.e., $2K_2 + (l - 2)K_1$.

Hint: The reasoning is the same as in item a.

c: Hint: Using the item b, construct a graph of order n with k components and maximal possible number of edges. Deleting the edges from it while keeping the number of the components constant, we obtain the required graphs.

1.2.24. a: The statement will become false. Hint: Suppose that the graph is disconnected and consider the maximal possible degree of a vertex in the component of the smallest order.

b: No, we cannot. Hint: Use the Exercise 1.2.23a.

c: It will become invalid. Hint: Use König's theorem and the Exercise 1.2.23a.

1.2.25. a: Hint: Use the theorem stating that any permutation of n numbers may be produced from any other one via an appropriate sequence of transpositions.

b: $n-1$. Hint: Use the theorem stating that the minimal number of transpositions whose product is equal to a given permutation s of the set $\{1, 2, \ldots, n\}$ is equal to its decrement $n - c$, where c is the number of independent cycles of s.

c: Each vertex is both central and peripherical. Hint: For any permutation $a_1, a_2, \ldots, a_n$ the decrement of the substitution

$$\begin{pmatrix} a_1 & a_2 & \ldots & a_{n-1} & a_n \\ a_2 & a_3 & \ldots & a_n & a_1 \end{pmatrix} = (a_1, a_2, \ldots, a_n)$$

is $n - 1$.

1.2.26. a: Solution: Let u be a central vertex of G. Then

$$d(u, w) \leq \max_{v \in VG} d(u, v) = e(u) = r(G)$$

for every vertex $w \in VG$. Consider any diametral (a, b)-path L. Since L is the shortest (a, b)-path, its length does not exceed the length of the walk composed of the shortest (a, u)-path and the shortest (u, b)-path.

b: Hint: Prove that a diametral (a, b)-path contains two adjacent vertices u, v such that $e(u) = d(u, a)$, $e(v) = d(v, b)$. Further, use the Exercise 1.2.26a.

1.2.28. a: Hint: Suppose that G is disconnected and consider the partition $VG = A \cup B$, where A is a connected domain of G, $B = VG \backslash A$.

c: Solution: Let L be a diametral (a,b)-path. Since its length is at least 3, the vertices a and b are nonadjacent. By the same reason, every $v \in VG$ is not adjacent either to a or to b. Hence $\overline{G}$ has the edge ab and each vertex of $\overline{G}$ is adjacent either to a or to b. This implies that $d(\overline{G}) \le 3$.

1.2.29. a: $n-1$, if $n>1$; b: $p+q-2$, if $p+q \ge 3$; c: 2; d: 1.

1.2.30. Hint: Consider a diametral path $(v_1, v_2, \dots, v_d, v_{d+1})$ of G and the set

$$A = VG \backslash \{v_1, v_2, \dots, v_d\}.$$

1.2.31. Hint: Use König's theorem.

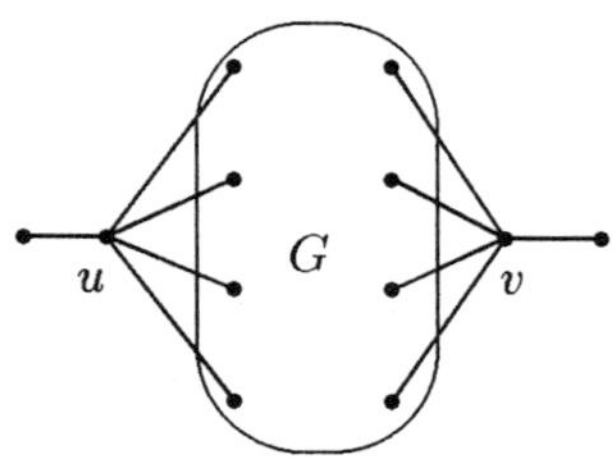

Figure A1.2.2: To Exercise 1.2.35

1.2.32. Solution: Let v be a vertex of degree $\Delta(G)$ and let $N(v)$ be its neighborhood. Consider $u \in N(v)$; clearly, $|N(u)| \ge \delta(G)$. Since G has no triangles, $N(v) \cap N(u) = \emptyset$. This implies the required inequality.

1.2.33. Hint: Use the Exercise 1.2.32.

1.2.35. Hint: See the graph at Fig. A1.2.2 in which u and v are adjacent to all vertices of G.

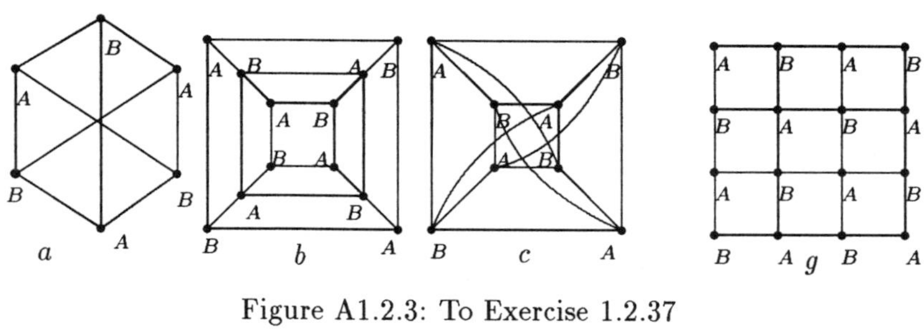

Figure A1.2.3: To Exercise 1.2.37

1.2.36. a: Hint: Let the vertex $\tilde{x} \in VK(G)$ correspond to the edge $x = uv$ of G. From the definition of $K(G)$ it follows that $\tilde{x}$ is adjacent only to $\tilde{u}$ and $\tilde{v}$ which correspond to the vertices $u$and v of G.

b: Hint: Use the Exercise 1.2.36a.

1.2.37. The graphs a, b, c, g are bipartite. At Fig. A1.2.3 the vertices of one part are marked by A and the vertices of the other one are marked by B.

1.2.38. a: $d(P_n) = n-1$; $r(P_n) = \begin{cases} (n-1)/2, & \text{if } n \text{ is odd,} \\ n/2-1, & \text{otherwise.} \end{cases}$

The center of P_n consists of a single vertex if n is odd and of two vertices if n is even. The endvertices are the peripherical vertices.

b: $d(C_n) = r(C_n) = \begin{cases} (n-1)/2, & \text{if } n \text{ is odd,} \\ n/2, & \text{otherwise.} \end{cases}$

Every vertex is both central and peripherical.

c: $d(K_n) = r(K_n) = 1 (n \geq 2)$; every vertex is both central and peripherical.

d: $d(K_{1,n}) = 2,\ r(K_{1,n}) = 1 (n \geq 2)$; every pendant vertex is peripherical. The center consists of a single vertex.

e: $d(K_{m,n}) = r(K_{m,n}) = 2 (n \geq 2, m \geq 2)$; every vertex is both central and peripherical.

1.2.39. a: $r(G) = 2,\ d(G) = 2$; very vertex is both central and peripherical.

b: $r(G) = 3,\ d(G) = 4$; the central and the peripherical vertices are shown at Fig. A1.2.4a. (At Fig. A1.2.4, the central vertices are marked by "*" and the peripherical ones are marked by "+".)

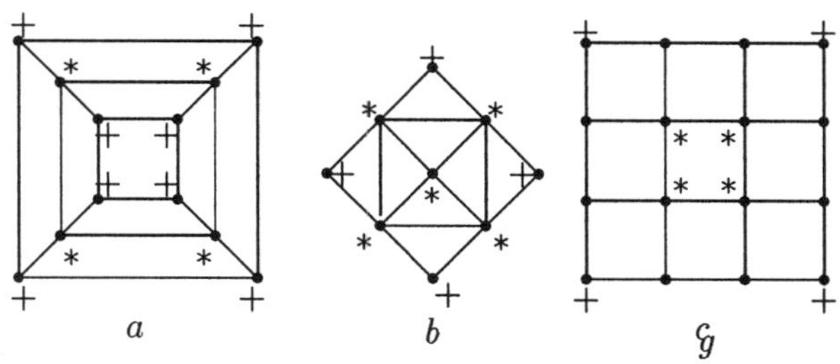

Figure A1.2.4: To Exercise 1.2.39

c: $r(G) = 2,\ d(G) = 2$; each vertex is both central and peripherical.

d: $r(G) = 3,\ d(G) = 3$; each vertex is both central and peripherical.

e: $r(G) = 2,\ d(G) = 2$; each vertex is both central and peripherical.

f: $r(G) = 2,\ d(G) = 3$; the central and the peripherical vertices are shown at Fig. A1.2.4b.

g: $r(G) = 4,\ d(G) = 6$; the central and the peripherical vertices are shown at Fig. A1.2.4c.

1.2.41. Solution: Let $G(n, U)$ be a circulant graph with $U = \{u_1, \ldots, u_k\}$. Let d denote the largest common divisor of $u_1, \ldots, u_k, n$. Vertices a, b are in the same component of G if and only if G has an (a, b)-path. In the considered situation this means that b is obtained from a by the addition of elements from U. In other words, there are integers $x_1, \ldots, x_k, y$ such that

$$u_1x_1 + \ldots + u_kx_k + yn = b - a.$$

But the above equation is solvable in integers if and only if $b - a$ is divisible by d. Therefore the number of components of G is equal to the number of residuals modulo d, i.e., it is equal to d.

1.3 Subgraphs and Hereditary Properties of Graphs. Reconstructibility

1.3.3. a: (1) Yes; (2) only C_8; (3) only C_5.
b: No.

1.3.5. a: $2^n - 1$. b: 2^m. c: $\sum_{i=1}^{n} C_n^2 2^{i(i-1)/2}$.

1.3.6. No.

1.3.7. (a) No. (b) Yes.

1.3.8. (a) O_n. (b) $2K_2$. (c) C_4. (d) K_n.

1.3.9. Graphs without cycles.

1.3.10. Hint: Consider the graph H_G described in Exercise 1.1.30c.

1.3.11. No.

1.3.12. Hint: Prove that such graph must be bipartite. Therefore it must have three pairwise nonadjacent vertices.

1.3.13. Let G be a graph with $\Delta(G) = \Delta$. We extend it as follows. For every vertex $v \in VG$ we add a new set S_v of $\Delta - \deg v$ pendant vertices adjacent to v (the sets S_v are pairwise disjoint). Further, we take a nonnegative integer t such that $|\bigcup_{v \in VG} S_v| + 2t \geq \Delta - 1$ and add $2t$ more new vertices $x_1, \ldots, x_t, y_1, \ldots, y_t$ together with edges x_iy_i, $i = 1, 2, \ldots, t$. The set of all vertices added to G is denoted by S_1, and the resulting graph is denoted by G_1. Let G_2 be a copy of G_1 (with VG_2 disjoint from VG_1) with $S_2 \subset VG_2$ being the corresponding copy of S_1. Finally, we construct a $(\Delta - 1)$-regular bipartite graph G_3 with parts S_1 and S_2 (see the answer to the Exercise 1.1.16a). Clearly, the graph $G_1 \cup G_2 \cup G_3$ is the required graph.

1.3.14. K_1 and the graphs shown at Fig. A1.3.1.

1.3.15. Hint: Consider the set of graphs which do not possess the given hereditary property.

1.3.16. Graphs whose components are complete graphs.

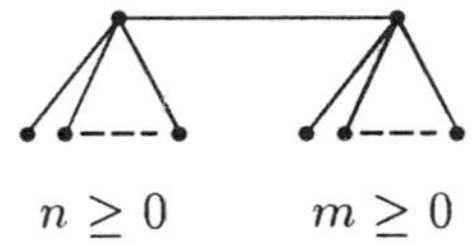

Figure A1.3.1: To Exercise 1.3.14

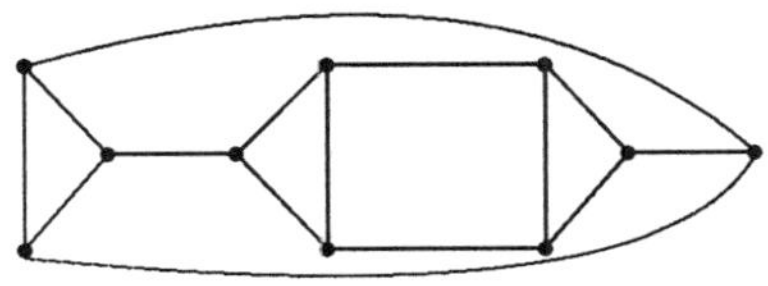

Figure A1.3.2: To Exercise 1.3.17

1.3.17. A graph shown at Fig. A1.3.2.

1.3.18. Suppose that the required regular graph G exists. Let w denote a central vertex of the star $K_{1,n}$ induced by the neighborhood of a vertex $v \in VG$. Since $n \geq 2$, there exists a vertex $x \in N(v)\backslash w$.

The vertex x is adjacent to no vertex from $N(v)\backslash w$, since $N(v)$ induces the star $K_{1,n}$. Moreover, it is adjacent to no vertex $y \notin N(v) \cup v$, since its neighborhood $N(x)$ must also induce the star $K_{1,n}$. Therefore $\deg x = 2 \leq \deg v$, which contradicts to the regularity of G.

1.3.19. a: K_4. b: $\overline{3K_2}$.

1.3.20. $\{H_1, H_2, \ldots\} = \{\overline{H}_1, \overline{H}_2, \ldots\}$.

1.3.21. (a) Yes; a minimal FIS is $\{K_2\}$. (b) Yes; a minimal FIS is $\{O_2\}$. (c) No. (d) No. (e) No. (f) Yes; a minimal FIS is $\{P_3\}$. (g) Yes; a minimal FIS is $\{K_{1,3}, C_n, n = 3, 4, \ldots\}$. (h) Yes; a minimal FIS is $\{C_3, C_4, P_4\}$. (i) No. (j) No. (k) Yes; a minimal FIS is

$$\{K_{1,3}, \overline{K_1 \cup P_3}, K_4, K_4 - e\}.$$

(l) No. (m) No. (n) Yes; a minimal FIS is $\{C_3, C_5, \ldots, C_{2k+1}, \ldots\}$. (o) Yes; a minimal FIS is $\{C_n, n = 3, 4, \ldots\}$. (p) No. (q) No.

1.3.22. a: No. b: Yes.

1.3.23. Hint: Sufficiency. Every simple cycle C_{2k+1} is either isomorphic to K_3 or contains the induced subgraph $\overline{P_3}$. Therefore a graph without induced K_3 and $\overline{P_3}$ is bipartite. Furthermore, every non-isolated vertex is adjacent to all vertices from the opposite part; otherwise one could readily find the induced $\overline{P_3}$.

1.3.24. Yes.

1.3.25. a: Yes. b: Yes.

1.3.26. a: Hint: Use the Exercise 1.3.25a.

b:Hint: Clearly, every graph from the class $[\mathcal{K}]$ has no induced subgraphs P_4, since $P_4 \notin [\mathcal{K}]$. Now we can demonstrate that every graph that has no induced P_4 is in $[\mathcal{K}]$. To this end one may apply the induction over the number of vertices. Let us assume the induction hypothesis (i.e., every graph of order $n > 1$ with no induced subgraphs P_4 is in $[\mathcal{K}]$) and consider a graph G of order $n+1$ with no induced subgraphs P_4. Consider $G - v$, $v \in VG$, and show that we may assume that this subgraph is disconnected. Using the fact that G has no induced subgraphs P_4, prove that v is adjacent to every vertex $w \in VG \backslash \{v\}$, and hence the graph is not connected.

1.3.27. Hint: First reconstruct the set of vertices of the graph together with their labels. Then reconstruct the edge set following the rule: the vertices with labels i, j are adjacent in the reconstruction if and only if they are adjacent in at least one card of the labelled deck. Substantiate this rule.

1.3.28. a: No. b: Yes.

1.3.29. a: Yes. b: No.

1.3.31. $\deg v = \sum_{u \in VG} m_u/(|G|-2) - m_v$, where $m_x = |EG_x|$.

1.3.32. The number of edges of G is $(\sum_{v \in VG} m(G_v))/(|G|-1)$, where $m(G_v)$ is the number of edges of the card G_v.

The number of components of G is $\min\{k(G_v) : v \in VG\}$, where $k(G_v)$ is the number of components of the card G_v.

1.3.33. Hint: Use the fact that every subgraph F of G is contained in each of $|G| - |F|$ cards G_v such that $v \notin VF$.

1.3.34. K_4.

1.3.35. No.

1.3.36. Yes.

1.3.40. Solution: Assume that $V = V(K_5) = \{1,2,3,4,5\}$ and $V(L(K_5)) = \{xy : x, y \in V,\ x \neq y\}$. The set of maximal cliques of $L(K_5)$ is separable into two classes. The first class consists of the four-element cliques:

$$Q_v = \{vx : x \in V,\ x \neq v\},\ v \in V.$$

The second class consists of the three-element cliques:

$$Q_{uvw} = \{uv, uw, vw\},\ u, v, w \in V,$$

where u, v, w are pairwise different. A clique Q_v consists of the edges of K_5 incident to the vertex v. A clique Q_{uvw} corresponds to the triangle in K_5 with the vertices u, v, w.

Thus, $|Q(L(K_5))| = 5 + C_5^3 = 15$.

The neighborhoods of the vertices of $Q(L(K_5))$ are the following.

$$N(Q_v) = \{Q_x,\ Q_{vxy} :\ x, y \in V,\ x \neq y,\ x \neq v,\ y \neq v\};$$

$$N(Q_{uvw}) = \{Q_x,\ Q_{yzt} :\ x = u, v, w,\ |\{y, z, t\} \cap \{u, v, w\}| = 2\};$$

$\deg Q_v = 10$; $\deg Q_{uvw} = 9$.

1.3.41. Hint: Consider the correspondence $v \mapsto Q_v$, where $v \in VG$ and Q_v is the clique of $L(G)$ consisting of all edges of G incident to v.

1.3.42. c: Solution: Necessity. Since $K_{1,3}$ is not a line graph (see Exercise 1.1.36a), then by item a) the graph G contains no induced subgraphs $K_{1,3}$. It is easily verified that there exists a single connected graph (see Fig. A1.3.3) such that $L(H) = K_4 - e$. Since H contains a triangle, by item b) the graph G contains no induced subgraphs $K_4 - e$.

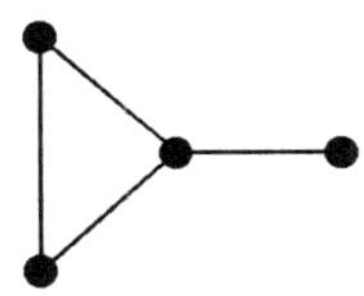

Figure A1.3.3: To Exercise 1.3.42

1.3.43. Hint: Assume that

$$VG = \{1, 2, \ldots n\},\ EG = \{e_1, e_2, \ldots e_m\}$$

and construct the sequences

$$\alpha^1 = (\alpha_1^1, \alpha_2^1, \ldots \alpha_m^1), \ldots, \alpha^n = (\alpha_1^n, \alpha_2^n, \ldots \alpha_m^n)$$

by setting $\alpha_j^i = 1$ if the vertex i is incident to the edge e_j and $\alpha_j^i = 0$ otherwise. To each α^i, $i = 1, \ldots, n$, we append the sequence $\beta^i = (\beta_1^i, \ldots \beta_{n^2-2m}^i)$, where $\beta_j^i = 1$ if

$$(i-1)n - (d_1 + \ldots + d_{i-1}) + 1 \leq j \leq in - (d_1 + \ldots + d_i)$$

and $\beta_j^i = 0$ otherwise. If $i = 1$ we assume that $d_1 + \ldots + d_{i-1} = 0$. Denote

$$\gamma^i = (\alpha_1^i, \ldots \alpha_m^i, \beta_1^i, \ldots \beta_{n^2-2m}^i).$$

Then for adjacent vertices i, j,

$$d(\gamma^i, \gamma^j) = d_i - 1 + d_j - 1 + n - d_i + n - d_j = 2n - 2,$$

and for nonadjacent vertices i, j,

$$d(\gamma^i, \gamma^j) = d_i + d_j + n - d_i + n - d_j = 2n.$$

To conclude the proof, consider the Hamming graph $H_{n^2-m, 2n-2}$.

1.4 Operations on Graphs

1.4.3. $G + H = \overline{\overline{G} \cup \overline{H}}$.

1.4.4. No.

1.4.5. No.

Hint: Consider the property specified by the forbidden induced subgraph $2K_2$.

1.4.6. The edge e belongs to no cycle of the graph G.

1.4.7. Hint: Use the fact that the distance $d(u, v)$ is odd.

1.4.8. $\lfloor n/2 \rfloor$.

1.4.9. The number k is equal to the diameter of the graph.

1.4.10. a: No. b: Yes.

1.4.11. $|EG| + |EH| + |VG||VH|$.

1.4.12. Hint: Prove that G has a dominating vertex.

1.4.13. See Fig. A1.4.1.

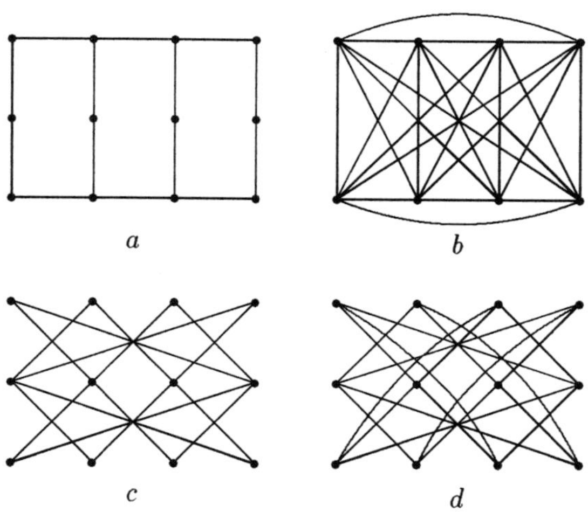

Figure A1.4.1: To Exercise 1.4.13

1.4.14. a: $n_1 m_2 + n_2 m_1$. b: $n_1 m_2 + n_2^2 m_1$. c: $2m_1 m_2$ d: $2(m_1 m_2 + m_1' m_2')$, where $m_i' = n_i(n_i - 1)/2 - m_i$, $i = 1, 2$.

1.4.15. Cartesian product, conjunction and modular product.

1.4.16. $d_i + d'_j$, $i = 1, \ldots, n,$, $j = 1, \ldots, n$.

1.4.18. Yes.

1.4.20. a: Hint: Let the vertex set of a simple cycle C_{2m+1} be $V = \{0, 1, \ldots, 2m\}$, with each vertex $i \in V$ being adjacent to $i + 1$ (addition is modulo $2m + 1$). Verify that the mapping $\varphi : V^2 \to V^2 : \varphi(i, j) = (i + j, 2m + 1 - i + j)$ is an isomorphism of the graphs $C_{2m+1} \times C_{2m+1}$ and $C_{2m+1} \wedge C_{2m+1}$ for any pair $(i, j) \in V^2$.

1.4.21. c: $\deg u \deg v + (|G| - \deg u - 1)(|G'| - 1)$.

1.4.23. a:Solution: Set $H_1 = C_5$. Suppose that the graph H_n have already been constructed. Set $H_{n+1} = C_5 \leftarrow H_n$. Clearly, H_n is a selfcomplementary graph of order 5^n for all $n > 0$.

1.4.26. $|Q_n| = 2^n$, $|EQ_n| = n2^{n-1}$.

1.4.27. a: 8. b: 14.

1.5 Matrices Associated with Graphs

1.5.4. a,b: Hint: Without loss of generality assume that $VG = VH = \{1, 2, \ldots, n\}$. $G \cong H$ implies that there exists a permutation of the set VG such that $ij \in EG \iff s(i)s(j) \in EH$. If $A = A(G)$ and $B = A(H)$ then the above condition may be rewritten as

$$B_{s(i)s(j)} = A_{ij},\ i, j = 1, 2, \ldots, n. \tag{A1.5.1}$$

1.5.6. a: Solution: Since a permutation s is a bijective mapping, for each $i \in X = \{1, 2, \ldots, n\}$ there exists exactly one $j \in X$ such that $s(j) = i$. Hence exactly one element of the i-th row of M_s, say, the j-th element, is equal to 1. A k-th column of M_s also contains a single 1, say, at the position $s(k)$. A determinant which has a single nonzero element in each row and column is equal to the product of its nonzero elements up to the sign.

b: Solution: For $M = M_s$,

$$(MA)_{ij} = \sum_{k=1}^{n} M_{ik} A_{kj}.$$

But $M_{ik} = 1$ for $i = s(k)$ and $M_{ik} = 0$ otherwise. Hence

$$(MA)_{ij} = M_{s(k)k} A_{kj} = A_{kj},\ k = s^{-1}(i).$$

d: Solution: Graphs G, H of order n are isomorphic if and only if there exists a permutation s of the set $\{1, 2, \ldots, n\}$ satisfying the equalities (A1.5.1). This system of equalities is equivalent to a single matrix equality

$$B = M_s A M_s^{-1}. \tag{A1.5.2}$$

In fact, (A1.5.2) is equivalent to $BM_s = M_s A$, i.e.,

$$(BM_s)_{pq} = (M_s A)_{pq}, \; p, q = 1, 2, \ldots, n. \tag{A1.5.3}$$

But

$$(BM_s)_{pq} = \sum_{k=1}^{n} B_{pk}(M_s)_{kq} = B_{pk}, \; k = s(q).$$

In the same way,

$$(M_s A)_{pq} = A_{kq}, \; k = s^{-1}(p).$$

Hence the system (A1.5.3) assumes the form

$$B_{ps(q)} = A_{s^{-1}(p)q}, \; p, q = 1, 2, \ldots, n.$$

Setting $s^{-1}(p) = i$, $q = j$, we obtain (A1.5.1).

1.5.8. a: Hint: If the highest coefficient of a polynomial of degree $n > 0$ is 1 then the coefficient at x^{n-1} is the sum of the roots of the polynomial with the opposite sign. On the other hand, the sum of the roots of the characteristic polynomial of a matrix is equal to the trace of the matrix.

b, c: Hint: The coefficient of the characteristic polynomial of an $n \times n$-matrix at x^{n-2} is equal to the sum of all second-order minors along the main diagonal, and the coefficient at x^{n-3} is the sum of all third-order minors along the main diagonal with the opposite sign.

1.5.9. $x^3(x-2)(x+2)$ is the characteristic polynomial for the first pair of graphs and $(x+1)^2(x-1)(x^3 - x^2 - 5x + 1)$ is the characteristic polynomial for the second pair of graphs.

1.5.10. Hint: Use induction with respect to k.

1.5.13. Solution: Let G be a connected graph of order $n > 1$. First, let us prove that

$$\text{rank } I(G) \geq n - 1. \tag{A1.5.4}$$

To this end, let us choose a spanning tree T_1 in G (see Section 2.2). Assign index 1 to a pendant vertex and to the edge incident to it. Delete these vertex and edge obtaining a tree T_2 with $n-1$ vertices. Assign index 2 to a pendant vertex of T_2 and to the edge incident to it. Proceed in the same way until we are left with a single vertex. Assign index n to it. Thus all vertices of G and $n-1$ its edges become indexed. The remaining edges (if any) are indexed by $n, n+1, \ldots$ in an arbitrary way. Under this indexing, the incidence matrix of G assumes the form

$$I(G) = \begin{bmatrix} n \left\{ \begin{matrix} 1 & 0 & \cdots & 0 & 0 & * & \cdots \\ * & 1 & \cdots & 0 & 0 & * & \cdots \\ & & \cdots & & & & \cdots \\ * & * & \cdots & 1 & 0 & * & \cdots \\ * & * & \cdots & * & 1 & * & \cdots \\ * & * & \cdots & * & 1 & * & \cdots \end{matrix} \right. \end{bmatrix},$$

and its upper left minor of order $n-1$ is equal to 1. Hence (A1.5.4) is proved.

Let now $G(A, B; E)$ be a bipartite graph, $|A| = p$. Let us rewrite the matrix $I(G)$ after indexing its vertices is such way that the indices of the vertices from part A become less than those for part B. If we add the sum of rows $2, 3, \ldots, p$ to row 1 and subtract the sum of the remaining rows from it, then the first row will contain only zeroes. Hence, rank $I(G) \leq n-1$, i.e., actually, rank $I(G) = n-1$.

Now let us consider the case when $G = C_n$ is a simple cycle of odd length. Let us index its vertices and edges according to their natural order:

$$v_1, e_1, v_2, e_2, \ldots, v_n, e_n.$$

Under this indexing,

$$I(C_n) = \begin{bmatrix} 1 & 0 & 0 & \cdots & 0 & 1 \\ 1 & 1 & 0 & \cdots & 0 & 0 \\ 0 & 1 & 1 & \cdots & 1 & 0 \\ & & & \cdots & & \\ 0 & 0 & 0 & \cdots & 1 & 1 \end{bmatrix},$$

$\det I(C_n) = 2$, rank $I(C_n) = n$.

Let now G be any connected non-bipartite graph. Then G has a simple cycle C_{p_1} of odd length p_1. If $VC_{p_1} \neq VG$, then either the induced subgraph $G_1 = G(VG \backslash VC_{p_1})$ is bipartite or it contains a simple cycle C_{p_2} of odd length p_2. In the latter case consider the induced subgraph

$$G_2 = G(VG \backslash V(C_{p_1} \cup C_{p_2})).$$

Iterating this process, we arrive at one of the following two partitions:

$$VG = VC_{p_1} \cup VC_{p_2} \cup \ldots \cup VC_{p_k}, \tag{A1.5.5}$$

$$VG = VC_{p_1} \cup VC_{p_2} \cup \ldots \cup VC_{p_k} \cup W, \tag{A1.5.6}$$

where $G(W)$ is a bipartite graph.

Suppose that we arrive at A1.5.5. If $I_q = I(C_{p_q})$, then under an appropriate indexing of the vertices and edges the matrix $I(G)$ assumes the form

$$I(G) = [AB],\ A = \text{diag}[I_1, I_2, \ldots, I_k];$$

rank $I(G) = p_1 + \ldots + p_k = n$.

If otherwise we arrive at A1.5.6, then similarly $I(G)$ assumes the form

$$I(G) = [AB],\ A = \text{diag}[I_1, I_2, \ldots, I_k, I],$$

where $I = I(G(W))$, with the additional feature that B has a column such that one of its 1's is at the position with index at most the sum $p_1 + \ldots + p_k$ and another 1 has the index larger than the sum. Therefore,

rank $I(G) = p_1 + \ldots + p_k + |W| - 1 = n$.

1.5.14. Hint: Index the vertices of a bipartite graph in such a way that the indices of the vertices of one part be less than those of the other part.

1.5.15. Solution: Consider the adjacency matrix A of the circulant graph $G = G_{n,U}$. It is convenient to index the rows and columns of this matrix by modulo n residuals: the row and the column that correspond to vertex i are indexed by $i \in \mathbf{Z}_n$. By the definition of circulant graph,

$$N(x) = \{x + u : u \in U\} \tag{A1.5.7}$$

for $x \in \mathbf{Z}_n$, where N(x) is the neighborhood of x. If $A_{ij} = 1$, i.e., $ij \in EG$, then $j = i + u$, $u \in U$. But then $j + 1 = (i + 1) + u$, i.e., $A_{i+1,j+1} = 1$. Thus the elements of A satisfy the conditions

$$A_{i+1,j+1} = A_{ij},\ i, j = 1, 2, \ldots, n, \tag{A1.5.8}$$

i.e., A is a circulant.

Conversely, let the vertices of G be indexed by elements of the additive group $\mathbf{Z}_n$ of modulo n residuals and let $A = A(G)$ be a circulant. Set

$$U = \{x \in \mathbf{Z}_n : A_{x+1,1} = 1\}.$$

For the elements of the circulant A the equalities (A1.5.8) are valid. Therefore (A1.5.7) is valid for each vertex $x \in \mathbf{Z}_n$, i.e., $G = G_{(n,U)}$.

1.5.16. a, b: Solution: Let $G_1 = G = G_{(n,U_1)}$ be a nonempty circulant graph and let n be prime. Then the adjacency matrix $A = A(G_1)$ is a circulant: $A = f_1(h)$ is a polynomial in h (see Exercise 1.5.15) of degree at most $n-1$. The characteristic polynomial of h is $x^n - 1$, its roots are the complex numbers $1, \epsilon, \epsilon^2, \ldots, \epsilon^{n-1}$, where $\epsilon^n = 1$, $\epsilon \neq 1$. Hence

$$f_1(1), f_1(\epsilon), f_1(\epsilon^2), \ldots, f_1(\epsilon^{n-1})$$

is the spectrum of $A = f_1(h)$.

If $G = G_{(n,U_2)}$ is another circulant graph with the incidence matrix $B = A(G_2)$, then similarly, $B = f_2(h)$ and

$$f_2(1), f_2(\epsilon), f_2(\epsilon^2), \ldots, f_2(\epsilon^{n-1})$$

is the spectrum of B.

Now, if the spectra of A and B are the same, then, in particular, $f_2(\epsilon) = f_1(\epsilon^k)$; hence ϵ is a root of the polynomial $p(x) = f_2(x) - f_1(x^k)$. But it is known that for prime n the polynomial $q(x) = 1 + x + x^2 + \ldots + x^{n-1}$ with root ϵ is irreducible over the field of rational numbers. Therefore $p(x)$ is divisible by $q(x)$.

It is easy to understand that $p(x)$ is also divisible by $x - 1$. In fact, the elements of the first column of the matrix $f_1(h)$ are the coefficients of the polynomial

$$f_1(x) = a_0 + a_1 x + a_2 x^2 + \ldots + a_{n-1} x^{n-1}.$$

Hence $a_0 = 0$ and $f_1(1)$ is the sum of the elements of a column, i.e., it is the degree of G_1. The same is true for $f_2(1)$. But by Theorem 1.5.3 the degree of a regular graph is uniquely specified by its characteristic polynomial, hence we have

$$\deg G_1 = \deg G_2, \; f_1(1) = f_2(1), p(1) = f_2(1) - f_1(1^k) = 0.$$

Thus we obtain

$$p(x) \equiv 0 \pmod{(x^n - 1)},$$
$$p(h) = f_2(h) - f_1(h^k) = 0,$$
$$f_2(h) = f_1(h^k).$$

The latter equality means that $U_2 = kU_1$ (see Exercise 1.5.15). $0 \notin U_2$, therefore $k \neq 0$. Further, see Exercise 1.1.26a.

1.6 Automorphism Group of Graph

1.6.2. Set $F = \langle(1,2),(3,4)\rangle$, $H = \langle(1,2),(3,4),(1,3),(2,4)\rangle$. Then $F \cong H$, but the groups F and H are not similar.

1.6.3. Let F be a subgroup of the symmetric group $S(X)$ and let H be a subgroup of $S(Y)$, $|Y| = k$. Suppose that $(h_i, \pi_i) \in F \operatorname{Wr} H$, $i = 1, 2$ and $(h_1, \pi_1) = (h_2, \pi_2)$. Then for any $x \in X$, $y \in Y$,

$$(h_1, \pi_1)(x, y) = (h_2, \pi_2)(x, y),$$

i.e., by the definition,

$$\pi_1(y)(x) = \pi_2(y)(x), \; h_1(y) = h_2(y),$$

i.e.,

$$\pi_1(y) = \pi_2(y), \; \pi_1 = \pi_2, \; h_1 = h_2.$$

Therefore $|F \text{ Wr } H| = |\Pi||H|$. Since the group Π is isomorphic to the direct product of k copies of the group F, we have $|\Pi| = |F|^k$, hence $|F \text{ Wr } H| = |F|_k|H|$.

1.6.4. a: Solution: Suppose that $(h_i, \pi_i) \in F \text{ Wr } H$, $i = 1, 2$. Consider the product $(h_1, \pi_1)(h_2, \pi_2)$. By formula (1.6.3), for any $(x, y) \in X \times Y$,

$$(h_1, \pi_1)(h_2, \pi_2)(x, y) = (h_1, \pi_1)((h_2, \pi_2)(x, y)) =$$

$$(h_1, \pi_1)(\pi_2(y)(x), h_2(y)) = (\pi_1 h_2(y)\pi_2(y)(x), h_1 h_2(y)).$$

But by formula (1.6.2),

$$\pi_1 h_2(y)\pi_2(y) = (\pi_1 h_2 + \pi_2)(y).$$

Thus,

$$(h_1, \pi_1)(h_2, \pi_2)(x, y) = ((\pi_1 h_2 + \pi_2)(y)(x),\ h_1 h_2(y)) = (h_1 h_2, \pi_1 h_2 + \pi_2)(x, y).$$

Hence we have proved that if we multiply the elements of the wreath product as mappings of $X \times Y$, then

$$(h_1, \pi_1)(h_2, \pi_2) = (h_1 h_2, \pi_1 h_2 + \pi_2). \tag{A1.6.1}$$

Further, let o denote the mapping $Y \to F$ such that $o(y) = e_X$ for all $y \in Y$. This mapping is the zero of the group Π. By (1.6.3),

$$(e_Y, o)(x, y) = (o(y)(x), e_Y(y)) = (e_X(x), y) = (x, y)$$

for any $(x, y) \in X \times Y$, i.e., $(e_Y, o) = e$ is the unit of the symmetric group $S(X \times Y)$.

For any pair (h, π), formula (A1.6.1) implies

$$(h, \pi)(h^{-1}, -\pi h^{-1}) = (e_Y, \pi h^{-1} - \pi h^{-1}) = (e_Y, o) = e.$$

Similarly,

$$(h^{-1}, -\pi h^{-1})(h, \pi) = e.$$

This proves that the wreath product $F \text{ Wr } H$ is a subgroup of the symmetric group $S(X \times Y)$.

b: Solution: Let F be a transitive subgroup of $S(X)$ and let H be a transitive subgroup of $S(Y)$. Let us prove the transitivity of the wreath product as a subgroup of the group $S(X \times Y)$. Take $a, b \in X$, $c, d \in Y$. F has a permutation f such that $f(a) = b$ and H has a permutation h such that $h(c) = d$. Consider $\pi : Y \to F$ such that $\pi(y) = f$ for all $y \in Y$. Then by (1.6.3),

$$(h, \pi)(a, c) = (\pi(c)(a), h(c)) = (f(a), d) = (b, d),$$

i.e., $F \text{ Wr } H$ is indeed transitive. Proof of the converse is similar.

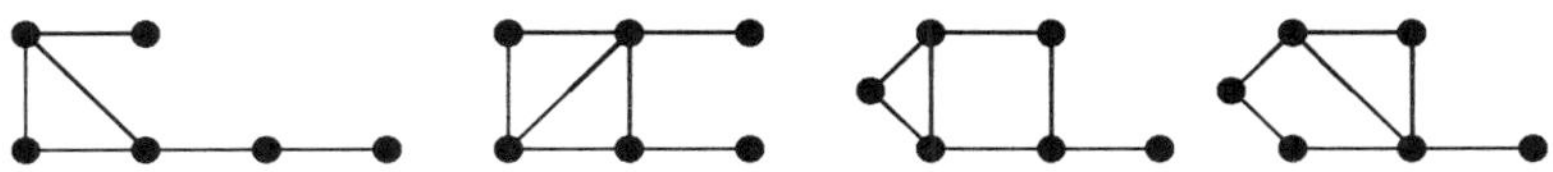

Figure A1.6.1: To Exercise 1.6.8

1.6.8. Graphs shown at Fig. A1.6.1 and their complements.

1.6.10. a, b: Aut $G = S_n$.

c: If $1, 2, \ldots, n$ are the vertices of the path P_n in the natural order, then Aut $P_n = \langle a \rangle$, where

$$a = \begin{cases} (1,n)(2,n-1)\ldots(k,k+1), & n = 2k; \\ (1,n)(2,n-1)\ldots(k,k+2), & n = 2k+1; \end{cases}$$

d: Hint: Let the vertices of C_n be indexed by elements of the additive group $\mathbf{Z}_n$ of modulo n residuals in the natural order along the cycle: $VC_n = \{0, 1, 2, \ldots, n-1\}$. For $x \in \mathbf{Z}_n$ the actions of permutations a, b will be defined by the following equalities: $a(x) = -x$, $b(x) = x + 1$. If $A = \langle a \rangle$, then

$$\text{Aut } C_n = \langle a, b \rangle = A \cup bA \cup \ldots \cup b^{n-1}A, \ |\text{Aut } C_n| = 2n.$$

1.6.11. Hint: Prove first that for $n > 3$ A_n is twice transitive, i.e., for any two pairs (i, j), (k, l) such that $i \neq j$, $k \neq l$ there exists a permutation $s \in A_n$ satisfying the conditions $s(i) = k$, $s(j) = l$.

1.6.12. a: Hint: If $\varphi \in$ Aut G and V is the connected domain of a graph G then $\varphi(V) = \{\varphi(v) : v \in V\}$ is also a connected domain.

b: Solution: Let $\{G_1, \ldots, G_k\}$ be the set of components of G and let them be isomorphic to each other. Set

$$G_1 = F, \ V_i = VG_i, \ i = 1, 2, \ldots, k,$$

and for each $i = 2, \ldots, k$ we pick an isomorphism $\varphi_i : V_1 \to V_i$ of G_1 and G_i. We shall write the elements $v \in V_1$ in the form of pairs $(v, 1)$:

$$V_1 = \{(v_1, 1), (v_2, 1), \ldots, (v_n, 1)\},$$

Similarly, we may assume that $\varphi_i(v, 1) = (v, i)$, i.e., we may write

$$V_i = \{(v_1, i), (v_2, i), \ldots, (v_n, i)\}, i = 1, 2, \ldots, k,$$

where $\{v_1, \ldots, v_n\} = VF$. Let us prove the inclusion

$$(\text{Aut } F) \text{ Wr } S_k \subseteq \text{Aut } G. \tag{A1.6.2}$$

Suppose that $\pi : \{1, 2, \ldots, k\} \to$ Aut F. Consider the element (e, π) of the wreath product, where e is the identity permutation of the set $\{1, 2, \ldots, k\}$.

By formula (1.6.3), $(e,\pi)(v,i) = (\pi_i)(v), i)$, hence $(e,\pi) \in \text{Aut } G$. Consider the element (h,o), where $h \in \text{Aut } F$ and $o : \{1,2,\ldots,k\} \to \text{Aut } F$ is such that $o(i) = e$, where e is the unit of Aut F. By formula (1.6.3), $(h,o)(v,i) = (v,h(i))$. Therefore $(h,o) \in \text{Aut } G$. But by formula (1.6.4) each element $(h,\pi) \in (\text{Aut } F) \text{ Wr } S_k$ may be represented as a product $(h,\pi) = (e,\pi)(h,o)$. Therefore the inclusion (A1.6.2) is proved.

Let now s be any automorphism of G. Define $\overline{s} \in S_k$ to be such that $\overline{s}(i) = j$ if $s(V_i) = V_j$, $i = 1,2,\ldots,k$, and consider the mapping $(\overline{s},o)$. We have previously demonstrated that it is in Aut G. Set $t = (\overline{s},o)^{-1}s$. Then $t \in \text{Aut } G$, $t(V_i) = V_i$, $i = 1,2,\ldots,k$. Hence, $t = (e,\pi)$, $\pi : \{1,2,\ldots,k\} \to \text{Aut } F$.

Thus, $s = (\overline{s},o)t \in (\text{Aut } F) \text{ Wr } S_k$, and together with (A1.6.2) this implies $\text{Aut } G = (\text{Aut } F) \text{ Wr } S_k$.

1.6.14. a: Under an appropriate indexing of vertices, $\text{Aut } G = (S_2 \text{ Wr } S_3) \times S_2$.

b: $\text{Aut } G = S_m \times S_n$.

c: $\text{Aut } G = S_m \text{ Wr } S_2$.

1.6.15. $\text{Aut } G = S_{m-1} \text{ Wr } S_2$.

1.6.18. b: Solution: Consider the mapping

$$\varphi : \text{Aut } G \to \text{Aut}_E G,\ f \mapsto f_E,\ f \in \text{Aut } G.$$

It is a surjective homomorphism of groups. If a and b are two isolated vertices of G or if the edge ab together with a,b is a component of G then the transposition (a,b) is an automorphism of G, and moreover if $f = (a,b)$ then f_E is the identity mapping, i.e., $f \in \text{Ker } \varphi$. Hence if one of the conditions (1), (2) is violated then

$$\text{Aut}_E G \cong \text{Aut } G/\text{Ker } \varphi \ncong G.$$

Assume now that both these conditions hold, $f \in \text{Ker } \varphi$ and $f(a) \neq a$ for some vertex $a \in VG$. Suppose that $x = ab \in EG$. Since $x = f_E(x) = f(a)f(b)$, we have $f(a) = b$, $f(b) = a$. There exists a third vertex $c \in VG$ adjacent to either a or b; assume, e.g., that $x_1 = ac \in EG$. Set $d = f(c)$. Then $f_E(x_1) = bd = x_1$, $b = c$. The resulting contradiction shows that $\text{Ker } \varphi = \langle e \rangle$, and hence φ is the isomorphism of groups.

c: Solution: For any vertex $v \in G$ let Q_v denote the clique of $L(G)$ that consists of the edges of G incident to v. If Q is the set of all such cliques then the mapping

$$\alpha : VG \to Q,\ v \mapsto Q_v,\ v \in VG$$

is the isomorphism of G onto the subgraph F of the clique graph $Q(L(G))$ induced by the vertex set Q (see Exercise 1.3.41). If G has no triangles then Q is the set of all maximal cliques of $L(G)$. If G has triangles then, by the assumption, it has no vertices of degree 3. Therefore the cardinalities of the cliques from the set Q are not equal to 3 and the cardinalities of the remaining maximal cliques of $L(G)$ are equal to 3.

Suppose now that $g \in$ Aut $L(G)$ and $\overline{g}$ is the corresponding automorphism of $Q(L(G))$, see Exercise 1.6.17. From the above discussion it follows that Q is invariant with respect to $\overline{g}$. If g' is the restriction of $\overline{g}$ onto Q (see Exercise 1.6.16) then

$$\{g(e)\} = g(Q_a) \cap g(Q_b) = Q_u \cap Q_v.$$

Consider now the automorphism $f_E \in \mathrm{Aut}_E G$. If

$$e = ab \in EG,\ g(Q_a) = Q_u,\ g(Q_b) = Q_v,$$

then

$$f(a) = \alpha^{-1}g'\alpha(a) = \alpha^{-1}g'(Q_a) = \alpha^{-1}g(Q_a) = \alpha^{-1}(Q_u) = u.$$

Similarly, $f(b) = v$. Hence $f_E(e) = f(a)f(b) = uv$. On the other hand, $\{e\} = Q_a \cap Q_b$. Therefore

$$\{g(e)\} = g(Q_a) \cap g(Q_b) = Q_u \cap Q_v.$$

But $Q_u \cap Q_v = \{uv\}$, hence $g(e) = uv = f_E(e)$. This proves that Aut $L(G) \subseteq \mathrm{Aut}_E G$, hence Aut $L(G) = \mathrm{Aut}_E G$.

Under the considered conditions $\mathrm{Aut}_E G \cong$ Aut G, see item b.

1.6.19. a: The group Aut $L(K_4)$ is similar to the wreath product S_2 Wr S_3, and $|$ Aut $L(K_4)| = 48$. Hint: $L(K_4) = \overline{3K_2}$.

b: Solution: The degrees of all vertices of K_5 are 4, hence (see Exercise 1.6.18c)

$$\mathrm{Aut}\ L(K_5) = \mathrm{Aut}_E K_5 \cong \mathrm{Aut}\ K_5 = S_5.$$

The symmetric group S_5 is generated by the cycle $a = (1,2,3,4,5)$ and the transposition $b = (1,2)$. Hence Aut $L(K_5) = \langle a_1, b_1\rangle$,

$$a_1 = (12,23,34,45,51)(13,24,35,41),\ b_1 = (13,23)(14,24)(15,25).$$

1.6.20. a: Solution: Let G be the Petersen graph. Indexing its vertices as shown at Fig. A1.1.4, we see that $G = \overline{L(K_5)}$ (see Exercise 1.1.40). Hence

$$\mathrm{Aut}\ G = \mathrm{Aut}\ \overline{L(K_5)} = \mathrm{Aut}\ L(K_5).$$

Further, see Exercise 1.6.19b.

1.6.22. $\mathrm{Aut}_E G$ coincides with the Klein four-group.

1.6.23. <u>Solution</u>: Assume that $V = \{1, 2, \ldots, n\}$ is the vertex set for all labelled graphs isomorphic to G. If G_1 is one of these graphs, then reindexing its vertices by means of a permutation $s \in S(V)$ we obtain the graph $s(G_1)$. The equality $s(G_1) = t(G_1)$, $s, t \in S(V)$, holds if and only if $t^{-1}s(G_1) = G_1$, i.e., $t^{-1}s \in \text{Aut } G_1$.

In other words, the permutations s, t must belong to the same left coset of $S(V)$ modulo the subgroup $\text{Aut } G_1$. Therefore the number of graphs isomorphic to G is equal to the number of these cosets, i.e., $n!/|\text{Aut } G_1|$. But since the groups $\text{Aut } G_1$ and $\text{Aut } G$ are similar, $|\text{Aut } G_1| = |\text{Aut } G|$.

1.6.25. Let G be a circulant graph of order n. Under an appropriate indexing of the vertices the adjacency matrix $A(G)$ coincides with some polynomial in h (see Exercise 1.5.15), where $s = (1, 2, \ldots, n)$ is the cyclic permutation of length n. Clearly, $hf(h) = f(h)h$, therefore $M_s \in \text{Aut } G$ (see Exercise 1.6.24).

Conversely, let $\text{Aut } G$ contain a cyclic permutation of length n. We may index the vertices of G in such a way that $s = \{1, 2, \ldots, n\}$. Under this indexing, the adjacency matrix $A(G)$ will commute with M_s, i.e., $M_s A(G) = A(G) M_s$. But the characteristic and minimal polynomials of M_s coincide, therefore all matrices commuting with M_s are polynomials in M_s of degree at most $n - 1$. Thus, $A(G) = f(M_s)$, i.e., it is a circulant, therefore G is a circulant graph.

1.6.26. b: <u>Hint</u>: Let $f : VF \to VG$ be an isomorphism of graphs F and G and let a be a vertex of maximal degree in F. Suppose further $VF \cap VG \neq \emptyset$ and let φ be the permutation of the set $VF \cup VG$ specified by the conditions

$$\varphi(x) = \begin{cases} f(x), & \text{if } x \in VF, \\ f^{-1}(x), & \text{if } x \in VG. \end{cases}$$

Then $b = f(a)$ and φ satisfies the assumptions of the exercise.

Conversely, if $\varphi \in \text{Aut } H$ and $\varphi(a) \neq a$ then $b = \varphi(a)$ is a vertex of maximal degree in G and the mapping

$$f : VF \to VG, \; x \mapsto \varphi(x), x \in VF$$

is an isomorphism of graphs.

Answers to Chapter 2:

Trees

2.1 Trees: Basic Notions

2.1.2. No.

2.1.3. Hint: The number of such trees is 16.

2.1.4. Hint: The number of such trees is 3 for $n = 5$, 6 for $n = 6$, 11 for $n = 7$, 23 for $n = 8$.

2.1.6. Stars.

2.1.8. $2, 3, \ldots, n-1$.

2.1.9. K_1 and P_4.

Solution: Let T be a selfcomplementary tree with n vertices. Then $|ET| = |E\overline{T}|$, i.e.,

$$n - 1 = n(n-1)/2 - (n-1),$$

hence $n = 1$ or $n = 4$. An immediate check gives the answer.

2.1.10. K_1 and K_2.

2.1.11. Hint: Use the Theorem 2.1.1.

2.1.12. Solution: For any n-vertex tree the handshake Lemma and the Theorem 2.1.1 imply

$$\sum_{i=1}^{n}(d_i - 2) = -2, \tag{A2.1.1}$$

where d_i is the degree of the i-th vertex. Let $k \geq 2$ be a fixed integer. Since $d_i \neq 2$ and the number of pendant vertices ($d_i = 1$) is equal to k, it follows that n is limited (otherwise the left side of (A2.1.1) could be arbitrarily large). Therefore $\mathcal{G}_k$ is a finite set.

2.1.13. $2/(2 - d_{mean})$.

2.1.14. $2 + \sum_i (d_i - 2)$, where the summation is over all branching vertices.

Hint: Use the equation (A2.1.1).

2.1.15. a: Solution: Let a tree have $n > 1$ vertices. Then by Theorem 2.1.1 it has $n - 1$ edges. Therefore by Handshake Lemma, $\sum_{i=1}^{n} d_i = 2n - 2$, where d_i is the degree of the i-th vertex. Since $d_i > 0$, the latter equality holds only if at least two right-hand terms are equal to 1, i.e., the tree has at least two pendant vertices.

b: Hint: Proof is by induction, based on the uniqueness of the path connecting any pair of vertices in a tree.

2.1.17. c: Solution: If $D = D(T)$ is the matrix of distances between the pendant vertices of a tree T then T is said to be a *realization* of the matrix D. We shall prove that all realizations of the matrix D are isomorphic using the induction over the number of pendant vertices. This is evident if the order k of D is 2, since every tree with two pendant vertices is a path (see Exercise 2.1.15b).

Suppose that $k > 2$ and let all realizations of $D(T')$ for any tree T' with $k-1$ pendant vertices be isomorphic, i.e., $D(T')$ has a unique realization. Let further T be a tree with pendant vertices $1, 2, \ldots, k$. Then T has branching vertices. Let a be the one closest to the vertex k and let A denote the simple (k, a)-path whose length is denoted by l. Deleting all vertices of $A - a$ from T, we obtain a tree T' whose matrix $D(T')$ may be obtained from $D(T)$ by the deletion of the last row and the last column.

Clearly, the tree T is uniquely constructed from T' if we know the length l and the branching vertex $a \in VT'$. Therefore the problem will be solved if we express l in terms of the elements of $D(T)$ and indicate the vertex a in T', since by the induction hypothesis, all realizations of the matrix $D(T')$ are isomorphic.

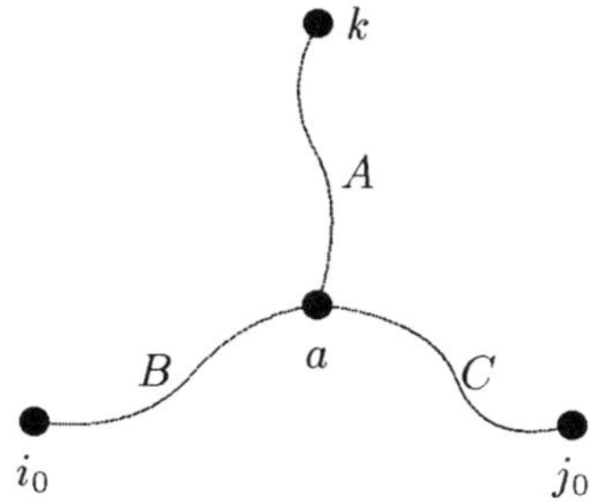

Figure A2.1.1: To Exercise 2.1.17, c

Let b and c be two different vertices of T' adjacent to a; $B = (a, b, \ldots, i_0)$, $C = (a, c, \ldots, j_0)$ are simple paths in T', with i_0, j_0 being pendant vertices, see Fig. A2.1.1. Then $A \cup B$ is the simple (k, i_0)-path, $A \cup C$ is the simple

(k, j_0)-path, $B \cup C$ is the simple (i_0, j_0)-path. Therefore if x and y are the lengths of paths B and C respectively then

$$D_{ki_0} = x + l, \; D_{kj_0} = y + l, \; D_{i_0 j_0} = x + y.$$

Therefore $l = (D_{ki_0} + D_{kj_0} - D_{i_0 j_0})/2$.

On the other hand, for arbitrary pendant vertices i, j, the simple (i,j)-, (i,k)-, and (j,k)-paths have a single common vertex t, see Fig. A2.1.2, with

$$d(k,t) = (D_{ki} + D_{kj} - D_{ij})/2.$$

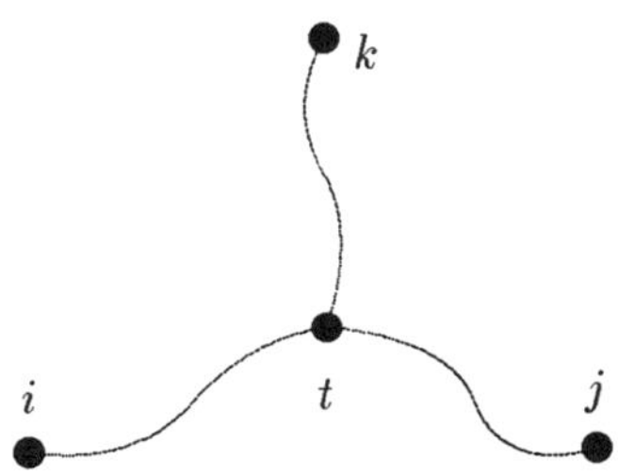

Figure A2.1.2: To Exercise 2.1.17, c

Therefore the length l of the simple path A may be determined as follows:

$$l = \min\left\{(D_{ki} + D_{kj} - D_{ij})/2 : \; 1 \le i < j \le k-1\right\}.$$

Further, if i_0, j_0 are the values of i, j which deliver the minimum, then the vertex a belongs to the simple (i_0, j_0)-path of T', and its location at this path is specified by the condition

$$d(i_0, a) = D_{i_0 k} - l.$$

d: Hint: Use of the solution for item c.

2.1.18. $\dfrac{nd - 2n + 2}{d - 1}$, where d is the degree of a nonpendant vertex.

Solution: The Handshake Lemma for a tree of order n implies $\sum_{i=1}^{n} d_i = 2(n-1)$, where d_i is the degree of the i-th vertex. Since all nonpendant vertices of T have the same degree d,

$$d(n-p) + p = 2(n-1),$$

where p is the number of pendant vertices in T, hence the formula.

2.1.19. Hint: Modify the solution of the Exercise 2.1.18.

2.1.20. a: Hint: Use the answer for the Exercise 2.1.14.

b: The bound is attained only for trees with vertex degrees either 1 or 3.

2.1.21. a: Solution: Let P be a diametral (a, b)-path of a tree T. Then a, b are pendant vertices of T (see Exercise 2.1.16). Consider any vertex c of T not in P. T has a unique (a, c)-path Q and the latter is the union $Q = S \cup R$, where S is the (a, d)-subpath of P and R is a (d, c)-path with $d \neq a$, see Fig. A2.1.3. Let x, y, z be the lengths of the paths $S, P \backslash S, R$ respectively

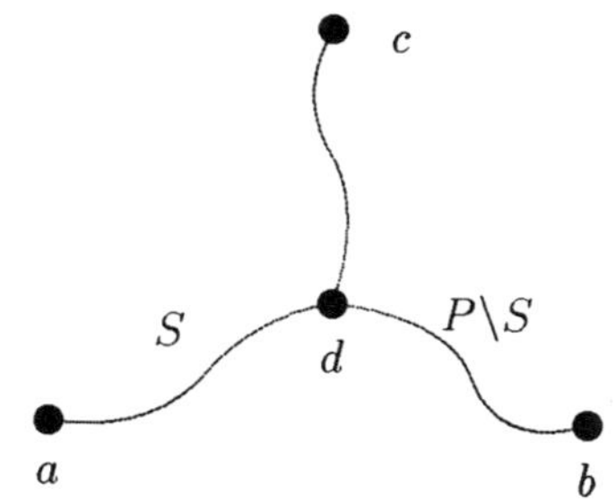

Figure A2.1.3: To Exercise 2.1.21a

and suppose that, e.g., $x \leq y$. Then we know the excentricities: $e(d) = y$, $e(c) = z + y$, and hence $e(d) < e(c)$, i.e., c is not central. This proves that the center of T is in P.

b: Hint: This follows from the item a.

c: Hint: Bearing in mind the item a, it is sufficient to demonstrate the Jordan theorem for the simple path P_n. In this case for odd n the center is a single vertex and for even n the center is two adjacent vertices.

d: Hint: When passing from T to T' (as well as at all similar steps of the algorithm), the excentricity of all remaining vertices is decreased by 1, therefore the centers of T and T' coincide.

e: Hint: See Exercises 2.1.21a,b, as well as the hint to the Exercise 2.1.21c.

2.1.25. The graphs that may be obtained from a star by replacing each edge by a path (all paths are of equal length).

Hint: First prove that the graph is a tree.

2.1.26. Hint: To do this, prove the inequality:

$$2r(T) - 1 \leq d(T) \leq 2r(T)$$

using the Exercises 2.1.21 and 1.2.26a.

2.1.27. K_1 and K_2.

2.1.28. 2 or 3.

2.1.29. Hint: Two paths $v_1, v_2, \ldots, v_k$ and $u_1, u_2, \ldots, u_k$ are said to be equal if and only if either $u_i = v_i$, $i = 1, 2, \ldots, k$ or $u_i = v_{k-i+1}$, $i = 1, 2, \ldots, k$.

Solution: Such tree has a simple path $P = v_1, v_2, \ldots, v_{2k-2}$ of length $2k - 3$. This path contains $k - 2$ subpaths of length k starting at the vertices $v_1, v_2, \ldots, v_{k-2}$. In addition, there are $n - 2k - 2$ vertices not lying on the path P. Each such vertex u is connected with P by a simple path Q_u. If Q_u is longer than $k - 1$ then the required path of length k is taken to be the segment of Q_u of length k starting at u. Otherwise the required path consists of Q_u plus a part of P. Hence there additionally exist $n - 2k + 2$ simple paths of length k. Hence the tree contains at least $k - 2 + n - 2k + 2 = n - k$ simple paths of length k.

2.1.30. a: Hint: See Table A2.1.1.

Table A2.1.1:

	1	2
1		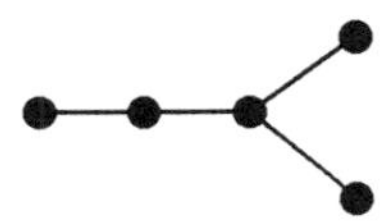
2	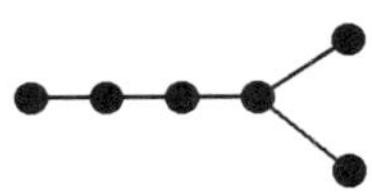	

b: Hint: See Fig. A2.1.4, where v_1 and u_1 are the central vertices and v_2, u_2 are the centroidal vertices.

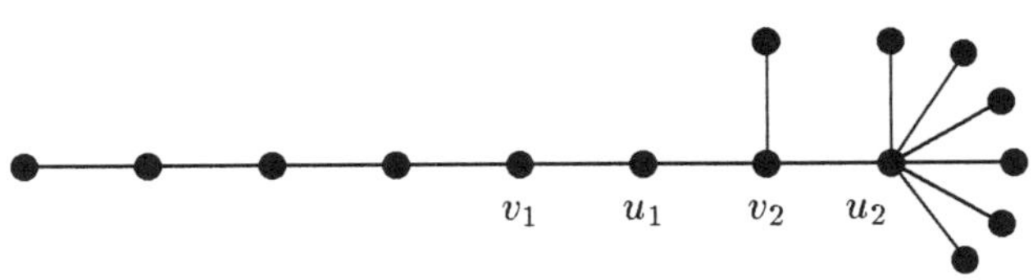

Figure A2.1.4: To Exercise 2.1.30b

c: Solution: Let v be a centroidal vertex of the tree T. The branches at this vertex are denoted by H_i, $i = 1, 2, \ldots, k = \deg v$, i.e., $T_v = \{H_1, H_2, \ldots, H_k\}$. A branch $H_i \in T_v$ such that $|VH_i| = w(v)$ will be called a ***branch weighted at*** v. Let H_1 be such branch, i.e., $|VH_1| = w(v)$, and let e_1 be the appending edge of the branch, i.e., $e_1 = (v, u)$. The branches at the vertex u are denoted by H'_i, $i = 1, 2, \ldots, p = \deg u$, where $e_1 \in EH'_1$, see Fig. A2.1.5. Let us

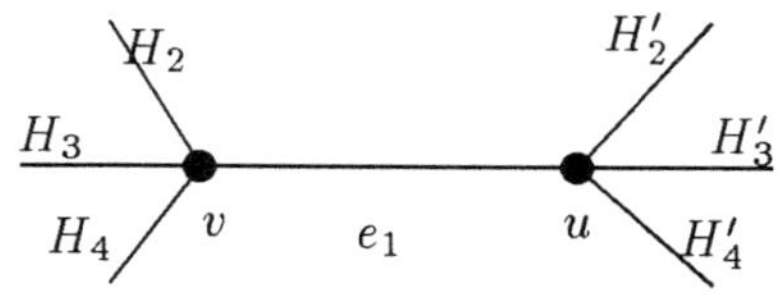

Figure A2.1.5: To Exercise 2.1.30c

demonstrate that H'_1 is the branch weighted at u. From Fig. A2.1.5 we see that

$$|EH_1| = \sum_{i \neq 1} |EH'_i| + 1.$$

Therefore if the branch weighted at u were a branch H'_j, $j \neq 1$, then it would be

$$|EH_1| \geq w(u) + 1$$

or $w(v) > w(u)$. But this would contradict to the fact that v is a centroidal vertex. Hence H'_1 is the single branch weighted at u.

Case 1: The vertex u is not centroidal, i.e., $w(u) > w(v)$. Let us demonstrate that in this case T has a single centroidal vertex (which is v). Indeed, Fig. A2.1.5 shows that no vertex $v' \in \bigcup_{i \neq 1} VH_i$, $v' \neq v$, can be centroidal since it has a branch that strictly contains H_1, i.e., $w(v') > w(v)$.

Similarly, every vertex $v'' \in \bigcup_{i \neq 1} VH'_i$ has a branch that strictly contains H'_1. Therefore $w(v'') > w(u)$. But in view of $w(u) > w(v)$ this means that v'' cannot be centroidal.

Case 2: The vertex u is centroidal. Repeating the reasoning similar to the previous case, we verify that no vertex other than u, v can be centroidal in T.

d: Hint: See the solution for the Exercise 2.1.30c. Basing on Fig. A2.1.2, the equality $w(v) + w(u) = |ET| + 1$ is evident, which implies the required statement.

e: Hint: See Exercise 2.1.30d.

2.1.31. Hint: a: The proofs of both the necessity and sufficiency are by contradiction.

b: To prove the necessity, verify that every subtree of an interval tree is also an interval tree, and the tree shown at Fig. 2.1.3 is not the interval one.

The sufficiency is easily proved by direct construction of the required bijection.

2.1.32. The tree of 7-th order shown at Fig. A2.1.6.

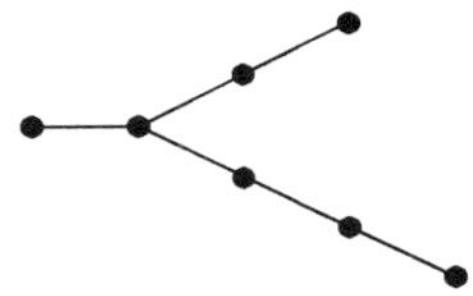

Figure A2.1.6: To Exercise 2.1.32

2.1.33. K_1 and K_2.

Hint: If a graph has the transitive automorphism group, it is regular. Next, see Exercise 2.1.10.

2.1.34. Let T be a tree of order n and let G be a graph with $\delta(G) \geq n-1$. We prove that G has a subgraph isomorphic to T. For $n = 2$ the statement is evident. Let us apply induction over n. Assume that $n > 2$ and the statement is true for trees with less vertices. If T' is a tree of order $n-1$, then by the induction hypothesis we may assume that T' is a subtree of G. For any $a \in VT'$ the following inequalities hold:

$$\deg_{T'} a \leq n-2, \ \deg_G a \geq n-1.$$

Hence there exists a vertex $b \in N_G(a)\backslash VT'$. Adding the vertex b and the edge ab to T', we obtain a subgraph of G which is a tree of order n. Since every tree of order n is obtained from some tree of order $n-1$ by adding a new vertex and a new edge, the required statement is proved.

2.1.35. a: Simple paths. b: Stars. c: Trees of diameter less than $k-1$. d: Simple paths or $K_{1,m}$. e: Trees with maximal degree at most 3.

2.1.36. Hint: See Exercise 1.5.13.

2.1.37. a: $(3,3,4,4,6,6)$.

b: $(3,3,4,4,6,6)$.

c: $(8,5,8,6,5,6)$.

d: $(6,6,7,7,7,8,8,9,9,8,12,13,13)$.

e: $(3,5,5,3,2,8,2,1,10,10,13,13,12,12,10)$.

2.1.38. Hint: Using induction over $n \geq 3$ prove the equivalent statement that

$$\{a_1, \ldots, a_{n-2}\}$$

is the set of all nonpendant vertices of the graph.

2.1.39. a: Hint: Use the previous exercise. b: See Fig. A2.1.7.

2.1.40. a: Hint: Proof may be carried out by contradiction using the algorithm of the reconstruction of a tree by its Prüfer code (see Exercise 2.1.39).

b: Hint: Use the algorithm of the reconstruction of a tree by its Prüfer code.

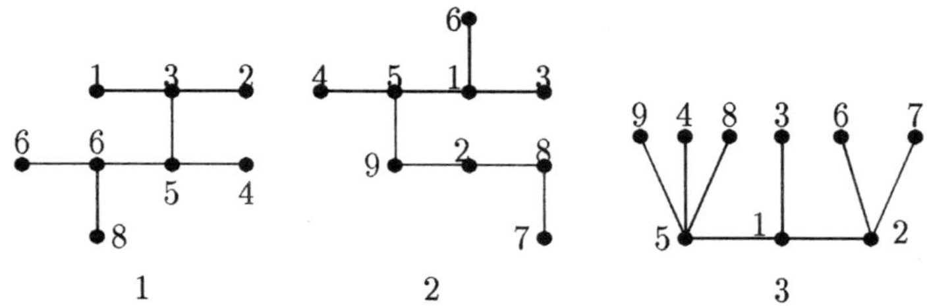

Figure A2.1.7: To Exercise 2.1.39

2.1.41. The star with at least three rays.

2.1.42. P_n, $n \geq 4$.

2.1.44. Hint: According to the Exercise 1.3.32, the family $P(G)$ allows to determine the number of edges of G and whether G is connected.

2.1.45. a: Solution: We shall use induction over k. Two subtrees surely satisfy our statement. Suppose that any $k-1 \geq 2$ pairwise intersecting trees have a common vertex. This, in particular, means that

$$V_1 = \bigcap_{i=1}^{k-1} VT_i \neq \emptyset,\ V_2 = \bigcap_{i=1}^{k} VT_i \neq \emptyset,\ V_3 = VT_1 \bigcap VT_2 \neq \emptyset.$$

It is sufficient to consider the case when there exist three different vertices

$$u \in V_1,\ u \notin V_2 \cup V_3,\quad v \in V_2,\ v \notin V_1 \cup V_3,\quad w \in V_3,\ w \notin V_1 \cup V_2,$$

otherwise the statement would obviously be true. For the graph T, let P_1 denote the (u, v)-path, let P_2 denote the (v, w)-path, and let P_3 denote the (v, w)-path. Since T is a tree,

$$V_4 = VP_1 \bigcap VP_2 \bigcap VP_3 \neq \emptyset$$

(otherwise T would contain a cycle). At the same time, every tree $T_i, i = 1, 2, \ldots, k$, contains at least two vertices from $\{u, v, w\}$. Therefore every tree $T_i, i = 1, 2, \ldots, k$, contains all vertices of one of the paths P_1, P_2, P_3. Therefore,

$$\emptyset \neq V_4 \subseteq \bigcap_{i=1}^{k} VT_i.$$

b: Hint: It is sufficient to prove the connectivity of the induced subgraph.

2.1.46. Solution: Let P be a path of length $e(x)$ starting at x. If P does not contain the vertices y and z then clearly

$$e(y) = e(x) + 1,\ e(z) = e(x) + 1,$$

and the statement is proved. If otherwise P contains one the vertices y, z, say, y, then

$$e(y) \geq e(x) - 1, \; e(z) = e(x) + 1,$$

and the statement is proved just the same.

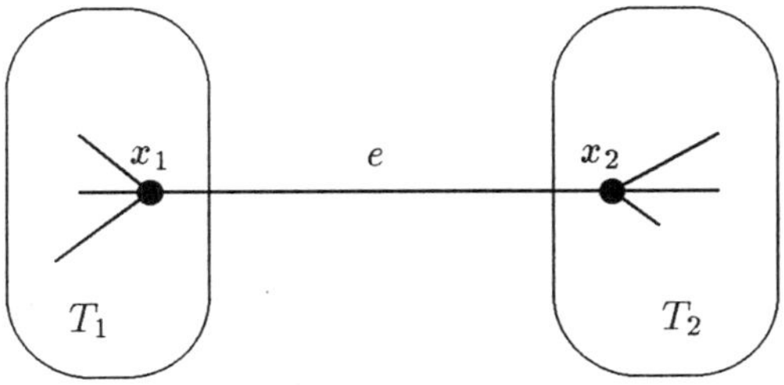

Figure A2.1.8: To Exercise 2.1.47

2.1.47. Solution: First of all let us determine the relation between $\sigma(x_1)$ and $\sigma(x_2)$ when x_1, x_2 are adjacent vertices of T. Denote $e = x_1x_2$; $T - e$ consists of two components T_1, T_2. Suppose that $|T_1| = n_1$, $|T_2| = n_2$, i.e., $n_1 + n_2 = |T|$. Fig. A2.1.8 shows us that

$$\sigma(x_1) = \sigma(x_2) - n_1 + n_2. \tag{A2.1.2}$$

Let now y and z be the vertices of T adjacent to x. Denote $e_1 = xy$, $e_2 = xz$; $T - e_1 - e_2$ consists of three components K_1, K_2, K_3,

$$y \in K_1, \; z \in K_2, \; x \in K_3,$$

$$|K_1| = k_1, \; |K_2| = k_2, \; |K_3| = k_3,$$

see Fig. A2.1.9. Clearly, $k_i > 1, i = 1, 2, 3$, $k_1 + k_2 + k_3 = |T|$. Bearing in

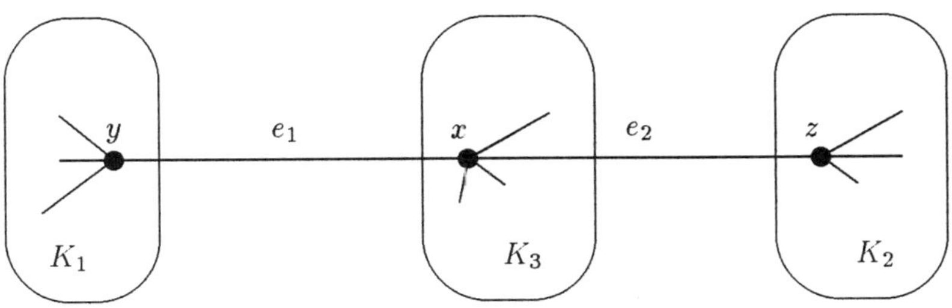

Figure A2.1.9: To Exercise 2.1.47

mind the relation A2.1.2, we may see from Fig. A2.1.9 that

$$\sigma(x) = \sigma(y) - k_2 - k_3 + k_1,$$

$$\sigma(x) = \sigma(z) - k_1 - k_3 + k_2,$$

hence

$$2\sigma(x) = \sigma(y) + \sigma(z) - 2k_3 < \sigma(y) + \sigma(z).$$

2.1.48. a: Solution: Suppose that T has two nonadjacent vertices u, v such that the value $\sigma(u) = \sigma(v)$ is minimal. Consider the (u, v)-path of T

$$x_1 = u, x_2, \ldots, x_{p-1}, x_p = v, \ p > 2.$$

Since $\sigma(u) \leq \sigma(v)$, strict convexity of the function $\sigma(x)$ (see Exercise 2.1.47) implies

$$\sigma(x_1) + \sigma(x_3) > 2\sigma(x_2) \geq \sigma(x_1) + \sigma(x_2),$$

$$\sigma(x_3) > \sigma(x_2) \geq \sigma(x_1),$$

$$\sigma(x_2) + \sigma(x_4) > 2\sigma(x_3) \geq \sigma(x_1) + \sigma(x_2),$$

$$\sigma(x_4) > \sigma(x_1),$$

and so on. Finally, we derive

$$\sigma(v) = \sigma(x_p) > \sigma(1) = \sigma(u),$$

which contradicts to our assumption.

b: Solution: Consider a tree T of order $m + n$:

$$T = P_m \cup \{e_1, e_2, \ldots, e_n\},$$

where P_m is the (v_1, v_m)-path $(v_1, v_2, \ldots, v_m)$ and $e_i = v_i u_i$, $i = 1, 2, \ldots, n$. If m is even then the center of T is the vertex $v_{m/2}$. At the same time if $n = C_m^2$, then v_1 is the barycenter. Therefore the distance between the center and the barycenter is $m/2 - 1$, i.e., it may be arbitrarily large.

2.1.49. a: A star graph. b: A path graph.

Solution: a: For any tree, the expression $\sum_{(x,y)} d(x, y)$ has exactly $n - 1$ terms equal to 1. The remaining terms are at least 2 and they are equal to 2 only for the star.

b: Let a be a pendant vertex of T. Then

$$\sum_{(x,y):\, x,y \in VT} d(x, y) = \sigma(a) + \sum_{(x,y):\, x,y \in V(T-a)} d(x, y), \qquad \text{(A2.1.3)}$$

where $\sigma(a)$ is defined in Exercise 2.1.47.

If T is a path then $\sigma(a) = 1 + 2 + \ldots + (n - 1)$. If T is an arbitrary tree and $e(a) = e$ (see Exercise 2.1.46) then for every $i = 1, 2, \ldots, e$ there exists a vertex at distance i from a. Therefore

$$\sigma(a) = 1 + 2 + \ldots + e + e_1 + e_2 + \ldots + e_{n-1-e},$$

where $e_k < e$, $k = 1, 2, \ldots, n - 1 - e$.

Clearly,

$$1 + 2 + \ldots + (n - 1) \geq 1 + 2 + \ldots + e + e_1 + e_2 + \ldots + e_{n-1-e},$$

where the equality takes place only if $e = n - 1$, i.e., when T is a path. Thus, the first term $\sigma(1)$ in (A2.1.3) is maximal if and only if T is a path. The maximality of the second term on the right side of (A2.1.3) in the case when T is a path is proved by induction over the number of vertices.

2.1.50.

$$t(n,k) = \frac{n}{k}(t(n-1,k) + (n-k)t(n-1,k-1),\ n \geq 3,\ k \geq 1.$$

Solution: Assume that the vertex set of the tree is $\{1, 2, \ldots, n\}$. Clearly,

$$t(2,2) = 1,\ t(n,0) = t(n,1) = t(n,n) = 0,\ n \geq 3.$$

Let $\mathcal{T}$ be the set of labelled trees of order n with k pendant vertices, $2 \leq k \leq n-1$, $n \geq 3$. Let $\mathcal{T}_i$ be the subset of $\mathcal{T}$ of trees whose i-th vertex is pendant. It is easily proved that

$$|\mathcal{T}_i| = kt(n-1,k) + (n-k)t(n-1,k-1).$$

But $\mathcal{T} = \bigcup_{i=1}^{n} \mathcal{T}_i$, and moreover, every tree from $\mathcal{T}$ belongs to exactly k sets $\mathcal{T}_i$. Therefore,

$$t(n,k) = |\mathcal{T}| = 1/k \sum_{i=1}^{n} |\mathcal{T}_i| = n/k(t(n-1,k) + (n-k)t(n-1,k-1)),$$

$$1 \leq k \leq n-1,\ n \geq 3.$$

2.1.51. Solution: Let T be a tree, $|T| = 2k+1$. Suppose that there exists an automorphism f for which no vertex of T is stationary. Choose a maximal subtree T' such that $f(VT') \cap VT' = \emptyset$. Since $|T| = 2k+1$, there exists a vertex $v \in VT$ such that $v \notin VT'$, $v \notin f(VT')$, but v is adjacent to a vertex of the tree T'. Therefore

$$f(V(T'+v)) \cap V(T'+v) = \emptyset,$$

contrary to the maximality of T'.

2.1.52. Hint: Let v be a vertex of T and let v' be a vertex maximally distant from v, i.e., $e(v) = d(v,v')$. To prove the required formula, it is sufficient to prove that all vertices of $C(T)$ belong to the (v,v')-path. To this end, use the Exercise 2.1.21.

2.1.53. Solution: Suppose the contrary, i.e., the graph $SI(T)$ has an induced cycle $C_k = (v_0, \ldots, v_{k-1}, v_0)$, $k \geq 4$. Let T_i denote the subtree of T that corresponds to v_i, $i = 1, 2, \ldots, k$. Consider $w_i \in VT_i \cap VT_{i+1}$, $i = 0, 1, \ldots, k-1$ (here and below the summation of indices is modulo k). Since C_k is an induced cycle and $k \geq 4$, all vertices $w_0, \ldots, w_{k-1}$ are pairwise different. Let Q_i denote the unique simple path connecting w_i and w_{i+1}, $i = 0, 1, \ldots, k-1$. Let further x_{i+1} denote the common vertex of Q_i and Q_{i+1} which is most distant from w_{i+1}, $i = 0, 1, \ldots, k-1$. It is easily seen that the union of the (x_i, x_{i+1})-subpaths Q_i constitutes an induced simple cycle in T, which is impossible.

2.2 Skeletons and Spanning Trees

2.2.1. Solution: Let an (n,m)-graph G have k components, $k>0$. Clearly,

$$m_i \geq n_i - 1,\ i = 1,2,\ldots,k,$$

where m_i is the number of edges of the i-th component, n_i is the number of vertices of G. Summing the inequalities, we have

$$m = \sum_{i=1}^{k} m_i \geq \sum_{i=1}^{k} (n_i - 1) = n - k,$$

i.e., $\nu(G) = m - n + k \geq 0$.

2.2.3. Hint: The solution immediately follows from the definition of the cyclic rank of graph.

2.2.4. Hint: The numbers of such skeletons are: a: 16; b: 1; c: 7; d: 1; e: 12.

2.2.5. a: 2; b: 3; c: 6; d: 1; e: 4; f: 2; g: 5.

2.2.6. a: 21; b: 15; c: 81; d: 64.

2.2.7. In the case when the graph is acyclic.

2.2.8. Solution: Choose an arbitrary vertex of the graph and complement the set of edges incident to it to obtain a skeleton of the graph. Deleting the edges of this skeleton from the graph, we obtain an isolated vertex.

2.2.9. Hint: Let G be a connected graph. Pick any vertex and label it by 0. Let v be the vertex with the smallest label i whose neighborhood contains unlabelled vertices $u_1, \ldots, u_k$. We label all these vertices by $i+1$ and select all edges $vu_1, \ldots, vu_k$. Repeat this until all vertices are labelled. The selected edges constitute a skeleton.

2.2.10. 1: Hint: See Theorem 2.1.1.

3: Hint: See Theorem 2.1.1.

2.2.11. Hint: See Theorem 2.1.1.

2.2.12. Hint: The minimal(maximal) weight of the skeleton is: a: 37(59); b: 28(60); c: 18(60).

2.2.14. Hint: See Theorem 2.1.1.

2.2.15. The number of skeletons of G is the product of the numbers of skeletons of G' and G''.

2.2.16. Solution: For any edge e from a cycle one may construct a skeleton that contains all edges of the cycle except of e.

2.2.17. No. Hint: See Exercise 2.2.17.

2.2.18. P_n, $n \geq 2$ and C_n, $n \geq 3$.

2.2.20. a: Hint: Use breadth-first search from v (see Exercise 2.2.9).

b: Hint: See Exercise 2.1.21. Apply Exercise 2.2.21a for every central vertex.

2.2.22. a: Yes. b: No.

2.2.23. Both statements are false. Hint: a: Consider the cycle C_4. b: Consider K_n, $n \geq 3$.

2.2.24. n^{n-2}.

2.2.25. $m^{n-1}n^{m-1}$.

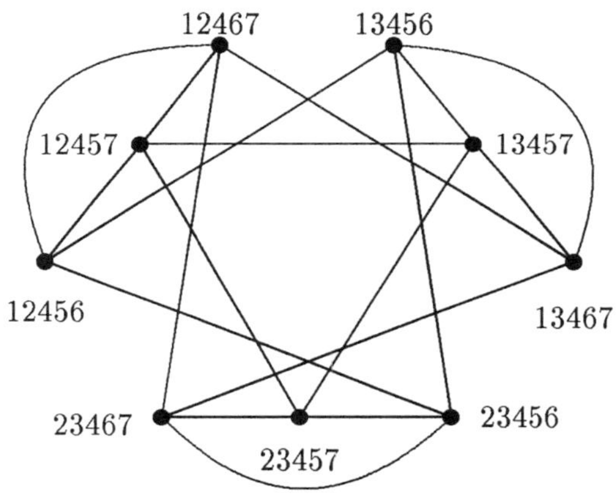

Figure A2.2.1: To Exercise 2.2.29a

2.2.27. Solution: Since the number of labelled trees of order n is n^{n-2} (Cayley Theorem), the total number of the outcomes is n^{n-1}. Let us count the number of favorable outcomes, i.e., when the picked vertex is pendant. Its label may be one of numbers $1, 2, \ldots, n$. It may be adjacent to any of the remaining $n-1$ vertices of the tree $T-v$. The number of labelled trees of order $n-1$ is $(n-1)^{n-3}$. Thus the number of favorable outcomes is $n(n-1)(n-1)^{n-3}$. Therefore

$$\lim_{n\to\infty} P_n = \lim_{n\to\infty} (\frac{n-1}{n})^{n-2} = 1/e.$$

2.2.28. a: Hint: Without loss of generality assume that the graph G is connected. If we delete an edge e from the skeleton T_1, it becomes a two-component graph with the vertex sets denoted by V' and V''. The tree T_2 has an edge $f = u'u''$ such that $u' \in V'$, $u'' \in V''$. It is easily seen that $(T_1 - e) + f$ is a skeleton of G.

b: Hint: Use item a of the exercise.

2.2.29. a: See Fig. A2.2.1.

b–e: Hint: See Exercise 2.2.29.

f: Connected graphs that have two edge-disjoint skeletons.

g: For the graph of order n shown an Fig. A2.2.2 this difference is $n-3$, i.e., it may be arbitrarily large.

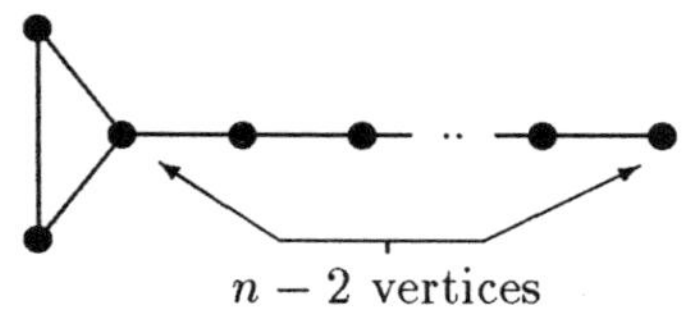

Figure A2.2.2: To Exercise 2.2.29g

h: $\nu^*(G)$, which is equal to $n-k$.

i: Graphs for which every component has two edge-disjoint skeletons.

2.2.30. See Fig. A2.2.3.

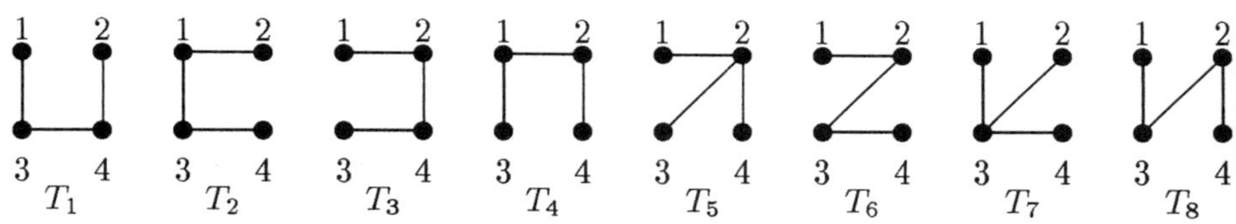

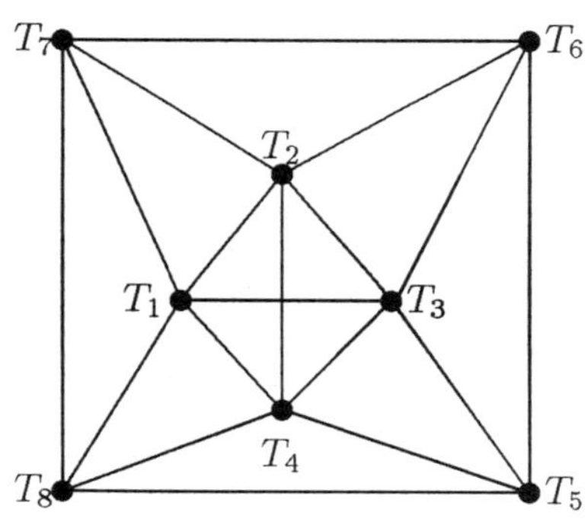

$S(G)$

Figure A2.2.3: To Exercise 2.2.30

2.2.31. Let (G, w) be a connected weighted graph and let T be its minimal spanning tree which cannot be constructed by Kruskal's algorithm. Let us order the list of weights of the edges of T by magnitude: $w_1 \leq w_2 \leq \ldots \leq w_m$ and reindex the edges of T correspondingly: $w(e_i) = w_i$, $i = 1, 2, \ldots, m$, with the additional requirement that among possible indexings we select the one for which an application of Kruskal's algorithm produces the maximal number of the first successive edges $e_1e_2, \dots e_k$ of this ordered edge list. By our supposition, $k < m$. Since G is connected, the $(k+1)$-st iteration of Kruskal's algorithm is possible. Let e'_{k+1} be the edge selected at this iteration. Clearly, $w_{k+1} > w(e'_{k+1})$, and therefore $e'_{k+1} \notin ET$. The graph $T + e'_{k+1}$ has a single cycle C. It contains the edge e'_{k+1} and some edge e_p with $p \geq k+1$. The graph $T' = T + e'_{k+1} - e_p$ is a tree such that $w(T') = w(T) + w(e'_{k+1}) - w_p < w(T)$. This contradiction proves the statement of the exercise.

2.2.32. Solution: Assume that

$$ET = \{e_1, e_2, \dots, e_m\},\ w(e_i) = c_i,\ i = 1, 2, \dots, m$$

and

$$ET' = \{e'_1, e'_2, \dots, e'_m\},\ w(e'_i) = d_i,\ i = 1, 2, \dots, m.$$

By Exercise 2.2.31, we may assume that $T = T_m$ is a spanning tree constructed by some run of Kruskal's algorithm. Suppose that there exists a number k such that $w(e_k) > w(e'_k)$. Let us choose the minimal possible k. Clearly, $k > 1$. Let T_{k-1} denote the spanning forest of G constructed at the $(k-1)$-th iteration of Kruskal's algorithm. Then because of $w(e_k) > w(e'_k)$, for every e'_j, $j = 1, 2, \dots, k$, either $e'_j \in ET_{k-1}$ or $T_{k-1} + e'_j$ contains a cycle, otherwise e_k would not have been selected at the k-th iteration. Therefore both ends of e'_j belong to the same component of T_{k-1}. Let S be a component of T_{k-1}, $|S| = p$. Since the edges of the set $E' = \{e'_1, \dots, e'_k\}$ produce no cycles, $E' \cap E(G(S))| \leq p - 1 = |ES|$ (recall that $G(S)$ denotes the subgraph of G induced by S). Summing these inequalities over all components S, we obtain a contradiction: $k - 1 = |ET_{k-1}| = \sum_S |ES| \geq |E'| = k$.

Answers to Chapter 3:

Independence and Coverings

3.1 Independent Vertex Sets and Cliques

3.1.1. a: 3; b: 6; c: 4; d: 4; e: 3; f: 5; g: 8.

3.1.2. a: 1; b: n; c: $\max\{n, m\}$; d: $n/2$ for even n, $(n+1)/2$ for odd n; e: $n/2$ for even n, $(n-1)/2$ for odd n; f: 4; g: 2^{n-1}.

Hint: Look for a pattern for small m, n and then proceed by induction.

3.1.3. Tetrahedron: 1; cube: 4; octahedron: 2; dodecahedron: 8; icosahedron: 3.

Hint: The dodecahedron graph D contains an independent set of cardinality 8, see Fig. A3.1.1, i.e., $\alpha_0(D) \geq 8$. On the other hand, every face of the

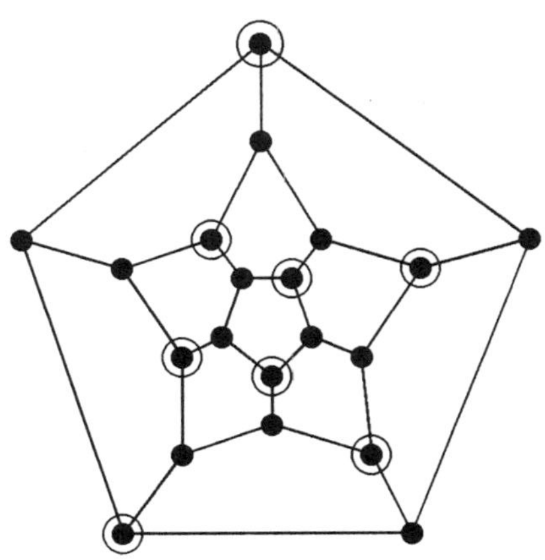

Figure A3.1.1: To Exercise 3.1.3

dodecahedron contains at most two independent vertices. Since every vertex belongs to 3 faces of the dodecahedron and there are 12 faces, we may write $3\alpha_0(D) \leq 2 \cdot 12$.

3.1.4. a: O_n; b: K_n.

3.1.5. No.

3.1.6.

$$\text{a: } \alpha_0(G) - 1 \leq \alpha_0(G - v) \leq \alpha_0(G).$$

$$\text{b: } \alpha_0(G) \leq \alpha_0(G - e) \leq \alpha_0(G) + 1.$$

$$\text{c: } \alpha_0(G) - 1 \leq \alpha_0(G + e) \leq \alpha_0(G).$$

Here v is the vertex being deleted, e is the edge being inserted or deleted.

3.1.7. No.

3.1.8. Hint: To proof the sufficiency, consider a cycle of odd length in the graph.

3.1.9. Hint: Take advantage of the fact that a tree is a bipartite graph.

3.1.10. Hint: Notice that either a pendant vertex or the one adjacent to it, but not both, belongs to a maximum cardinality independent set.

3.1.11. Hint: Use the Exercise 3.1.10.

3.1.12. No. See, e.g., Fig. A3.1.2

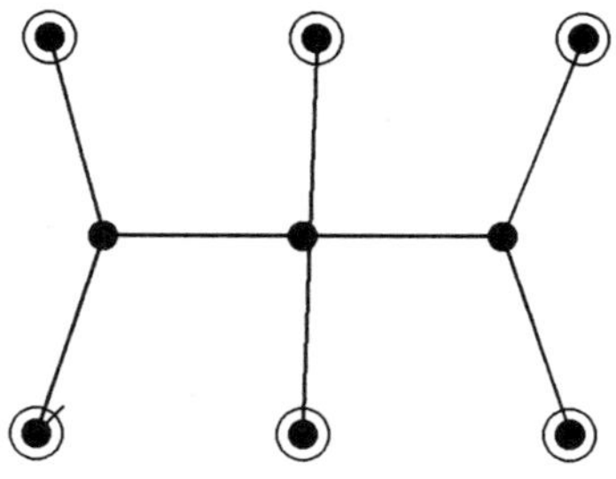

Figure A3.1.2: To Exercise 3.1.12

3.1.13. a: 1; b: 1; c: $\min\{m, n\}$; d,e: $\lceil n/3 \rceil$; f: 3; g: 2^{n-1}.

3.1.14. Hint: Consider the stars $K_{1,n}$, $n \geq 2$.

3.1.15. Hint: c: If X_1, X_2 are independent sets of graphs G and H respectively then $X_1 \times X_2$ is an independent set of the graph $G \times H$.

3.1.16. Hint: a: A proof is by induction over the number of vertices of graph. Clearly, (3.1.1) holds for $n = 2$. Suppose that $|VG| = n > 2$ and the inequality holds for graphs of order less than n. Let G' denote the graph obtained from G by the deletion of a vertex x of minimal degree and of all vertices of its neighborhood. If I' is a maximum cardinality independent set of vertices of G', then the set $I = I' \cup \{x\}$ is independent in G. To prove (3.1.1), it is sufficient to prove that

$$S_1 = \sum_{v \in VG} (1 + \deg_G v)^{-1} \leq S_2 = \sum_{v \in VG'} (1 + \deg_{G'} v)^{-1} + 1.$$

b: Consider the complete graphs.

c: Consider an algorithm such that at each iteration a vertex of minimal degree is put into the required independent set X and then deleted from the graph together with its neighborhood.

3.1.17. Hint: Use formula (3.1.1) and the Cauchy-Bunyakowsky inequality

$$(\sum_{i=1}^{n} a_i^{1/2} b_i^{1/2})^2 \leq \left(\sum_{i=1}^{n} a_i\right)\left(\sum_{i=1}^{n} b_i\right),$$

where $a_i, b_i \geq 0$, assuming $a_i = 1 + \deg v_i$, $b_i = (1 + \deg v_i)^{-1}$ for all $v_i \in VG$.

3.1.18. Solution: Since the set I is maximal, every vertex of $VG \backslash I$ is adjacent to at least one vertex from I, i.e., $|VG \backslash I| \leq \rho(I)$. Adding $|I|$ to both sides, we arrive at the required inequality.

3.1.19. Hint: Use the previous exercise and the fact that $\Delta(G)|I| \geq \rho(I)$.

3.1.20. Solution: Let

$$P_1 = (a_1, \ldots, b_1),\ P_2 = (a_2, \ldots, b_2),\ \ldots,\ P_l = (a_l, \ldots, b_l)$$

be a required partition of VG into the minimal number of paths such that $l(G) \geq \alpha_0(G) + 1$. Then among $a_1, \ldots, a_l$ a pair a_i, a_j of adjacent vertices exists. Linking P_i, P_j into a path $P = (b_j, \ldots, a_j, a_i, \ldots, b_i)$, we obtain the partition of the vertex set into less number of paths, contrary to the assumption.

3.1.21. Solution: Every vertex of a maximum cardinality independent set of G is incident to at least $\delta(G)$ edges. Hence $|EG| \geq \alpha_0(G)\delta(G)$.

3.1.22. Hint: Denoting the independence number of G by α, reindex the vertices of G in such a way that its adjacency matrix assumes the form $A(G) = A = \begin{bmatrix} 0_\alpha & D \\ B & C \end{bmatrix}$, where 0_α denotes the zero $\alpha \times \alpha$-matrix.

Prove that rank $A \leq 2(n - \alpha)$. Then notice that $n = p^0 + p^+ + p^-$ and rank $A = p^+ + p^-$.

3.1.23. Solution: From one hand, $\alpha_0(G') \geq \alpha_0(G)$, since G is an induced subgraph of G'. From the other hand, if an independent vertex set I of G' contains u', it does not contain u, since u is adjacent to u'. Since $G \cong G' - u$, replacing u' by u, we obtain an independent vertex set of G, i.e., $\alpha_0(G') \leq \alpha_0(G)$. Finally, if an independent vertex set I of G' does not contain u, it is also independent in G.

3.1.24. Solution: a: Let u, v be adjacent in $G^m_{n,\alpha}$ with $\deg v > \deg u$. Deleting v from the graph and splitting u (see Exercise 3.1.23), we obtain a graph G' of order n with less edges.

From the other hand,

$$\alpha = \alpha_0(G^m_{n,\alpha}) \geq \alpha_0(G^m_{n,\alpha} - v) = \alpha_0(G'),$$

which is a contradiction.

b: Let u, v be nonadjacent in $G^m_{n,\alpha}$ with $\deg v > \deg u + 1$. Deleting v from the graph and splitting u (see Exercise 3.1.23), we obtain a graph G' of order n with less edges and with $\alpha_0(G') \leq \alpha$, which is a contradiction.

c: Assume that a vertex u is adjacent to v and w and v is not adjacent to w. By item a of the exercise, $\deg v = \deg u = \deg w$. Delete v, w from $G^m_{n,\alpha}$ denoting $G' = G^m_{n,\alpha} - v - w$. Splitting u in G' and then splitting it again in the resulting graph, we obtain a graph G'' of order n with less edges and with $\alpha_0(G'') \leq \alpha$. The resulting contradiction means that v, w are adjacent to each other. This implies that every component of $G^m_{n,\alpha}$ is a complete graph.

d: Take $G = G^m_{n,\alpha}$; k is the number of its components. From the above it follows that

$$G = qK_{p+1} \cup (k-q)K_p, \ q \leq k = \alpha_0(G) \leq \alpha. \tag{A3.1.1}$$

Without loss of generality assume that

$$q < k, \tag{A3.1.2}$$

for otherwise

$$G = kK_{p+1} = q'K_{p'+1} \cup (k-q')K_{p'}, \ q' = 0, \ p' = p+1,$$

and since $n = kp' = \alpha p' + 0$, the required relation holds.

By (A3.1.1),

$$n = kp + q. \tag{A3.1.3}$$

(A3.1.2) and (A3.1.3) together imply that p is the quotient and q is the remainder of the division of n by k. It remains to prove that $k = \alpha$.

Suppose the contrary, i.e., $k < \alpha$. Suppose that the division of n by $k+1$ gives the quotient p_1 and the remainder q_1 and

$$G_1 = q_1 K_{p_1+1} \cup (k+1-q_1)K_{p_1}.$$

Clearly,

$$p_1 \le p,\ n = (k+1)p_1 + q_1,\ \alpha_0(G_1) = k+1 \le \alpha. \tag{A3.1.4}$$

Let us evaluate the difference $t = |EG| - |EG_1|$. We have

$$2|EG| = q(p+1)p + (k-q)p(p-1) = p(2q + k(p-1)).$$

Similarly, $2|EG_1| = p_1(2q_1 + (k+1)(p_1 - 1))$. Taking into account the relations (A3.1.3), (A3.1.4), we obtain

$$q_1 = k(p-p_1) + q - p_1,\ 2|EG_1| = p_1(k(2p - p_1 - 1) - p_1 - 1 + 2q).$$

Therefore,

$$2t = 2q(p-p_1) + k(p-p_1)(p-p_1-1) + p_1^2 + p_1. \tag{A3.1.5}$$

Since $k+1 \le n$, it follows that $p_1 > 0$, and (A3.1.5) implies that $t > 0$. The latter conclusion contradicts to the definition of G, and this contradiction proves that $k = \alpha$.

3.1.25.

$$m = C^2_{p+1}q + C^2_p(\alpha - q).$$

Hint: Use the previous Exercise.

3.1.26. Hint: Use the previous two Exercises.

3.1.27. Solution: Denote $\alpha = \alpha_0(G)$. Since $n/\alpha - 1 < p \le n/\alpha$, the previous exercise and the fact that $n - \alpha \ge 0$ together imply that

$$m \ge p(n - \alpha/2 - \alpha p/2) > (n/\alpha - 1)(n - \alpha/2 - \alpha(n/\alpha)/2) =$$

$$(n/\alpha - 1)(n/2 - \alpha/2) = \frac{n^2}{2\alpha} - n + \frac{\alpha}{2} > \frac{n^2}{2\alpha} - n.$$

This implies $n + m > \frac{n^2}{2\alpha}$, or $\alpha_0(G) = \alpha > \dfrac{n^2}{2(n+m)}$.

3.1.28. a: Solution: The sufficiency is evident. Let us prove the necessity. Assume that G has a component K which is not a complete graph. Then $\alpha_0(K) \ge 2$. Since $\alpha_0(H) = \alpha_0(H^p)$, on may choose a maximum cardinality independent set N of the graph H which is a maximum cardinality independent set of H^p. Let us choose two vertices u, v from N such that the distance $d_H(u,v)$ is minimal possible. Since u, v are not adjacent in H^p, $d_H(u,v) \ge p+1$. Consider a vertex w from a shortest (u,v)-path such that $d_H(u,w) = 2$, see Fig. A3.1.3; $w \notin N$, since $d(u,w) < p+1$. In addition, w must be adjacent to some vertex z from N, otherwise $N \cup \{w\}$ would be a larger independent set. But the (u,w,z)-path is shorter than the chosen (u,v)-path, contrary to the choice of u, v. Therefore the assumption that K is not a complete graph is false.

b: For example, P_4.

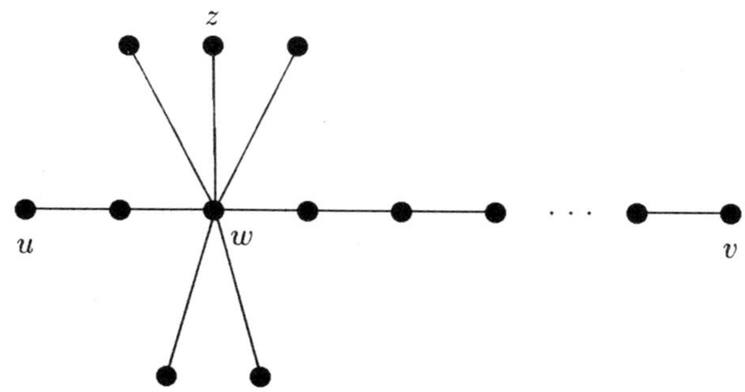

Figure A3.1.3: To Exercise 3.1.28

3.1.30. a: Hint: Delete edges from the graph successively until the required graph is obtained.

b: Solution: Assume that $\alpha_0(G) = \alpha_0$ and $EG = \{e_1, e_2, \ldots, e_m\}$. Let us construct a graph $\tilde{H}$ as follows:

(1)$V\tilde{H} = VG \cup V_1 \cup \ldots \cup V_m$, $|V_i| = \alpha_0 - 1$.

(2)$EG \subseteq E\tilde{H}$.

(3) Every vertex $v \in V_j$, $j = 1, 2, \ldots, m$, is adjacent to all vertices of $\tilde{H}$ with the exception of the vertices from V_j and the ends of e_j. Clearly, $\alpha_0(\tilde{H}) = \alpha_0$.

Delete an edge $e_j = ab \in EG$ from $\tilde{H}$. Since the set $V_j \cup a \cup b$ is independent, $\alpha_0(\tilde{H} - e_j) = \alpha_0 + 1$.

Therefore to obtain an α_0-critical graph H, it is sufficient to delete "redundant" edges from $E\tilde{H} \setminus EG$.

3.1.31. Tetrahedron, icosahedron.

3.1.32. a: O_n; b: K_1; c: K_n; d: P_{n-1}; e: K_1, if $n = 3$; C_n, if $n > 3$; f: $K_m \times K_n$; g: the line graph of the Petersen graph.

3.1.33. Hint: Consider a graph induced by the maximal clique.

3.1.34. $G = \overline{G^m_{n,\varphi}}$ with $m = C^2_{p+1}r + C^2_p(\varphi - r)$, where p is the quotient and r is the remainder of the division of n by φ. $|EG| = C^2_n - m$.

Hint: See the Exercises 3.1.24, 3.1.25.

3.1.35. Solution: Assume that $VG = \{1, 2, \ldots, n\}$. Let G' have an induced subgraph H isomorphic to G and let

$$\varphi : VG \to VH : i \mapsto v_i, \; i = 1, 2, \ldots, n,$$

be the corresponding isomorphism. By the definition of the modular product it immediately follows that the vertices $(1, v_1), (2, v_2), \ldots, (n, v_n)$ are pairwise adjacent in $G \diamond G'$ and hence $\varphi(G \diamond G') \geq n$.

Conversely, suppose that $\varphi(G \diamond G') \geq n$ and let C be a clique in $G \diamond G'$ with n vertices. Suppose that $C = \{(1, v_1), (2, v_2), \ldots, (n, v_n)\}$. The vertices $v_i, i = 1, 2, \ldots, n$, are pairwise distinct. If we set $H = G'(v_1, v_2, \ldots, v_n)$ then the mapping $i \mapsto v_i$ will clearly be an isomorphism of G and H.

3.1.36. Hint: The solution is similar to that of the previous exercise.

3.1.37. b: Hint: Prove that

$$x_1 + x_2 + \ldots + x_n \leq n$$

is a separating inequality for the graph O_n.

c: Hint: Prove that

$$x_1 + x_2 + \ldots + x_n \leq 1$$

is a separating inequality for the graph K_n.

d: Hint: Let G be a threshold graph of order n, $G \neq O_n$, let

$$\alpha_1 x_1 + \alpha_2 x_2 + \ldots + \alpha_n x_n \leq \beta \tag{A3.1.6}$$

be a separating inequality for G and let H be a graph obtained by the addition of an isolated vertex v_{n+1} to G.

Then

$$\alpha_1 x_1 + \alpha_2 x_2 + \ldots + \alpha_n x_n + \alpha_{n+1} x_{n+1} \leq \beta,$$

where $\alpha_{n+1} = 0$, is a separating inequality for H.

e: Hint: Let G be a threshold graph with a separating inequality (A3.1.6). Bearing in mind item c, we assume that $G \neq O_n$. Let b be the minimum among the sums $\alpha_1 x_1 + \ldots + \alpha_n x_n$, $x_i \in \{0, 1\}$, $i = 1, \ldots, n$, greater than β. Then

$$\alpha'_1 x_1 + \alpha'_2 x_2 + \ldots + \alpha'_n x_n \leq b$$

where $\alpha'_i = \alpha_i + (b - \beta)/n$ is also a separating inequality for G.

f: Hint: Let G be a threshold graph with a separating inequality (A3.1.6) in which all coefficients are positive. If $H = G + v$, where v is a new dominating vertex then

$$\alpha_1 x_1 + \alpha_2 x_2 + \ldots + \alpha_n x_n + \beta x_{n+1} \leq \beta$$

is a separating inequality for H.

g: Hint: The separating inequality for the graph G shown at Fig. 3.1.2 is convenient to look for in the form

$$\alpha(x_1 + x_2) + \beta(x_3 + x_4) + \gamma(x_5 + x_6) \leq \delta.$$

It must be:

$$\alpha, \beta, \gamma \leq \delta; \quad 2\alpha, \alpha + \beta, \alpha + \gamma, 2\beta > \delta; \quad \beta + 2\gamma \leq \delta.$$

We may take $\alpha = 5$, $\beta = 3$, $\gamma = 1$, $\delta = 5$.

h: Hint: Let G be a threshold graph with a separating inequality (A3.1.6) and let H be a subgraph of G induced by the set of vertices $\{v_{i_1}, \ldots, v_{i_k}\}$. Consider the inequality

$$\alpha_{i_1} x_{i_1} + \ldots + \alpha_{i_k} x_{i_k} \leq \delta, \tag{A3.1.7}$$

where δ is the maximum of the sums $\alpha_{i_1} x_{i_1} + \ldots + \alpha_{i_k} x_{i_k}$ obtained by the substitution of the characteristic vectors of the independent sets of vertices of H into the left side of (A3.1.7).

i: Hint: Suppose that the graph $G = 2K_2$ is a threshold one with $EG = \{(v_1, v_2), (v_3, v_4)\}$ and with a threshold inequality $\alpha_1 x_1 + \alpha_2 x_2 + \alpha_3 x_3 + \alpha_4 x_4 \leq \beta$. Then

$$\alpha_1 + \alpha_3 \leq \beta, \quad \alpha_2 + \alpha_4 \leq \beta, \quad \alpha_1 + \alpha_2 > \beta, \quad \alpha_3 + \alpha_4 > \beta.$$

Summing the first two inequalities and then the second two ones, we obtain a contradiction.

In a similar way one may prove that P_4 and C_4 are not threshold graphs.

j: Hint: Use items h), i) of this exercise.

k: Hint: Use items h), j) of this exercise.

3.1.38. Hint: There are 8 such graphs.

3.1.39. a: Hint: The graphs at Figs. 1.2.4a,b,c,d,g have no triangles, therefore $cc(G)$ for them is equal to the number of edges. For graph e the answer is 8; for graph f it is 6.

3.1.41. See Fig. A3.1.4.

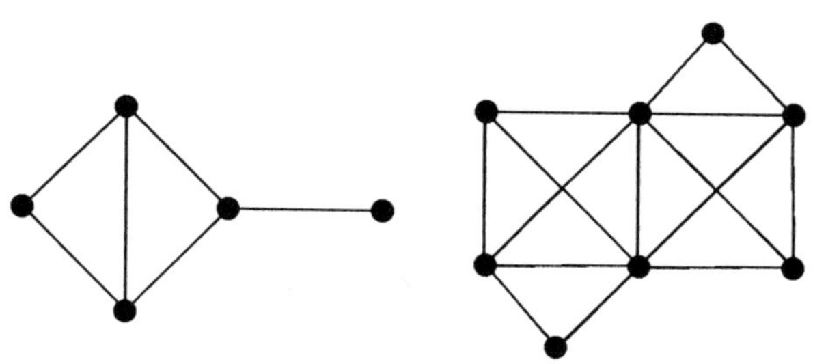

Figure A3.1.4: To Exercise 3.1.41

3.1.43. a: $G_A = G_B$. b: Does not follow. Suppose for example that

$$A = \begin{bmatrix} 1 & 1 & 1 & 0 \\ 0 & 0 & 1 & 1 \end{bmatrix}, \quad B = \begin{bmatrix} 1 & 1 & 0 & 0 \\ 1 & 0 & 1 & 0 \\ 0 & 1 & 1 & 0 \\ 0 & 0 & 1 & 1 \end{bmatrix}.$$

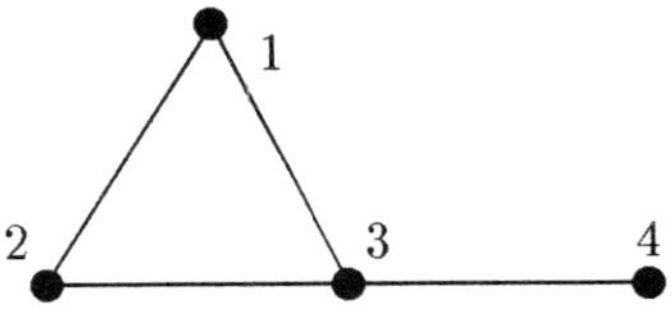

Figure A3.1.5: To Exercise 3.1.43

Since $G_A = G_B$ (see Fig. A3.1.5), the sets of integer points of the polyhedra $P(A)$, $P(B)$ coincide. But for $x = (1/2, 1/2, 1/2, 1/2)$ it is easily seen that $x \in P_B$, and $x \notin P_A$.

3.1.44. Hint: Use the Exercise 3.1.40

3.1.45. Solution: Sufficiency: Let $Q = (Q_i, i = 1, \dots, p)$ be a covering of a graph G by cliques satisfying the conditions 1 and 2. Without loss of generality one may assume that every vertex v belongs to exactly two cliques. Otherwise one may add a single-vertex clique $\{v\}$ to the covering Q.

Consider the intersection graph $\Omega = \Omega(Q)$ of the collection Q, where $V\Omega = Q$ and $Q_iQ_j \in E\Omega$ if and only if $Q_i \cap Q_j \neq \emptyset$. Let us prove that $L(\Omega) \cong G$.

Recall that $VL(\Omega) = E\Omega$. Consider the mapping $\varphi : VG \to VL(\Omega)$ such that $\varphi(v) = Q_iQ_j$, where $v \in Q_i$, $v \in Q_j$. The bijectivity of the mapping follows from the conditions 1 and 2 of the exercise.

Let us prove that the mapping preserves the adjacency relation. Suppose that $vw \in EG$ for $v, w \in VG$. This means that v, w belong to the same clique Q_i of G. Therefore $\varphi(v) = Q_iQ_\alpha$, $\varphi(v) = Q_iQ_\beta$. The edges Q_iQ_α and Q_iQ_β have a common endpoint in Ω, therefore the corresponding vertices of $L(\Omega)$ are adjacent. The converse implication about the adjacency may be proved in a similar way.

Necessity: Suppose that $G = L(H)$, where H is a graph. Let $Q(v)$ denote the set of edges of H incident to a vertex $v \in VH$. Then the collection $(Q_v : v \in VH)$ is a covering of G by cliques satisfying the conditions 1 and 2.

3.2 Coverings

3.2.1. a: 3, 3; b: 6, 6; c: 4, 4; d: 6, 5; e: 3, 3; f: 4, 5; g: 8, 8.

3.2.2. a: $\beta_0 = \lfloor n/2 \rfloor$; $\beta_1 = \lfloor (n+1)/2 \rfloor$.

b: $\beta_0 = \lfloor (n+1)/2 \rfloor$; $\beta_1 = \lfloor (n+1)/2 \rfloor$.

c: $\beta_0 = n - 1$; $\beta_1 = \lfloor (n+1)/2 \rfloor$.

d: $\beta_0 = 1$; $\beta_1 = n$.

e: $\beta_0 = \min\{m, n\}$; $\beta_1 = \max\{m, n\}$.

f: $\beta_0 = 6$; $\beta_1 = 5$.

3.2.4. No. Hint: Use the formula (3.2.1).

3.2.5. Stars $K_{1,n}$.

3.2.6. Hint: If $e = uv$ is a pendant edge of the tree and v is a pendant vertex then the replacement of v by u does not increase the cardinality of the covering.

3.2.7. Hint: Use the previous exercise.

3.2.10. For example, paths with even number of vertices and stars.

3.2.11. b: Hint: Use item a) of the exercise.

3.2.12. Hint: Use the fact that the set $\overline{U}$ is independent if and only if the graph has no edge with both ends in $\overline{U}$, i.e., at least one end of every edge is in U.

3.2.13. Hint: Use the previous exercise.

3.2.14. Solution: Since $\alpha_0(K_{m,n}) = \max\{m,n\}$ and $\beta_0(K_{m,n}) = \min\{m,n\}$, it follows that $\alpha_0(K_{m,n})\beta_0(K_{m,n}) = mn = q$.

Suppose that $G = (X, Y; E)$ is a non-complete bipartite graph. If $\alpha_0(G) = |X|$ then $\beta_0(G) = |Y|$, and $q < |X||Y| = \alpha_0(G)\beta_0(G)$.

Suppose now that $X = A \cup B$, $Y = C \cup D$, $A \cap B = \emptyset$, $C \cap D = \emptyset$, $|A| \neq 0$, $|C| \neq 0$, where $A \cup C$ is a maximum cardinality independent set of G. Then the Exercise 3.2.12 implies that $B \cup D$ is a minimum cardinality covering of G.

Since X and Y are independent sets of G, we have

$$\begin{array}{l} |A| + |B| \leq |A| + |C| \text{ and } |B| \leq |C|, \\ |C| + |D| \leq |A| + |C| \text{ and } |D| \leq |A|. \end{array} \tag{A3.2.1}$$

Since the vertices of the sets A and C do not adjacent to each other, we have $q \leq |B||C| + |A||D| + |B||D|$.

Now we use the relations (A3.2.1): $q \leq |B||C| + |A||D| + |B||D| \leq |B||C| + |A||D| + |C||D| < |B||C| + |A||D| + |C||D| + |A||B| = (|A| + |C|)(|B| + |D|) = \alpha_0(G)\beta_0(G)$.

3.2.15. Hint: Consider a vertex v of degree δ. Prove that a covering of edges incident to the vertex set $v \cup N(v)$ contains at least δ vertices.

3.2.17. Hint: Use the fact that if an edge covering H contains the edges ab, bc, cd, then $H - bc$ is also a covering.

3.2.18. Solution: If M is a minimum cardinality vertex covering of a graph G containing v then the set $M \backslash \{v\}$ covers $G - v$. Hence $\beta_0(G - v) \leq |M| - 1 = \beta_0(G) - 1$. Therefore v is a critical vertex.

Conversely, let v be a critical vertex of G. Consider a minimum cardinality vertex covering M' of the graph $G - v$. Since the set $M' \cup \{v\}$ is a vertex covering of G and the number of vertices in it is only by 1 more than in M', then it is of minimal cardinality.

3.2.19. Hint: Use the fact that the deletion of an end of an edge leads to the deletion of the edge itself.

3.2.20. Hint: Use the Exercise 3.2.19.

3.2.22. The statement is false. Hint: Consider, e.g., $K_{1,n}$, $n \geq 2$.

3.3 Dominating Sets

3.3.1. a: 2; b: 3; c: 2; d: 3; e: 2; f: 2; g: 4.

3.3.2. a: $\lceil n/3 \rceil$; b: $\lceil n/3 \rceil$; c: 1; d: 1; e: 2; f: 3.

3.3.5. See Fig. A3.3.1.

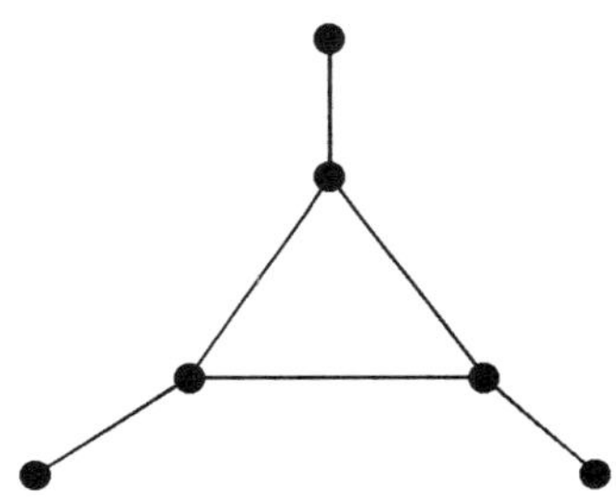

Figure A3.3.1: To Exercise 3.3.5

3.3.6. a: Solution: Let I be a maximal independent set of vertices of a graph G. Then every $v \in VG \backslash I$ is adjacent to a vertex of I, for otherwise the set $I \cup \{v\}$ would have also been independent. Hence I is a dominating set.

The proof of the converse statement is similar.

b: Hint: Use item a) of the exercise.

3.3.7. Hint: Use the Exercise 3.3.6a.

3.3.8. Hint: To prove the left inequality, notice that

$$|D|(\Delta(G)+1) \geq |VG|$$

for a minimum cardinality dominating set D.

To prove the right inequality, notice that $(VG \backslash N(v))$ is a dominating set, where v is a vertex of maximal degree in G.

3.3.9. Hint: Proof is by contradiction.

3.3.10. Hint: We may assume that G is connected. Consider any its spanning tree T and a vertex $v \in VT$. If D is the set of vertices at even distances from v in T, then $VG \backslash D$ is the set of vertices at odd distances from v in T, and D and $VG \backslash D$ are dominating sets.

3.3.11. Hint: Use the previous exercise.

3.3.12. Hint: Prove that if for some vertex $v_0 \in D$ there are no edges adjacent to it whose second end is in $VG \backslash D$ then D is not minimal.

3.3.13. Hint: Use the previous exercise.

3.3.14. Hint: Use the fact that either a pendant vertex or the vertex adjacent to it, but not both, is in a minimum cardinality dominating set.

3.3.15. Hint: Use the previous exercise.

3.3.16. Hint: Construct a graph H satisfying the conditions

(1) $VH = VG \cup V_1$; $VG \cap V_1 = \emptyset$; $|V_1| = |VG|$;

(2) $EG \subseteq EH$;

(3) Every vertex of VG is adjacent to exactly one vertex of V_1 and vice versa.

3.3.18. Hint: Use the Exercise 3.3.8.

3.3.19. Hint: d: See Fig. A3.3.2.

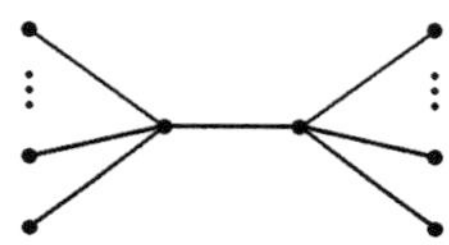

Figure A3.3.2: To Exercise 3.3.19

3.3.20. a: 3; b: 3; c: 4; d: 4; e: 2; f: 3; g: 4.

3.3.21. a: $\lceil n/3 \rceil$; b: $\lceil n/3 \rceil$; c: 1; d: 1; e: $\min\{m, n\}$; f: 3.

3.3.23. a: See, e.g., Fig. A3.3.2. b: K_3.

3.4 Matchings

3.4.1. a: 3; b: 6; c: 4; d: 5; e: 3; f: 4; g: 8.

Hint: Each of these graphs has a perfect matching, with the exception of the graph shown at Fig. 1.2.4f.

3.4.2. a: $\lfloor n/2 \rfloor$; b: $\lfloor n/2 \rfloor$; c: $\lfloor n/2 \rfloor$; d: 1; e: $\min\{m, n\}$; f: 5.

3.4.5. Hint: For Q_n, use the induction.

For K_{2n}, the required 1-factors are, e.g., the sets of edges

$$X_i = \{v_i v_{2n}\} \cup \{v_{i+j} v_{i-j} : j = 1, 2, \ldots, n-1\}, \ i = 1, 2, \ldots, 2n-1,$$

where the summation of the indices of vertices is modulo $2n - 1$.

For the Petersen graph, notice that every its 1-factor contains either two edges of the external cycle or no such edges.

3.4.6. Solution: Set $\alpha_1 = \alpha_1(G)$, $\beta_1 = \beta_1(G)$. In order to prove $\alpha_1 + \beta_1 = n$, we shall prove the inequalities

$$\alpha_1 + \beta_1 \leq n, \ \alpha_1 + \beta_1 \geq n.$$

Let M be a maximum cardinality matching in G. Consider the subset V' of all unsaturated vertices. Clearly, either V' is an independent set or it is empty (otherwise M would not be of maximal cardinality). In addition, $V' = |G| - 2\alpha_1$. If $V' \neq \emptyset$, then for every $v \in V'$ we pick in G an edge incident to it. The set of picked edges is denoted by E'. If $V' = \emptyset$, we set $E' = \emptyset$.

Since G has no isolated vertices and V' is independent,

$$|E'| = |V'| = |G| - 2\alpha_1.$$

Clearly, $E' \cup M$ is an edge covering, therefore

$$\beta_1 \leq |E' \cup M| = |E'| + |M| = (|G| - 2\alpha_1) + \alpha_1 = |G| - \alpha_1.$$

Let now P be a minimum cardinality edge covering of G. Consider a subgraph $G' = G(P)$ induced by the edges of the covering P. Clearly, G' has no cycles. Let t denote the number of the components of G and let k_i be the number of edges in the i-th component, $i = 1, 2, \ldots, t$. Selecting an edge from each component, we obtain a matching P' of cardinality t. Therefore $t \leq \alpha_1$. Since G has no isolated vertices,

$$|G| = \sum_{i=1}^{t}(k_i + 1) = \sum_{i=1}^{t} k_i + t = |P| + t = \beta_1 + t \leq \beta_1 + \alpha_1.$$

3.4.8. Solution: The proof is by induction over the number of vertices. For K_2 the statement is true. Let it be true for graphs of orders less than p. Then for our purposes it is sufficient to prove that if a graph G of order p has no induced subgraphs isomorphic to $K_{1,3}$ then it has two adjacent vertices whose deletion leaves the graph connected.

Let T be a spanning tree of G and let $D = (a, b, c, \ldots)$ be a diametral path of T.

If a vertex b is adjacent in T to the vertices a and c only, then $G - a - b$ is connected.

Suppose now that $R = N_T(b) \backslash \{a, c\} \neq \emptyset$. Since D is a diametral path, every $r \in R$ is pendant in T. Therefore if a is adjacent in G to some $r \in R$ then $G - a - b$ is connected.

Let a be adjacent to no vertex from R. If $|R| > 1$, then since G has no induced subgraphs isomorphic to $K_{1,3}$, the vertices of R induce a complete graph. Therefore any two vertices of R are the required ones.

If $R = \{r\}$, then by the same reason r and c are adjacent in G, so a, b are the required ones.

3.4.9. Hint: Notice that every vertex is incident to at most one edge of every matching.

3.4.10. Hint: To prove the sufficiency, suppose the existence of a matching M_1 such that $|M_1| > |M|$. Next, consider the graph induced by the edges of $M_1 \oplus M$ and use the previous exercise.

3.4.11. Use the Exercise 3.4.9.

3.4.12. Solution: Necessity: Let a tree T have a perfect matching M. Consider any vertex v and the vertices $u, w_1, w_2, \ldots, w_k$ adjacent to it, where the edge vu is in M. Let $\tilde{T}, T_1, T_2, \ldots, T_k$ denote the components of $T - v$ that contain the vertices
$u, w_1, w_2, \ldots, w_k$ respectively. The graphs $T_1, T_2, \ldots, T_k$ have perfect matchings, for otherwise T would also have no perfect matchings. Therefore the orders of $T_1, T_2, \ldots, T_k$ are even. By the same reason, $\tilde{T} \cup \{u\}$ has a perfect matching. Hence $\tilde{T}$ is of odd order.

Sufficiency: We shall proceed by induction. For K_2 the statement is true. Let it be true for trees with less than n vertices.

Let u be a pendant vertex and let v be adjacent to it. One of the components of $T - v$ is the vertex u, the remaining ones are denoted by $T_1, T_2, \ldots, T_k$. From the assumption of the exercise for T it follows that the orders of $T_1, T_2, \ldots, T_k$ are even.

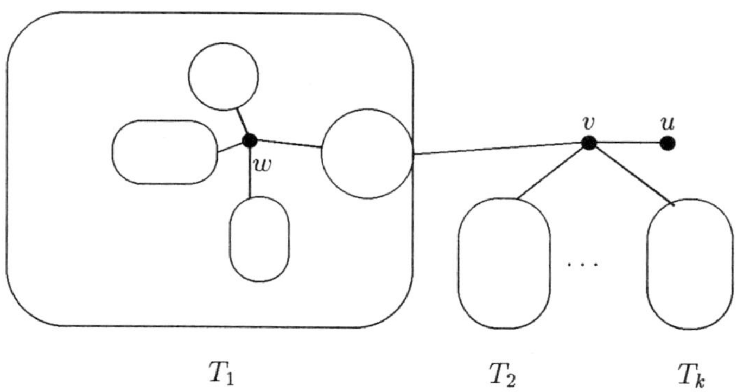

Figure A3.4.1: To Exercise 3.4.12

Let us prove that every T_i satisfies the assumption of the exercise. Refer to Fig. A3.4.1 to keep the track of the proof. If $w \in VT_1$ then one of components of $T_1 - w$ is obtained from a component of $T - w$ by the deletion of $u, v, VT_2, \ldots, VT_k$. Therefore the orders of these components are of the same parity. The remaining components of these graphs coincide.

Proofs for the remaining subtrees T_i are similar.

By the induction hypothesis every tree $T_1, T_2, \ldots, T_k$ has a perfect matching. Their union plus the edge uv is a perfect matching in T.

3.4.14. Hint: Notice first that $\delta(G) = |G| - |V_1|$, where V_1 is the largest part, and furthermore, if $\lfloor |G|/2 \rfloor < \delta(G)$ then $|V_1| < \lfloor |G|/2 \rfloor$, otherwise $|V_1| \geq \lfloor |G|/2 \rfloor$.

3.4.15. a: Solution: If M is a perfect matching then the inequality is evident. Suppose that a maximal matching M is not perfect and V_1 is the set of vertices not saturated by M. Clearly, V_1 is an independent set. Therefore every vertex $v \in V_1$ may be adjacent only to vertices saturated by the matching, i.e., $\delta(G) \leq \deg v \leq 2|M|$.

b: Solution: Suppose that there exists a maximal matching M_0 such that $|M_0| < \alpha_1(G)/2$. For every edge $e \in M_0$, consider a graph $T(e)$ induced by an edge e and edges adjacent to it. Since M_0 is a maximal matching, $\bigcup_{e \in M_0} T(e) = G$. On the other hand, since every graph $T(e)$ contains at most two independent edges, $\alpha_1(G) = \alpha_1(\bigcup_{e \in M_0} T(e)) \leq 2|M_0| < \alpha_1(G)$, which is a contradiction.

3.4.16. a: 3, 3; b: 4, 4; c: 4, 4; d: 4, 4; e: 3, 3; f: 2, 2, g: 6, 6.

3.4.17. a: $\lceil (n-1)/3 \rceil$, $\lceil (n-1)/3 \rceil$; b: $\lceil n/3 \rceil$, $\lceil n/3 \rceil$; c: $\lfloor n/2 \rfloor$, $\lfloor n/2 \rfloor$; d: 1, 1; e: $\min\{m,n\}$, $\min\{m,n\}$; f: 3, 3.

3.4.18. Hint: Notice that if edges uv, vw, wz are in a covering M then $M \backslash vw$ is also a covering.

3.4.19. Hint: The proof may be carried out by contradiction. Among all weak coverings, pick the one with maximal number of independent edges and suppose that this covering has adjacent edges. Such edges will induce stars (see the previous Exercise). Analyze the properties of the covering in the neighborhood of such star T.

Verify that every pendant vertex of T either belongs to a pendant edge of the graph or at distance 2 from it there is a vertex incident to an edge e of the covering, $e \notin T$. If either it belongs to a pendant edge or the mentioned distance is 2, then it is easy to construct another covering with less number of edges.

If the mentioned distance is 1, construct a sequence of edges alternately belonging/not belonging to the covering. Since the number of vertices of a graph is finite, this sequence terminates either at a pendant vertex or at an already traversed vertex. Switching the edges of the covering in an appropriate way, one may obtain another covering with less number of edges.

3.5 Matchings in Bipartite Graphs

3.5.1. Solution: Consider a bipartite graph

$$G' = (X_1 \cup X_2, Y_1 \cup Y_2; M_1 \cup M_2).$$

Since every vertex of G' is of degree 1 or 2, every its component is either a path or cycle with edges alternately in M_1 and M_2, or an edge that belongs to both M_1 and M_2.

The degree in G' of every vertex $y \in Y_2 \backslash Y_1$ is 1. Therefore every $y \in Y_2 \backslash Y_1$ belongs to a component which is a (y, x)-path P_y, where $x \in X_2 \backslash X_1$ or $x \in Y_1 \backslash Y_2$.

In the first case, the last edge of P_y belongs to M_2. Therefore $M_1 \oplus P_y$ is a matching from $X_1 \cup \{x\}$ into $Y_1 \cup \{y\}$ (recall that $A \oplus B = (A \cup B) \backslash (A \cap B)$). In the second case, the last edge of P_y belongs to M_1. Therefore $M_1 \oplus P_y$ is a matching from X_1 into $(Y_1 \backslash \{x\}) \cup \{y\}$.

In both cases the matching $M_1 \oplus P_y$ saturates $Y_1 \cap Y_2$. Therefore the matching $M_1' = M_1 \oplus P_y$ saturates all vertices of the set X_1, the vertex $y \in Y_2 \backslash Y_1$, and the vertices of $Y_1 \cap Y_2$.

Performing this operation several times, we obtain the required matching.

3.5.2. Hint: Let X_1 denote the subset of X of vertices of maximal degree. Consider the graph induced by $X_1 \cup N(X_1)$. Prove that it satisfies the assumptions of Theorem 3.5.1, i.e., there exists a matching M_1 saturating the vertices of X_1. Next, let Y_1 denote the subset of Y of vertices of maximal degree. Prove that there exists a matching M_2 saturating all vertices of Y_1. Now you may use the previous exercise.

3.5.3. Hint: Use the previous exercise.

3.5.4. Solution: Let us construct a new bipartite graph $G' = (X, Y', E')$ by adding $|X| - t$ new vertices to the part Y and connecting each of them with every vertex of X. Clearly, the existence of a matching of cardinality t in G is equivalent to the existence of a matching M from X into Y' in G'. A necessary condition for the existence of such M is $|A| \leq |N_{G'}(A)|$. The latter inequality is equivalent to the assumption of the theorem, since $|N_{G'}(A)| = |N_G(A)| + |X| - t$.

3.5.5. Hint: Use the Theorem 3.5.3.

3.5.6. Hint: The sufficiency may be proved by contradiction considering a cycle of odd length in the graph.

3.5.7. Hint: A binary matrix may be regarded as a reduced adjacency matrix of a bipartite graph; the ones of the matrix correspond to the edges of the bipartite graph. The independence of two ones of the matrix means that the corresponding two edges are not adjacent.

3.5.8. Hint: Use Corollary 3.5.2.

3.5.9. Hint: Consider a bipartite graph $G = (X, Y, E)$, where the vertices of X correspond to the elements of S and the vertices of Y correspond to the elements of P, and $x_i y_j \in E$ if and only if $p_j \in S_i$. Now you may use the Theorem 3.5.1.

3.5.10. Hint: Use the Exercise 3.5.4.

3.5.11. Hint: Use the Hall theorem replacing the collection S by the collection

$$S' = (S_{11}, S_{12}, S_{1k_1}, S_{21}, S_{22}, \ldots S_{2k_2}, \ldots, S_{m1}, S_{m2}, \ldots S_{mk_m},$$

where

$$S_{ij} = S_i,\ i = 1, 2, \ldots, m,\ j = 1, 2, \ldots, k_t,\ t = 1, 2, \ldots, m.$$

Answers to Chapter 4: Connectivity

4.1 Biconnected Graphs and Biconnected Components

4.1.3. d: Hint: Exercises 4.1.3b,c and 2.1.15 imply that every skeleton of the graph G is a simple path.

4.1.5. Hint: Use Exercise 4.1.1a and the Handshake Lemma.

4.1.6. A necessary and sufficient condition is that the graph G must have a bridge.

4.1.7. Hint: Use Exercise 4.1.2.

4.1.8. Hint: Use item 2) of Theorem 4.1.1.

4.1.9. See Fig. A4.1.1.

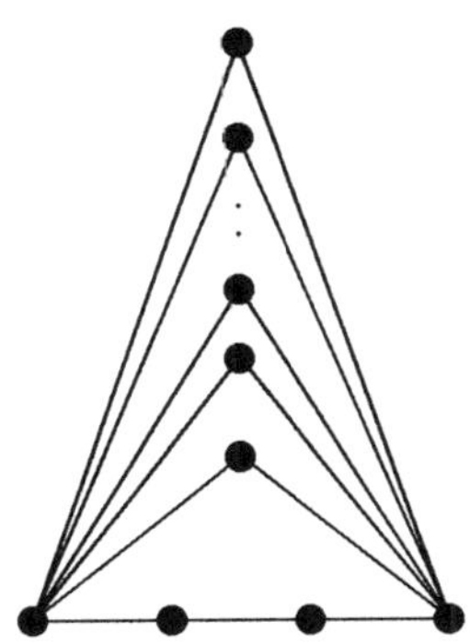

Figure A4.1.1: To Exercise 4.1.9

4.1.10. Hint: By item 6) of Theorem 4.1.1, for any two vertices a, b the graph G has a simple (a, b)-path H_1 containing v. Prove that H_1 contains a subpath that is the path required for the exercise.

4.1.11. a: See Fig. A4.1.2.

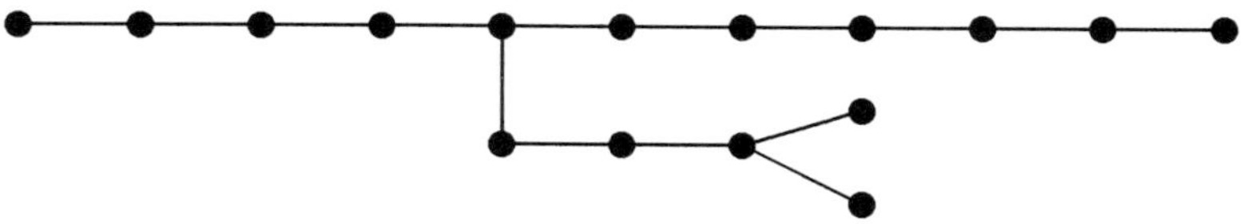

Figure A4.1.2: To Exercise 4.1.11

b: Solution: Clearly, the graph $bc(G)$ is connected. Therefore it is sufficient to prove that it is acyclic. Suppose that it contains a simple cycle

$$(c_{j_1}, B_{j_1}, c_{j_2}, B_{j_2}, \ldots, c_{j_k}, B_{j_k}, c_{j_1}).$$

Each of blocks B_{j_i}, $i = 2, \ldots, k$, contains a simple $(c_{j_i}, c_{j_{i+1}})$-path, and the union of these paths is a simple (c_{j_1}, c_{j_2})-path P of G. By Exercise 4.1.1b, the graph $B_{j_1} \cup P$ is biconnected, hence B_{j_1} is not a block of G, which is a contradiction.

4.1.12. Solution: Let B and C be the sets of blocks and cutpoints of the graph G respectively and let E be the set of edges of $bc(G)$. Since $bc(G)$ is a tree (Exercise 4.1.11b), $|B| = |E| - |C| + 1$.

On the other hand,

$$1 + \sum_{v \in VG} (b(v) - 1) = 1 + \sum_{v \in C} (b(v) - 1) = 1 + \sum_{v \in C} b(v) - |C| = 1 + |E| - |C|.$$

The two equalities imply $|B| = 1 + \sum_{v \in VG}(b(v) - 1)$.

4.1.13. Solution: Proof is by contradiction. Suppose that the graph $G_v = G - v$ has cutpoints for every $v \in VG$. Let B_v denote a block of G_v with maximal number of vertices; denote $t_v = |B_v|$. Choose a vertex v maximizing t_v. Let e be an edge of G with both ends in B_v. By Theorem 4.1.1, G has a simple cycle passing through e and v. This implies that G has two simple paths

$$h_1 = (v, x_1, x_2, \ldots, x_p),\ h_2 = (v, y_1, y_2, \ldots, y_q)$$

such that

$$Vh_1 \cap Vh_2 = v,\ Vh_1 \cap VB_v = x_p,\ Vh_2 \cap VB_v = y_q.$$

Set $H = Vh_1 \cup Vh_2 \cup VB_v$. If we suppose that G has a vertex $z \notin H$ then this would imply that $G_z = G - z$ has a block B' such that $H \subseteq VB'$, i.e., $t_z > t_v$, contrary to the choice of the vertex v. Therefore $VG = H$. This implies that $\max\{p, q\} > 1$; otherwise it would be $G - v = B_v$. To be specific, assume that $p > 1$.

Consider the vertex x_{p-1} of the path h_1. By the assumptions of the exercise, $\deg x_{p-1} \geq 3$. Hence G has a vertex w adjacent to x_{p-1} such that $wx_{p-1} \notin Eh_1$. There are four possibilities for the location of w :

$$w \in VG\backslash(Vh_1 \cup Vh_2 \cup VB_v),\ w \in Vh_1\backslash\{x_p, x_{p-2}\},\ w \in Vh_2\backslash\{v\},\ w \in VB_v.$$

The first and second cases contradict to the choice of v; the third and fourth cases contradict to the fact that B_v is a block of G_v.

4.1.14. Hint: Use the induction over the number of blocks of the graph.

4.1.15. Hint: This follows from the equality $G - x = G - x - e$.

4.1.16. Solution: Assume the contrary. Let B be a block of $G - e$ containing a, b, where $ab = e$, and let x be a vertex of $(G-e)\backslash B$. Since G is biconnected, by Theorem 4.1.1 it contains a simple cycle containing both e and x. Now the Exercise 4.1.1b readily leads to the contradiction with the assumption that B is a block of $G - e$.

4.1.18. Solution: Proof is by contradiction. Let blocks B_1, B_2 and vertices a, b be such that $a, b \in VB_1 \cap VB_2$. It is sufficient to consider the case when B_1, B_2 are distinct from K_2. For an arbitrary vertex $x \in VB_2\backslash VB_1$, by Theorem 4.1.1 B_2 has a simple (a, b)-path containing x. It has a subpath passing through x and having exactly two vertices in common with B_1. Now Exercise 4.1.1b contradicts to the fact that B_1 is a block.

4.1.19. Hint: Use the previous exercise.

4.1.20. Hint: Use the Exercises 4.1.19, 4.1.17.

4.1.21. Solution: The sufficiency readily follows from Theorem 4.1.1, item 2. Let us prove the necessity. Assume the contrary. Then by Exercise 4.1.17 the cycle C entirely lies in a block of $G - e$. On the other hand, Exercise 4.1.16 implies that a, b belong to different blocks of $G - e$.

4.1.22. Hint: Use the previous exercise.

4.1.24. Solution: Proof is by contradiction. For $|G| = 3$ the statement is evident. Consider a graph G of minimal order which is critically biconnected and has no vertices of degree 2. Since $\delta(G) > 2$, the graph $G - e$, $e \in EG$, has a block B which is not K_2. By Exercise 4.1.23, B is a critically biconnected graph, contrary to the minimality assumption for G.

4.1.25. Solution: Proof is by induction over $|G|$. For $|G| = 4$ the statement is evident. Assume that $|G| > 4$. For $e \in EG$ let $G_1, G_2, \ldots, G_m$ denote the blocks of $G - e$ distinct from K_2 and let $R_1, R_2, \ldots, R_k$ denote the two-vertex blocks of $G - e$. By Exercises 4.1.22, 4.1.23, each of blocks G_i is a critically biconnected graph and $|G_i| \geq 4$. Clearly, $G - e$ has $k + m - 1$ cutpoints and each cutpoint belongs to exactly two blocks of $G - e$. Therefore

$$\sum_{i=1}^{m} |G_i| + \sum_{i=1}^{k} |R_i| = |G| + k + m - 1.$$

By the induction hypothesis, $|EG_i| \le 2|G_i| - 4$, $i = 1, 2, \ldots, m$. In addition, obviously, $|ER_j| = 1 = 2|R_j| - 3$, $j = 1, 2, \ldots, k$. These relations imply

$$|EG| = \sum_{i=1}^{m} |EG_i| + \sum_{j=1}^{k} |ER_j| + 1 \le$$

$$2\sum_{i=1}^{m} |G_i| - 4m + 2\sum_{j=1}^{k} |R_j| - 3k + 1 = 2|G| - 2m - k - 1 \le 2|G| - 4.$$

The last inequality is always valid, if we observe that if $m = 0$ then $k \ge 3$, since G is not the triangle, and if $m = 1$ then $k \ge 1$, since $G - e$ has at least two blocks.

4.1.26. No; see Fig. A4.1.3.

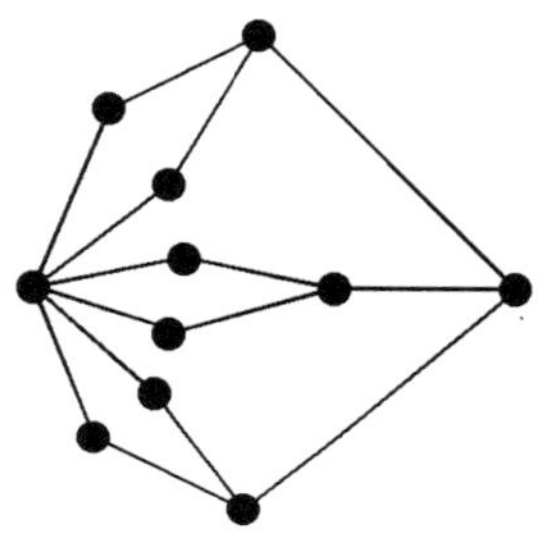

Figure A4.1.3: To Exercise 4.1.26

4.1.28. Hint: Use the Exercise 4.1.27.

4.1.29. Hint: Use the Exercise 4.1.11b.

4.1.30. Solution: Let a be a vertex of G. It is easily seen that $\deg_G a < |G| - 1$. Therefore there exists $b \in VG$ not adjacent to a. The hypothesis of the exercise implies that b is a cutpoint of $G - a$. Let s, t be two vertices of $G - a$ that are not its cutpoints and belong to its different pendant blocks (their existence follows from Exercise 4.1.29). Clearly, s, t are not adjacent. Suppose that $\deg_G a > 2$. Then G has a vertex $v \ne s, t$ adjacent to a. This means that $G - s - t$ is connected, contrary to the assumptions of the exercise. This contradiction implies that $\deg_G a = 2$, i.e., G is a simple cycle.

4.1.31. a: Hint: Use the Exercise 4.1.20.

b: Solution: : The following proof is by contradiction. Suppose that H is a maximal complete subgraph of $B(G)$ and $VB(G) \ne VH$. The connectedness of $B(G)$ and the choice of the set VH together imply the existence of vertices v_i, v_j, v_k such that $v_i \in VB(G) \backslash VH$, $v_j, v_k \in VH$, $v_i v_j \in EB(G)$, $v_i v_k \notin EB(G)$. This means that the block B_j contains two cutpoints of G, i.e., it is not a pendant block. But then according to item a of the exercise $B(G) - v_j$ is disconnected, contrary to the assumption that $B(G)$ is biconnected.

4.2 k-connectivity

4.2.1. $\kappa(G) = 3$, $\lambda(G) = 4$.

4.2.2. A subgraph induced by the set of vertices 1, 2, 3, 10.

4.2.3. One edge.

4.2.4. Hint: Use the Handshake Lemma and Theorem 4.2.2.

4.2.5. Solution: Proof is by contradiction. Let the subset $S \subset VG$ be such that $|S| \leq k-1$ and the graph $G - S$ is disconnected. Let V' denote the vertex set of a component of $G - S$ of smallest size. If $x \in V'$, then clearly $\deg x \leq |V'| + |S| - 1$. This together with the inequality $|V'| \leq (|G| - |S|)/2$ implies

$$\deg_G x \leq (|G| - |S| - 2)/2 \leq (|G| + k - 3)/2,$$

which contradicts to the assumption of the exercise.

4.2.6. A possible solution is the union of two copies of K_{n-2} with $n-4$ common vertices.

4.2.7. Hint: Use the exercise 4.2.5.

4.2.8. A possible solution is the union of two copies of K_{2k-2} with $k-4$ common vertices.

4.2.10. Hint: Use the exercise 4.2.9.

4.2.11. Solution: Let us take $S \subset EG$ with $|S| = \lambda(G)$ such that the graph $G - S$ is disconnected. Then $G - S$ has exactly two components G_1, G_2 (see Exercise 4.2.10). Without loss of generality assume that $|G_2| \geq |G_1| = t$. By the condition, $1 \leq t \leq \delta$. Let us estimate $|S|$:

$$|S| \geq \sum_{v \in VG_1} (\deg v - t + 1) \geq t(\delta + 1) - t^2 = (\delta - t)(t - 1) + \delta \geq \delta.$$

4.2.12. Hint: Consider, e.g., the graph $G = G_1 \cup G_2 \cup G_3$, where

$$G_1 = K_{n_1},\ G_2 = K_{n_2},\ VG_1 \cap VG_2 = \emptyset,\ n_1 + n_2 = n,$$

$$|n_1 - n_2| \leq 1;\ EG_3 = \{ab\},\ a \in VG_1,\ b \in VG_2.$$

4.2.13. Solution: Let $S \subset VG$ be such that $|S| = \kappa(G)$ and $G - S$ is disconnected. Let G' denote a component of $G - S$ of the smallest size. Clearly, $|G'| \leq (|G| - |S|)/2$. If $v \in VG'$ then

$$\delta(G) \leq \deg_G v \leq |G'| - 1 + |S| \leq (|G| + |S|)/2 - 1.$$

This implies that

$$2\delta(G) - |G| + 2 \leq |S| = \kappa(G).$$

4.2.14. Solution: Without loss of generality assume that G is connected. In the cases $\kappa(G) = 1$ and $\kappa(G) = 3$ the equality $\kappa(G) = \lambda(G)$ follows from Exercise 4.1.4 and Theorem 4.2.2. Consider the case $\kappa(G) = 2$. By Theorem 4.2.2 it is sufficient to prove that the graph has two edges e_1, e_2 such that $G - e_1 - e_2$ is disconnected. By the assumption of the exercise, G has two vertices a, b such that $G - a - b$ is disconnected, i.e., both $G - a$ and $G - b$ have a cutpoint. Hence by Exercise 4.1.4 $G - a$ has a bridge e_1 and $G - b$ has a bridge e_2. It is easily seen that $G - e_1 - e_2$ is disconnected.

4.2.15. Hint: This follows from the Theorem 4.2.2 and the Handshake Lemma.

4.2.16. Hint: Among the sets of edges incident to a cutpoint and belonging to the same block, consider the one of the smallest size.

4.2.17. Solution: Suppose that $S \subset EG$ is such that $|S| = \lambda(G)$ and $G - S$ is disconnected. By Exercise 4.2.10, $G - S$ has exactly two components G_1, G_2. Let $V_1 \subseteq VG_1$ and $V_2 \subseteq VG_2$ denote the subsets of vertices incident to edges from S. Clearly, $|V_1| \leq \lambda(G)$ and $|V_2| \leq \lambda(G)$. Moreover, the condition $d(G) = 2$ implies either $V_1 = VG_1$ or $V_2 = VG_2$; say, $V_1 = VG_1$ holds. Consider $v \in V_1$ and let $T_1 \subseteq V_1$, $T_2 \subseteq V_2$ be the sets of vertices adjacent to v. If $v \in V_1$, then let $m(v)$ and $m'(v)$ denote the sets of edges incident to v that belong to S and not belong to S respectively. Clearly, $\deg_G v = |m(v)| + |m'(v)|$ and $|m(v)| \geq 1$ for $v \in V_1$, and $S = \bigcup_{v \in V_1} m(v)$. Let us select a vertex $v_0 \in V_1$ maximizing $|m(v)|$. Then the above relations produce:

$$\lambda(G) = |S| \geq |V_1||m(v_0)| \geq |V_1| - 1 + |m(v_0)| \geq |m'(v_0)| + |m(v_0)| = \deg_G v_0,$$

i.e., $\lambda(G) \geq \delta(G)$. This together with the Theorem 4.2.2 delivers the required equality.

4.2.19. Hint: Use the Handshake Lemma and the Exercise 4.2.15.

4.2.20. One graph, the wheel W_4.

4.2.21. $k = 1, 2$.

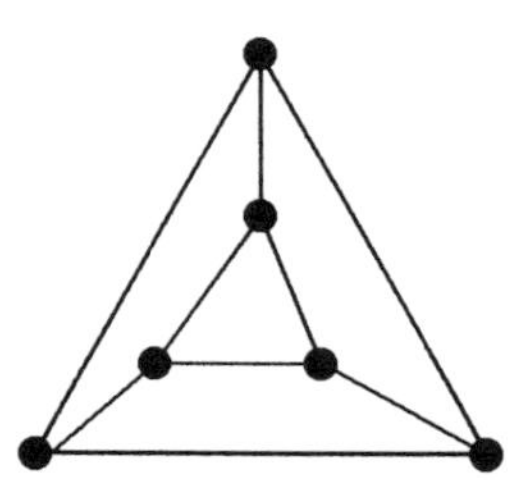

Figure A4.2.1: To Exercise 4.2.21

4.2.22. No; see Fig. A4.2.1

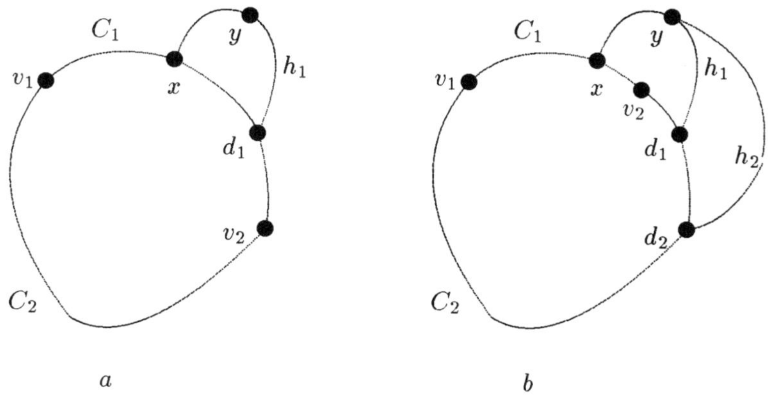

Figure A4.2.2: To Exercise 4.2.22

4.2.23. Solution: Let v_1, v_2, v_3 be three arbitrary vertices of a triconnected graph G. By Theorem 4.1.1, it has a simple cycle containing the vertices v_1, v_2. Let S denote the set of all vertices v of G such that G has a simple cycle containing v_1, v_2, v. Suppose that $S \neq V$. Since G is connected, it has an edge $e = xy$ such that $x \in S$, $y \in V \backslash S$. By the choice of S, G has a simple cycle C passing through v_1, v_2, x. Let C_1, C_2 denote the two simple (v_1, v_2)-paths such that $C_1 \cup C_2 = C$ and $x \in VC_1$. Since $G - x$ is biconnected, by Exercise 4.1.10 it has a simple (y, d_1)-path h_1 and a simple (y, d_2)-path h_2 such that

$$d_1, d_2 \in VC - \{x\},\ d_1 \neq d_2,\ Vh_1 \cap Vh_2 = \{y\}.$$

Consider two cases: (1) one of d_1, d_2, say, d_1 is in VC_1; (2) $d_1, d_2 \in VC_2$.

It is easily seen that in both cases G contains a simple cycle passing through v_1, v_2, y. For the first case its structure is shown at Fig. A4.2.2a, for the second case its structure is shown at Fig. A4.2.2b. This is the contradiction with the choice of S, y. Therefore, $C = V$, i.e., $v_3 \in S$.

4.2.24. No; see the graph $K_{3,4}$.

4.2.25. Let $G(A, B; E)$ be a bipartite graph with $\delta(G) = \delta$. It easy to show that the assumptions of the exercise imply that G is connected. Suppose that $S \subset EG$ is such that $|S| = \lambda(G)$ and $G - S$ is disconnected. $G - S$ has two connected components, G_1 and G_2 (see Exercise 4.2.10). Without loss of generality assume that $|G_1| \leq |G_2|$. Set $X_1 = VG_1 \cap A$, $X_2 = VG_1 \cap B$. Clearly, $|X_1| + |X_2| = |G_1| \leq |G|/2$. It is easily seen that if $X_1 = \emptyset$ or $X_2 = \emptyset$ then $|G_1| = 1$, and then, clearly, $\delta(G) = \lambda(G)$. Therefore we assume that $X_1 \neq \emptyset$ and $X_2 \neq \emptyset$, and hence $|G_1| \geq 2$. Let us bound the value $|S|$ from the below:

$$|S| \geq \sum_{v \in X_1} (\deg v - |X_2|) + \sum_{v \in X_2} (\deg v - |X_1|) \geq |X_1|(\delta - |X_2|) + |X_2|(\delta - |X_1|) =$$

$$\delta(|X_1|+|X_2|)-2|X_1||X_2| \geq \delta(|X_1|+|X_2|)-(|X_1|+|X_2|)^2/2 =$$

$$\delta|G_1|-|G_1|^2/2 = \delta + (|G_1|-1)(\delta - |G_1|^2/(2|G_1|-2)) \geq$$

$$\delta + (|G_1|-1)(\delta - |G|/4 - 1) \geq \delta.$$

This and Theorem 4.2.1 imply $\lambda(G) = \delta(G)$.

4.2.26. Hint: See the solution for the Exercise 4.2.11.

4.2.27. No. A counterexample is shown at Fig. A4.2.3.

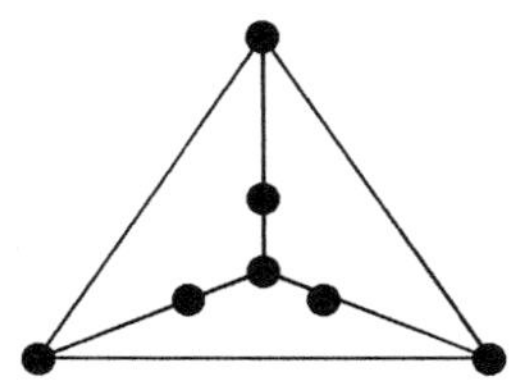

Figure A4.2.3: To Exercise 4.2.27

4.2.28. C_n.

4.2.29. $K_{n,n+1}$.

4.2.30. $K_{3,3}$.

4.2.31. $K_{3,4}$.

4.3 Cycles and Cuts

4.3.1. $K_{3,4}$.

4.3.2. Solution: Let C be a simple cycle of the graph G such that $|C| = g(G)$. The case $g(G) \leq 4$ is obvious. Assume that $g(G) \geq 5$ and consider the set T of vertices from $VG \backslash VC$ adjacent to the vertices of C. Since $\delta(G) \geq 3$ and cycle C has no diagonals, every vertex of C is adjacent to a vertex from T. Furthermore, it is easily seen that every vertex of T is adjacent to a single vertex of C, i.e., $|C| \leq |T|$, and hence $2|C| \leq |C| + |T| \leq |G|$.

4.3.3. Solution: Proof is by contradiction. Let C be a simple cycle of G, $C \neq G$. If $|C| = |G|$ then C must have a diagonal, and in this case G has a cycle of even length. Assume that $|C| < |G|$ and consider $v \in VG \backslash VC$. By Exercise 4.1.10, G has a simple (a,b)-path h passing through v such that $Vh \cap VC = \{a, b\}$. It is easily seen that $h \cup C$ has a cycle of even length.

4.3.4. a: Solution: Let $h = (v_1, v_2, \ldots, v_p)$ be a simple path of maximal length in G and let v_k be a vertex of h incident to v_1 with the largest index k. The maximality of the path h implies that v_1 is adjacent only to the vertices of h. Therefore $k \geq \delta(G) + 1$. Hence the simple cycle $(v_1, v_2, \ldots, v_k, v_1)$ is of length at least $\delta(G) + 1$.

4.3.5. Hint: See the Exercise 4.2.10.

4.3.6. Solution: Let E_1, E_2 denote the sets of edges of a cycle C with equicolored and differently colored ends respectively. Let us assign the weight $w(v) = 0$ to the vertices of the first color and the weight $w(v) = 1$ to the vertices of the second color. An edge $e = uv$ obtains the weight $w(e) = w(u) + w(v)$. Clearly, $\sum_{e \in E_2} w(e) = |E_2|$ and $t = \sum_{e \in E_1} w(e)$ is even. On the other hand,

$$\sum_{uv \in EC} w(uv) = \sum_{uv \in EC} (w(u) + w(v)) = 2 \sum_{x \in VC} w(x), \quad \sum_{e \in EC} w(e) = t + |E_2|.$$

Therefore $|E_2|$ is even.

4.3.7. Hint: Use the Exercises 4.3.5, 4.3.6.

4.3.8. Hint: Use the Exercise 4.3.5.

4.3.9. Ten cuts. Hint: To find them, use the Exercise 4.3.8.

4.3.10. Hint: Use the Exercise 4.3.8.

4.3.13. Hint: Use the previous exercise.

4.3.14. No.

4.3.16. Hint: Use the previous exercise.

4.3.17. Solution: Suppose that $k(G - (V_1, V_2)) = 2$ and let X be a proper subset of the separator set (V_1, V_2). The ends of an edge $e \in (V_1, V_2) \backslash X$ belong to different components of $G - (V_1, V_2)$, therefore adding e to $G - (V_1, V_2)$ we obtain a connected graph. Hence X is not separating, and therefore (V_1, V_2) is a cut of G. The converse is proved as in the Exercise 4.3.5.

4.3.18. Solution: Let C_1, C_2 denote the vertex sets of paths produced from C by the deletions of the edges e_1 and e_2 respectively. Among the separator sets of form

$$X = (V_1, V_2),\ C_1 \subseteq V_1,\ C_2 \subseteq V_2$$

we select a set $X^o = (V_1^o, V_2^o)$ such that $G - X^o$ has the minimal number of components. Suppose that $k(G - X^o) \geq 3$. Let G_1, G_2 denote the components of $G - X^o$ containing C_1, C_2 respectively. G has an edge connecting a vertex of some component $G_3 \neq G_1, G_2$ with a vertex of G_1 or G_2, say with that of G_1. Therefore $VG_3 \subset V_2^o$. Consider the separator set $\overline{X} = (V_1^o \cup VG_3, V_2^o \backslash VG_3)$. Clearly, $C_1 \subseteq V_1^o \cup VG_3$, $C_2 \subset V_2^o \backslash VG_3$ and $k(G - \overline{X}) < k(G - X^o)$, which is a contradiction. Therefore $k(G - X^o) = 2$, and by Exercise 4.3.17, X^o is a cut.

Answers to Chapter 5:

Matroids

5.1 Independence Systems

5.1.1. a: No; b: No; c: Yes, $\mathcal{B}(S) = \{\{1,2\}, \{1,3\}\}$, $C(S) = \{\{4\}, \{2,3\}\}$.

5.1.2. Hint: a: $I(S)$ is the set of all subsets of elements $B \in \mathcal{B}$.

b: A subset $X \subseteq E$ is independent if and only if no element of $\mathcal{B}$ is a subset of X.

5.1.5. If S_1, S_2 are independence systems with the same support then $\mathcal{C}(S_1 \cap S_2)$ is the set of all minimal under inclusion elements of the set $\mathcal{C}(S_1) \cup \mathcal{C}(S_2)$.

5.1.6. b: The maximal independent sets of vertices of G.

c: The edges of G.

5.1.8. Nine; four graphical ones.

5.1.9. b: The characteristic vectors of lower ones of the function f.

c: Hint: Let (E, I) be an independence system of order n with E assumed to be $\{1, 2, \ldots, n\}$. Let us define the Boolean function f in n variables by setting
$f(x_1, \ldots, x_n) = 0$ if and only if the subset $U \subseteq E$ with the characteristic vector $x_U = (x_1, \ldots, x_n)$ is independent.

d: $f \equiv 0$ and the monotonic Boolean functions for which the number of nonzero components of every lower one is equal to 2.

Hint: See the Exercise 5.1.7.

5.1.10. a: $\mathcal{C}(S) = \{\{1\}, \{2\}, \ldots, \{n\}\}$, $\mathcal{B}(S) = \{\emptyset\}$.

b: $\mathcal{C}(S) = \{\{1\}, \{4\}, \{2,3\}\}$, $\mathcal{B}(S) = \{\{2\}, \{3\}\}$.

5.1.11. b: Solution: Let $S = (E, I)$ be an independence system of order n, with E assumed to be $\{1, 2, \ldots, n\}$. If S is a free system then it is specified by the single inequality

$$x_1 + x_2 + \ldots + x_n \leq n.$$

If S is not free, take $C \in \mathcal{C}(S)$ and let $x_C = (a_{c_1}, \ldots, a_{c_n})$ be the characteristic vector of the cycle C, with k_C denoting the number of its nonzero components. Then the independence system S is specified by the system of inequalities

$$a_{c_1}x_1 + a_{c_2}x_2 + \ldots + a_{c_n}x_n \leq k_C - 1, \quad C \in \mathcal{C}(S).$$

c: The systems of inequalities of form (5.1.1) with $a_{ij} \in \{0,1\}$, $b_i = 1$.

5.1.12. b: The pairs of adjacent edges of G.

c: The maximal matchings of G.

5.2 Matroids

5.2.3. The support for all matroids is $\{1,2,3\}$. See the table below. There are three graphical independence systems and two graphic matroids.

ρ	Bases	Cycles
0	$\emptyset$	$\{1\},\{2\},\{3\}$
1	$\{1\}$	$\{2\},\{3\}$
1	$\{1\},\{2\}$	$\{3\},\{1,2\}$
1	$\{1\},\{2\},\{3\}$	$\{1,2\},\{1,3\},\{2,3\}$
2	$\{1,2\}$	$\{3\}$
2	$\{1,2\},\{1,3\}$	$\{2,3\}$
2	$\{1,2\},\{1,3\},\{2,3\}$	$\{1,2,3\}$
3	$\{1,2,3\}$	No

5.2.5. b:Hint: Use the Proposition 5.2.5.

5.2.6. a: Solution:

The set C' as an antichain, since it is a subset of the antichain C, i.e., Axiom C.1 holds for C'. Therefore the pair $M' = (E, C')$ is an independence system with the cycle set C' (Exercise 5.1.2b). By Theorem 5.2.4, M' is a matroid if C' satisfies Axiom C.2. Take $X, Y \in C'$, $X \neq Y$, $e \in X \cap Y$. Since $X, Y \in C$, the set $Z = (X \cup Y)\backslash\{e\}$ contains a cycle U of the matroid M. In the considered situation $|X| = |Y| = |Z| = 2$, $|U| \leq 2$. Therefore $U \in C'$, i.e., C' satisfies Axiom C.2.

b: A counterexample is the matroid with the set of cycles

$$\{\{1,2,3\},\{1,4,5\},\{2,3,4,5\}\}.$$

5.2.7. b: Solution: Recalling that an application of elementary operations to rows of a matrix preserves linear relations among the columns, we transform the matrix A as follows:

$$A = \begin{bmatrix} 1 & 2 & 3 & 1 \\ 0 & 1 & 1 & 2 \\ 1 & 1 & 0 & 1 \end{bmatrix} \to \begin{bmatrix} 1 & 2 & 3 & 1 \\ 0 & 1 & 1 & 2 \\ 0 & -1 & -3 & 0 \end{bmatrix} \to \begin{bmatrix} 1 & 2 & 3 & 1 \\ 0 & 1 & 1 & 2 \\ 0 & 0 & -2 & 2 \end{bmatrix} \to$$

$$\begin{bmatrix} 1 & 2 & 3 & 1 \\ 0 & 1 & 1 & 2 \\ 0 & 0 & 1 & -1 \end{bmatrix} \to \begin{bmatrix} 1 & 0 & 1 & -3 \\ 0 & 1 & 1 & 2 \\ 0 & 0 & 1 & -1 \end{bmatrix} \to \begin{bmatrix} 1 & 0 & 0 & -2 \\ 0 & 1 & 0 & 3 \\ 0 & 0 & 1 & -1 \end{bmatrix}.$$

It is seen now that rank $A = 3$ and every three columns are linearly independent. Thus, the three-element sets of columns of A are the bases of the matroid M and the single-element sets of columns of A are the cobases of M. The set of all columns is the single cycle. The two-element sets of columns are the cocycles.

5.2.10. b: Hint: A possible covering is the set of cycles of the considered independence system plus the set of all elements contained in no cycle.

d: Hint: For matroids of order three, use the table from the answer to Exercise 5.2.3. When passing to matroids of higher orders you must take into account that if M is the matroid of the partition $E = E_1 \cup \ldots \cup E_k$ corresponding to the choice $m = (m_1, \ldots, m_k)$, then the cycles of M are the $(m_i + 1)$-element subsets of E_i, $i = 1, 2, \ldots, k$. Let now M be the matroid with the support $\{1, 2, 3, 4\}$ and the cycle set $C = \{\{1, 2, 3\}, \{1, 2, 4\}, \{3, 4\}\}$. If it were a partition matroid then all three cycles should belong to the same part E_i of the partition and hence the number of elements in them should be the same. Therefore M is not a partition matroid.

5.2.11. a: Hint: If E is a finite nonempty set, $S = \{E_i : i = 1, \ldots, k\}$ is a covering of E, $M = (E, \mathcal{B})$ is the transversal matroid whose bases are the transversals of S, M_i, $i = 1, \ldots, k$ are uniform matroids of rank 1 with the supports E_i respectively, then $M = M_1 \cup \ldots \cup M_k$.

e: The matroid with the support $\{1, 2, 3, 4\}$ and with the set of bases

$$\{\{1, 2\}, \{1, 3\}, \{1, 4\}, \{2, 3\}, \{2, 4\}\}$$

is transversal, but it is not a partition matroid.

5.2.12. b: Hint: Prove that if M is a transversal matroid of order n then to every its independent set there corresponds a matching in the complete bipartite graph $K_{n,n}$. The number of possible sets of matchings is at most 2^{n^2}.

Solution: Let $M = (E, I)$ be a transversal matroid of order n. Without loss of generality let us assume that M is the matroid of transversals of a collection $F = (E_i : i = 1, 2, \ldots, k)$ of nonempty subsets of E, where F is a covering of E. Thus, the independent sets of M are the partial transversals of F plus the empty set. Assume that $k > n$ first.

Let us define a bipartite graph G_F with parts $A = \{1, 2, \ldots, k\}$ and $B = E$ as follows: $N(i) = E_i$, $i = 1, 2, \ldots, k$. Clearly, the mapping $F \mapsto G_F$ is a bijection between the set of coverings of E of length k and the set of bipartite graphs with parts A and B without isolated vertices. It is also clear that a subset $X \subseteq E$ is independent with respect to M if and only if G_F has a matching that covers X.

Let now $B = \{b_1, \ldots, b_\rho\}$ be a base of M and C be a matching that covers it. Without loss of generality assume that $C = \{1b_1, 2b_2 \ldots, \rho b_\rho\}$. The graph G_F has no edges of type xy, $x \in \{\rho+1, \ldots, k\}$, $y \notin N(1) \cup \ldots \cup N(\rho)$, otherwise C would not be a maximal matching and hence B would not be a base. Therefore the subset $F' = (E_i : i = 1, \ldots, \rho)$ is also a covering of E.

Let $M' = (E, I')$ be the matroid of transversals of the family F'. Let us prove that $M = M'$, i.e., $I = I'$. By the construction, $I' \subseteq I$. Suppose that $X = \{e_\alpha : \alpha = 1, \ldots, m\} \in I$, $D = \{i_\alpha e_\alpha : \alpha = 1, \ldots, m\}$ is a matching of G_F that covers X such that $i_\alpha \leq \rho$ if and only if $\alpha \leq q < m$. Set $X = X_1 \cup X_2$, where $X_1 = \{e_\alpha : \alpha \leq q\}$, $X_2 = X \backslash X_1$. The matching D is partitioned accordingly: $D = D_1 \cup D_2$, $i_\alpha e_\alpha \in D_1$ if and only if $e_\alpha \in X_1$.

Clearly, every edge $xy \in D_2$ is incident to a vertex $b_i \in B$. Replacing every edge xy of D_2 by the corresponding edge $ib_i \in C$ we obtain a matching of $G_{F'}$. Hence we have proved that $x \in I'$, $I \subseteq I'$, $I = I'$, $M = M'$.

Thus, we may assume that $k \leq n$ in the original collection F. Therefore the matroid M is specified by a bipartite graph G_F with parts of sizes at most n, i.e., by a subset of edges of $K_{n,n}$. The number of such subsets is 2^{n^2}.

5.2.13. Hint: Consider the dual matroid M^* for the matroid M of transversals of the collection of sets $\{\{1\}, \{2, 3\}\}$.

5.2.14. Hint: Use the axioms of cycles for the matroid.

5.2.15. b: Use Exercises 5.2.14 and 5.1.6.

5.2.16. Hint: Consider the line graph $L(G)$ and use the Exercise 5.2.14.

5.2.17. a: The only bases of the matroid $M(G)$ are the sets of edges of skeletons of the graph G. The sets of edges of the simple cycles of G are the cycles of the matroid. The cuts of G are the cocycles; $\rho(M(G)) = \nu^*(G)$ (the cocyclic rank of G); $\rho^*(M(G)) = \nu(G)$ (the cyclic rank of G).

b: Hint: See the Exercise 5.2.17a.

5.2.18. No.

5.2.19. The forests and the simple cycles.

5.2.20. Hint: Use the proposition 5.2.6.

5.2.21. Hint: The cuts of G are the cocycles of the matroid $M(G)$.

5.2.22. Hint: Use the proposition 5.2.9.

5.2.23. Hint: See the Exercise 4.3.19.

5.2.24. a: The rows with numbers 1, 2, 3, 4, 6. The sum of its elements is 41.

b: The rows with numbers 1, 2, 3, 4, 5. The sum of its elements is 35.

5.2.25. a: Solution: Let $E = \{e_1, \ldots, e_n\}$ be the set of all tasks with the same execution time t and let t_i be the deadline of the i-th task e_i. Let I be the set of sets of tasks that may be performed within their deadlines. Let us prove that $M = (E, I)$ is a matroid.

The validity of Axioms $\mathbf{I}_1$ and $\mathbf{I}_2$ is evident. Let us prove the validity of Axiom $\mathbf{I}_3$. Take $X, Y \in I$, $|X| < |Y|$. Without loss of the generality assume that $Y = \{e_1, \ldots, e_k\}$, $k \le n$ and the tasks in the list Y are written in the order of the possible execution. The execution time for Y is kt and for X it is at most $(k-1)t$. Therefore if $e_k \notin X$ then $X \cup e_k \in \mathcal{I}$, since $t_k \le kt$. If $e_k \in X$ then consider the task e_{k-1}. We have $t_{k-1} \le (k-1)t$. Therefore if $e_{k-1} \notin X$ then $X \cup e_{k-1} \in I$, since the task e_{k-1} may be executed within the time interval $[(k-2)t, (k-1)t]$ and the task e_k within the interval $[(k-1)t, kt]$. If further e_{k-1} is also in X, we consider e_{k-2} just in the same way, and so on. Clearly, eventually we will find a task e_i such that $e_i \notin X$ and $e_j \in X$, $j = i+1, \ldots, k$. Therefore $X \cup e_i \in I$, since the task e_i may be executed within the time interval $[(i-1)t, it]$, and each task e_j, $j = i+1, \ldots, k$ within the interval $[(j-1)t, jt]$.

b: The tasks must be performed in the nondecreasing order of their deadlines.

5.3 Binary Matroids

5.3.2. a: Solution: This matroid is not binary. In fact, the two-dimensional linear space over $\mathbf{Z}_2$ has only three nonzero vectors: $a, b, a+b$, where $\{a, b\}$ is a basis. Hence it has no four elements such that any two of them are linearly independent.

b: Yes.

5.3.3. Hint: The matrix

$$\begin{bmatrix} 1 & 0 & 0 & 0 \\ 1 & 1 & 0 & 1 \\ 0 & 1 & 1 & 0 \\ 0 & 0 & 1 & 1 \end{bmatrix}$$

is a representation of the considered matroid over $\mathbf{Z}_2$.

b: $\{2, 3, 4\}$ is the single cycle.

c: Yes.

5.3.6. Solution: Let us transform the matrix $I(G)$ by elementary transformations of rows over $\mathbf{Z}_2$.

$$\begin{bmatrix} 1&0&1&0&0&0\\ 1&1&0&1&0&0\\ 0&1&0&0&1&0\\ 0&0&0&1&1&1\\ 0&0&1&0&0&1 \end{bmatrix} \to \begin{bmatrix} 1&0&1&0&0&0\\ 0&1&1&1&0&0\\ 0&1&0&0&1&0\\ 0&0&0&1&1&1\\ 0&0&1&0&0&1 \end{bmatrix} \to \begin{bmatrix} 1&0&1&0&0&0\\ 0&1&1&1&0&0\\ 0&0&1&1&1&0\\ 0&0&0&1&1&1\\ 0&0&1&0&0&1 \end{bmatrix} \to$$

$$\begin{bmatrix} 1&0&1&0&0&0\\ 0&1&1&1&0&0\\ 0&0&1&1&1&0\\ 0&0&0&1&1&1\\ 0&0&0&1&1&1 \end{bmatrix} \to \begin{bmatrix} 1&0&1&0&0&0\\ 0&1&1&1&0&0\\ 0&0&1&1&1&0\\ 0&0&0&1&1&1\\ 0&0&0&0&0&0 \end{bmatrix} \to \begin{bmatrix} 1&0&1&0&0&0\\ 0&1&1&1&0&0\\ 0&0&1&0&0&1\\ 0&0&0&1&1&1\\ 0&0&0&0&0&0 \end{bmatrix} \to$$

$$\begin{bmatrix} 1&0&1&0&0&0\\ 0&1&0&0&1&0\\ 0&0&1&0&0&1\\ 0&0&0&1&1&1\\ 0&0&0&0&0&0 \end{bmatrix} \to \begin{bmatrix} 1&0&0&0&0&1\\ 0&1&0&0&1&0\\ 0&0&1&0&0&1\\ 0&0&0&1&1&1\\ 0&0&0&0&0&0 \end{bmatrix}.$$

Let e_i be the i-th column of $I(G)$, $i = 1, \ldots, 6$. The final form of the matrix implies that $\{e_1, e_2, e_3, e_4\}$ is a column basis,

$$e_5 = e_2 + e_4, \quad e_6 = e_1 + e_3 + e_4.$$

If now e_i is the i-th edge of G then $\{e_1, e_2, e_3, e_4\}$ is a skeleton of G and

$$C_{e_5} = \{e_2, e_4, e_5\}, \; C_{e_6} = \{e_1, e_3, e_4, e_6\}$$

is the cycle basis with respect to this skeleton. To obtain the cut basis, we apply the Theorem 5.3.2:

$$C^*_{e_1} = \{e_1, e_6\}, \; C^*_{e_2} = \{e_2, e_5\}, \; C^*_{e_3} = \{e_3, e_6\}, \; C^*_{e_4} = \{e_4, e_5, e_6\}.$$

Answers to Chapter 6:

Planarity

6.1 Embeddings of Graphs. Euler Formula

6.1.1. Hint: Using the Theorem 6.1.3 we embed a plane graph G onto the sphere in such a way that the north pole falls into the chosen internal face. Next, consider the stereographic projection G' of G onto the plane tangent to the sphere at the south pole. Clearly, the chosen face of G will be mapped into the external face of G'.

6.1.2. See Fig. A6.1.1.

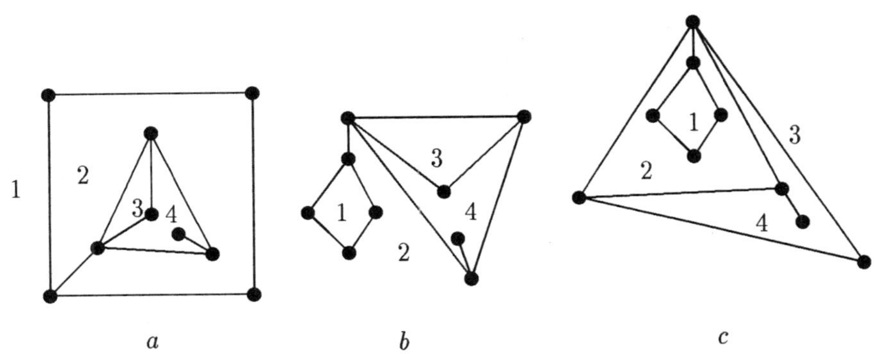

Figure A6.1.1: To Exercise 6.1.2

6.1.4. No.

Hint: Prove by contradiction that $K_{3,3}$ is nonplanar using the Euler formula and noticing that if the bipartite graph $K_{3,3}$ were planar, every its face would have been bounded by at least 4 edges.

6.1.5. See Fig. A6.1.2.

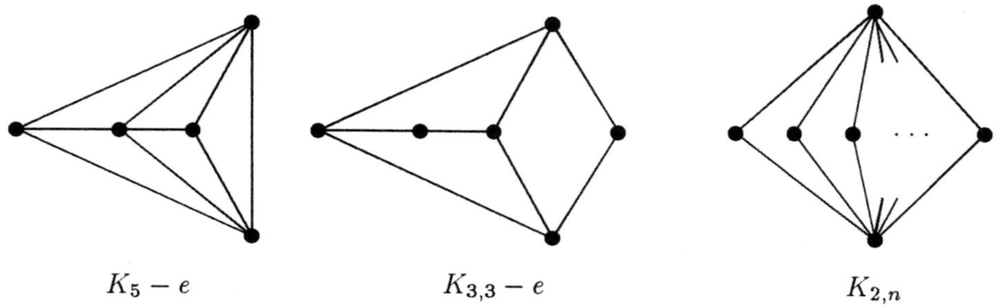

Figure A6.1.2: To Exercise 6.1.5

6.1.7. Hint: Apply the Euler formula to each component counting the external face only once.

6.1.8. To one or two faces.

Hint: If the edge is the bridge then the point belongs to one face, otherwise it belongs to two faces.

6.1.9. It may belong to any number $1, \ldots, d$ of faces.

6.1.10. Hint: See Theorem 5.3.2.

6.1.11. Hint: Use the Exercise 6.1.10.

6.1.12. Hint: Use the Euler formula and the inequality $hf \leq 2m$.

6.1.13. Hint: Use the Euler formula and the inequality $4f \leq 2m$.

6.1.14. Hint: Notice that a bipartite graph is triangle-free (see König's theorem on bipartite graphs). Next, use the previous exercise.

6.1.15. Hint: Use the Exercise 6.1.13 noticing that the n-cube Q_n is triangle-free (see Exercise 1.4.28 and König's theorem on bipartite graphs).

6.1.16. Solution: A possible proof is by contradiction. By Corollary 6.1.5, $m \leq 3n - 6$ for any planar (n, m)-graph G. Supposing that the degree of each vertex of G is at least 6 we would have $6n \leq 2m$, and hence the contradictive inequality $3n \leq 3n - 6$.

6.1.17. Solution: Let G be a 6-connected planar graph. Then $\delta(G) \geq \kappa(G) \geq 6$. But G has vertices with degrees less than 5 (see Exercise 6.1.16).

6.1.18. Solution: Let G be a 5-connected planar (n, m)-graph. Then $\delta(G) \geq \kappa(G) \geq 5$ and this implies $5 \leq 2m/n \leq 2(3n - 6)/n = 6 - 12/n$, i.e., $n \geq 12$.

An example of a 5-connected planar graph with minimal number of vertices $n = 12$ is the icosahedron graph, Fig. 1.1.2.

6.1.19. For $r = 0, 1, \ldots, 5$.

Hint: Such graphs are K_1, K_2, K_3, K_4, the graph from Fig. A6.1.3 and the icosahedron graph. For $r \geq 6$ there are no r-regular planar graphs, see Exercise 6.1.16.

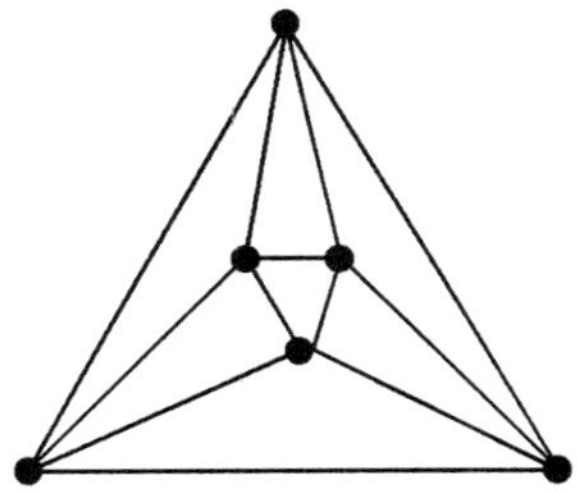

Figure A6.1.3: To Exercise 6.1.19

6.1.20. Hint: Use the Euler formula and the Handshake Lemma.

6.1.21. Hint: The Handshake Lemma, the Euler formula, and the equality $\sum_{i\geq 3} if_i = 2m$, where m is the number of edges of the graph, together imply $\sum_{i\geq 3}(6-i)f_i = 12$. This equality is impossible if $f_3 = f_4 = f_5 = 0$.

6.1.22. Hint: Similarly to the solution of Exercise 6.1.21, noticing that the degrees of the vertices are at least 3 ($3n \geq 2m$), one may prove

$$12 \leq 3f_3 + 2f_4 + f_5 - f_7 - 2f_8 - 3f_9 - \ldots.$$

If $f_3 = f_4 = 0$ then

$$12 \leq f_5 - f_7 - 2f_8 - 3f_9 - \ldots \leq f_5,$$

which contradicts to the inequality $f = \sum_{i\geq 3} f_i < 12$. Therefore $f_3 + f_4 \geq 1$.

6.1.23. Solution: Sufficiency: Since $4f = 2m$, the Euler formula implies $m = 2n - 4$ (see Exercise 6.1.13).

Necessity: Since $m = 2n - 4$, the Euler formula implies $2f = m$. Because of the equalities

$$\sum_{i\geq 4} if_i = 2m, \ \sum_{i\geq 4} f_i = f,$$

we deduce $f_i = 0$, $i \geq 5$ and $f_4 = m/2 = f$.

6.1.24. Hint: Proof is by induction basing on the Exercises 6.1.14, 6.1.23, and on the fact that bipartite graphs are triangle-free. During the inductive step, a new vertex is placed inside any face and is connected with two opposite vertices of the face.

6.1.25. Hint: Use the Exercises 6.1.24, 6.1.23.

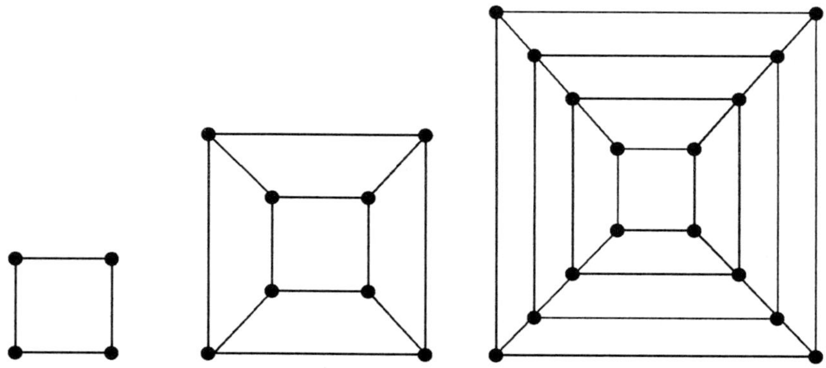

Figure A6.1.4: To Exercise 6.1.26

6.1.26. Hint: Proof is by induction basing on the Exercises 6.1.15, 6.1.23, and on the fact that Q_n is triangle-free. See Fig. A6.1.4, where plane subgraphs of Q_2, Q_3, Q_4 with maximal number of edges are shown.

6.1.27. Hint: Use the Exercises 6.1.26, 6.1.23.

6.1.28. a: Hint: It follows from the equality $5f_5 + 6f_6 = 2m$ (see Exercise 6.1.21 for notations).

b: Hint: Use the Exercise 6.1.21.

6.1.29. For $n = 3$ and for even $n \geq 4$.

6.1.31. Hint: Consider all faces as possible outer faces.

6.1.33. G_1.

6.1.34. For no $n \geq 3$ the wheel W_n is outerplanar.

Hint: Use the Exercise 6.1.30.

6.1.35. Solution: a: It is sufficient to prove the inequality for a maximal outerplanar
(n, m)-graph G. Let us construct a planar graph G' by adding a new vertex adjacent to all vertices of G. Clearly, $n' = n(G') = n + 1$, $m' = m(G') = m + n$. By Corollary 6.1.5, $3n' - 6 \geq m'$. Therefore $m \leq 2n - 3$.

b: Adding the equality $3(f - 1) = 3 + 3m - 3n$ (that follows from the Euler formula) to the inequality $m \leq 2n - 3$, we obtain the required relation.

6.1.36. For $n = 3$.

Hint: The graph $C_4^2 = K_4$ is not outerplanar, see the Exercise 6.1.31. For any $n \geq 5$ the graph G_n^2 has $m = 2n$ edges, which contradicts to the inequality of the Exercise 6.1.35a.

6.1.37. No.

Hint: A counterexample is K_4.

6.1.38. Solution: The graph G^2 has a subgraph K_4, therefore (see Exercises 6.1.31, 6.1.32) G^2 is not outerplanar.

6.1.39. Solution: The sufficiency is evident. The necessity my be proved by contradiction. Suppose that a connected graph G is neither the path graph nor K_3. Then two cases are possible.

Case 1: G is the simple cycle C_n, $n \geq 4$. Then G^2 is not outerplanar, see Exercise 6.1.36.

Case 2: G has a subgraph $K_{1,3}$. Then G^2 is not outerplanar by Exercise 6.1.38.

6.1.40. Hint: Since no internal face is a triangle, $4(f-1)+n \leq 2m$. Now apply the Euler formula.

6.1.41. Solution: First of all notice that every internal face of a maximal outerplanar graph is triangular.

a: Proof is by contradiction. Suppose that $\deg v \geq 3$ for each vertex v of the graph. Then the number of internal faces (triangles) is n. Therefore noticing that the external face is bounded by an n-cycle ($f_n = 1$), we have

$$\sum_{i \geq 3} i f_i = 3n + n = 4n = 2m,$$

where f_i is the number of faces bounded by i edges. This implies $m = 2n$, $f = n+1$, which contradicts to the Euler formula.

b: Proof is by induction over $n \geq 3$. For $n = 3$ the statement is evident. Assume that it is true for $n = k \geq 3$ and consider a graph G with $n = k+1$ vertices and q internal faces. Then $G - v$, where v is a vertex of degree 2 (see item a), has $q-1$ internal faces, i.e., $q-1 = k-2$, or $q = k-1 = n-2$.

6.1.42. See Fig. A6.1.5.

6.1.43. Hint: a: Use the Euler formula and the Exercise 6.1.41.

b: A possible proof is by induction over the number of vertices taking into account that the vertices of degree 2 are not adjacent to each other if $n \geq 4$.

c: Use items a, b of the exercise and the Handshake Lemma.

d: Noticing that every maximal outerplanar graph has at least one vertex of degree 2 (see Exercise 6.1.41a), a simple proof is by induction over the number of vertices.

6.1.44. Hint: It is sufficient to place the vertices in the space in such a way that no four lie in a common plane.

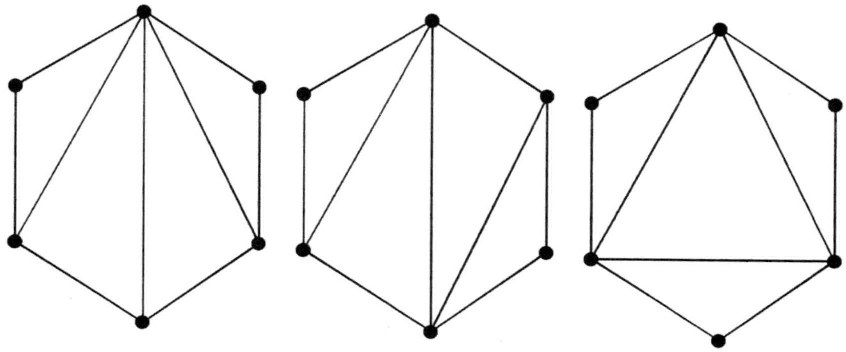

Figure A6.1.5: To Exercise 6.1.42

6.2 Plane Triangulation

6.2.1. Hint: Use Theorem 6.2.1.

6.2.2. $2n - 4$.

Hint: Use the Euler formula and Corollary 6.2.3.

6.2.3. Solution: Let G be a plane graph satisfying the conditions of the theorem. Adding the edges, convert G into a maximal plane graph G' with all vertices of degrees at least 3 (see Proposition 6.2.4). Then by the Corollary 6.2.4 and the Handshake Lemma we have

$$2(3n-6) = 2m = 3n_3 + 4n_4 + 4n_5 + \ldots \geq 3(n_3 + n_4 + n_5) + 6\sum_{i\geq 6} n_i,$$

where m is the number of edges of G', n is the number of its vertices and n_i is the number of vertices of degree i. This inequality implies the required $n_3 + n_4 + n_5 \geq 4$, if we notice that $\sum_{i\geq 3} n_i = n$.

6.2.4. Hint: For such (n, m)-graph the following equalities are valid:

$$5n = 2m, \quad m = 3n - 6, \quad n + f - m = 2,$$

where f is the number of faces of the graph. Hence $m = 30$, $n = 12$, $f = 20$. The uniqueness follows from the construction of a 5-regular triangulation.

6.2.5. $3n - 6$.

Hint: See Corollaries 6.1.5 and 6.2.3.

6.2.6. See Fig. A6.2.1.

6.2.7. No.

Solution: Suppose that such triangulation exists. Deleting the edge connecting the vertices of odd degree, we obtain the graph G' whose all faces are

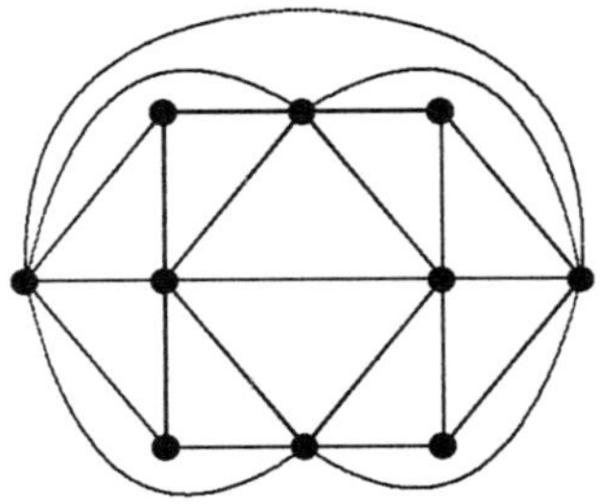

Figure A6.2.1: To Exercise 6.2.6

triangular with the exception of one, bounded by four edges. All vertices of G' are of even degree. Therefore the faces of G' may be colored red and blue in such a way that any neighboring faces are colored differently (see Exercise 9.4.6). Assume that the unique 4-face is colored red. Let r, b denote the numbers of faces G' colored red and blue respectively. Since every edge of G' separates red and blue faces, $3b = 4 + 3(r-1)$. But this equation has no integer solutions; hence such triangulation cannot exist.

6.2.8. Solution: Let the colors be numbered $1, 2, 3$. Each face may be characterized by the triple (i, j, k) of colors of its vertices. Let us transform the triangulation G into another triangulation G' as follows. We place a new vertex inside each monocolored face, i.e., a face of type (i, i, i), $i = 1, 2, 3$, color this vertex by a color $j \neq i$, and connect it with all vertices of this face, see Fig. A6.2.2. This operation preserves the number of tricolored faces (and the total number of faces remains even, see Exercise 6.2.2). Clearly, G' has no monocolored faces, i.e., it has only bi- and tricolored faces.

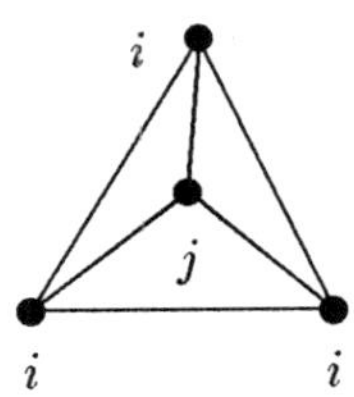

Figure A6.2.2: To Exercise 6.2.8

Every bicolored face of G' has a monocolored edge. At the same time, every monocolored edge belongs to two bicolored faces, see Fig. A6.2.3, where $k \neq i$, $j \neq i$. This implies that the number of bicolored faces of G' is even, the same for the number of tricolored faces of G' and hence of G.

6.3 Planarity Criteria

6.3.1. The Petersen graph contains a subgraph homeomorphic to $K_{3,3}$ with the

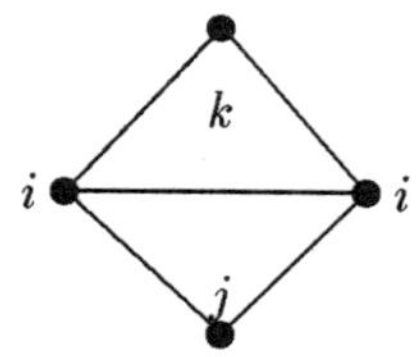

Figure A6.2.3: To Exercise 6.2.8

parts $\{1,2,3\}$ and $\{4,5,6\}$, see Fig. A6.3.1.

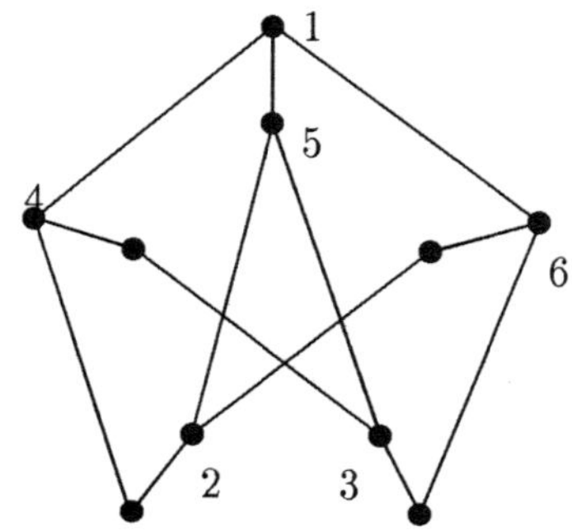

Figure A6.3.1: To Exercise 6.3.1

6.3.2. Hint: Find a subgraph isomorphic to $K_{3,3}$ in the n-permutation graph.

6.3.3. $n = 2, 3, \ldots, 6$.

Hint: G_7 is nonplanar since it contains a subgraph homeomorphic to K_5. Further, if $s < n$ then G_s is the subgraph of G_n.

6.3.4. 14. Hint: Construct all ten graphs of sixth order that contain the subgraph $K_{3,3}$. Further, let G be a unique graph of sixth order homeomorphic to K_5. Construct all graphs containing G as a spanning subgraph and verify that exactly two of them (having the vertex of degree two) were not constructed previously. Finally, it is easy to find two six-vertex graphs containing the subgraph K_5 but neither $K_{3,3}$ nor G.

6.3.5. All graphs are nonplanar.

Hint: All of them contain subgraphs homeomorphic to $K_{3,3}$.

6.3.6. a: For every $n \geq 2$. b: For $n = 2$ only.

Hint: a: A plane embedding may be constructed inductively.

b: For $n = 3$ it contains a subgraph $K_{3,3}$. For $n > 3$ it contains a subgraph homeomorphic to $K_{3,3}$.

6.3.7. The graph H_n is planar for every even $n \geq 4$. For any odd $n \geq 3$ H_n is nonplanar: $H_3 = K_{3,3}$, and H_{2k+1}, $k > 1$, contains a subgraph contractible to $K_{3,3}$.

6.3.8. Hint: If $n \leq 5$, the statement is evident. Assume that $n = 6$ and the graph G is nonplanar. Then three cases are possible: 1) $G = K_{3,3}$; 2) G has a subgraph K_5; 3) G is obtained from K_5 by a subdivision of an edge. In all these cases $\overline{G}$ is planar.

The case $n = 7$ may be analyzed in a similar way.

6.3.9. $n = 8$.

Hint: The graph $K_5 \cup O_3$ has the property. See also the previous exercise.

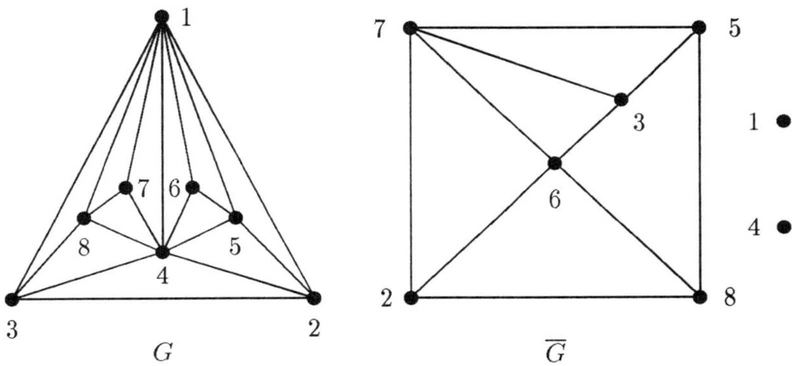

Figure A6.3.2: To Exercise 6.3.10a

6.3.10. a: See Fig. A6.3.2. b: The graph $K_5 \cup O_3$. c: K_8.

6.3.11. For any one.

Hint: The graph $\overline{G}$ must have 10 edges. A possible proof is by contradiction: enumerate all possible cases of nonplanar graphs $\overline{G}$ of 8th order with 10 edges ($K_{3,3} \cup P_2$, $K_5 \cup O_3$, $(K_{3,3} + e) \cup O_2$, etc.) and verify that for no case the graph $G = \overline{\overline{G}}$ may be a planar triangulation.

6.3.12. Solution: Clearly, $|EG| + |E\overline{G}| = n(n-1)/2$. If both $G, \overline{G}$ were planar then it would have been $|EG| \leq 3n - 6$, $|E\overline{G}| \leq 3n - 6$, i.e., $n(n-1)/2 \leq 6n - 12$, which is impossible for $n \geq 11$.

6.3.14. P_n and P_{n+k}, C_n and C_{n+k}.

6.3.15. Not always. At Fig. A6.3.3 G_1, G_2 are homeomorphic but $L(G_1), L(G_2)$ are not homeomorphic.

6.3.17. Not always; see Fig. A6.3.4.

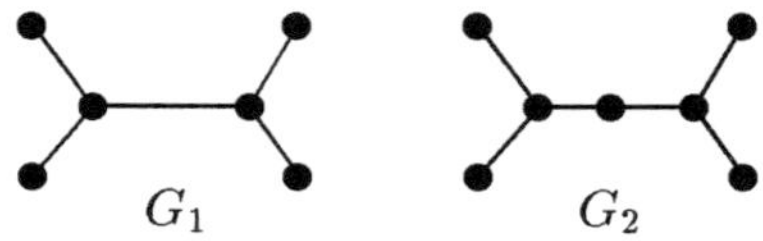

Figure A6.3.3: To Exercise 6.3.15

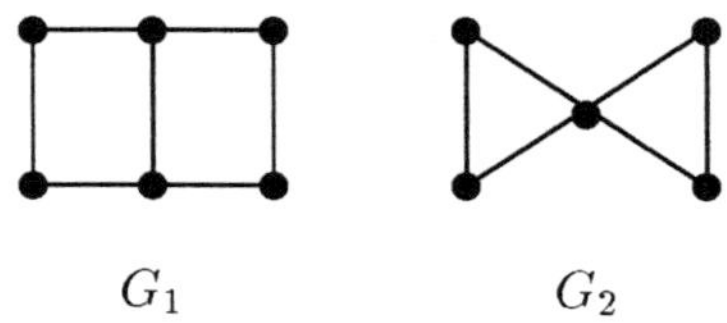

Figure A6.3.4: To Exercise 6.3.17

6.3.18. No. Hint: A counterexample is the Petersen graph: it is contractible to K_5 but it has no subgraph homeomorphic to K_5 since the vertex degrees of the Petersen graph are all equal to 3.

6.3.19. Hint: A possible proof is by contradiction using the Wagner criterion.

6.3.20. Hint: Use the previous exercise.

6.3.21. A counterexample is the graph from Fig. A6.3.5.

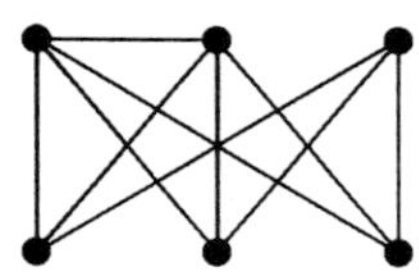

Figure A6.3.5: To Exercise 6.3.21

6.3.22. Hint: All vertices of $L(K_5)$ are of degree six, which is impossible for a planar graph, see Exercise 6.1.16. Verify that $L(K_{3,3})$ contains a subgraph homeomorphic to K_5.

6.3.24. Solution: By the Kuratowski criterion, the graph G contains a subgraph H homeomorphic either to K_5 or to $K_{3,3}$. Therefore by Exercise 6.3.23 $L(H)$ is contractible either to $L(K_5)$ or to $L(K_{3,3})$, which are both nonplanar (see Exercise 6.3.22). Therefore by Exercise 6.3.20 $L(H)$ is nonplanar, hence $L(G)$ is nonplanar.

6.3.25. Not always.

Hint: If, for example, a planar graph G has a vertex of degree 5, then $L(G)$ contains a subgraph K_5.

6.3.26. $r = 1, 2, 3$.

Hint: If a graph G is four-regular then $L(G)$ is nonplanar, since it is six-regular, cf. Exercise 6.1.19. Similarly for $r = 5$.

6.3.27. Hint:

1) The planarity of G follows from the Exercise 6.3.24.

2) Proof is by contradiction, see the hint for Exercise 6.3.25.

3) If some degree-4 vertex were not a cutpoint of G then $L(G)$ would have been nonplanar since it would contain a subgraph contractible to K_5.

6.3.28. Hint: Otherwise G^2 would contain a subgraph K_5.

6.3.29. Hint: Let G' be obtained from a graph G by the addition of a vertex adjacent to all vertices of G. It is easily seen (see Exercise 6.1.30) that G is outerplanar if and only if G' is planar. Further, use the Kuratowski criterion.

6.3.30. Graphs homeomorphic to K_5 and $K_{3,3}$.

Hint: Use the Kuratowski criterion.

6.3.31. K_5 and all 6-vertex graphs G that have the following two properties:

1) $|\{v \in VG : \deg v = 3\}| \geq 3$;

2) G has a subgraph $K_{3,3}$.

Hint: Use the Wagner criterion.

6.4 Duality and Planarity

6.4.3. Hint: Use induction over the number of vertices of G^*, i.e., over the number of faces of G.

6.4.4. Hint: A cutpoint of G^* is a vertex v^* lying inside a face of G that contains boundaries of at least two blocks of G.

6.4.5. Hint: Proof is by contradiction using the previous exercise and the Theorem 6.4.2.

6.4.6. Hint: See the previous exercise.

6.4.7. Their geometric duals are nonisomorphic.

6.4.8. Hint: The graphs G^* and G^{**} are connected.

6.4.10. When the graph is bridgeless and any two faces have at most one common edge.

6.4.11. Hint: Use the Euler formula and the Proposition 6.4.1.

6.4.13. Three.

Solution: Let G be a plane selfdual $(6, m)$-graph with f faces. Then by the Euler formula and the Proposition 6.4.1 we have

$$n = n^* = f = f^* = 6, \quad m = m^* = 10.$$

Since the degree of each vertex of G is at least 3 (see Exercise 6.4.10) and at most 5, we have

$$n_3 + n_4 + n_5 = 6, \quad 3n_3 + 4n_4 + 5n_5 = 20,$$

where n_i is the number of vertices of degree i in G. Therefore only the following cases are possible.

Case 1: $n_3 = 4$, $n_4 = 2$. In addition, due to the selfduality of G, $f_3 = 4, f_4 = 2$, where f_i is the number of faces of G bounded by i edges. Furthermore, two vertices of G of degree 4 may be either adjacent or nonadjacent. In both cases a selfdual graph exists, see Fig. A6.4.1.

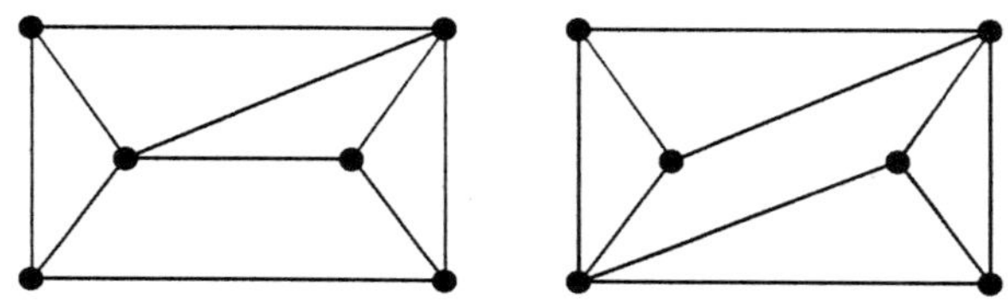

Figure A6.4.1: To Exercise 6.4.13

Case 2: $n_3 = 5$, $n_5 = 1$, with $f_3 = 5$, $f_5 = 1$, i.e., G is the wheel W_5.

6.4.14. Hint: The geometric dual of such graph would contain a subgraph K_5.

6.4.15. Hint: Let T be a spanning tree of G. Prove that the edges of G^* which do not cross the edges of T constitute a spanning tree of G^*.

Solution: Let T^* be the subgraph of G^* constituted by the edges that do not cross the edges of the spanning tree T. Clearly, T^* is acyclic, otherwise the vertices of T inside the cycle of T^* could not be connected with the vertices outside the cycle, i.e., T would have been disconnected. Assume further that $n = |G|$, $n^* = |G^*|$, $m = m(G)$, $m^* = m(G^*)$, f is the number of faces of G. Since $m(T) = n - 1$, the Proposition 6.4.1 and the Euler formula imply

$$m(T^*) = m^* - n + 1 = m - n + 1 = f - 1 = n^* - 1.$$

In addition, clearly, $|T^*| = n^*$. Hence T^* is the spanning tree of G^* (see Exercise 2.2.10). Thus a bijection $T \to T^*$ between the spanning trees of G and G^* is established.

6.4.16. Solution: In the biconnected graph G^* (see Exercise 6.4.5) the boundary of each face has even number of edges. Therefore (see Exercise 6.1.11) the length of any simple cycle in G^* is even. By König's theorem about bipartite graphs, G^* is bipartite.

6.4.17. Hint: Consider two embeddings of the graph in the plane with nonisomorphic geometric duals.

6.4.18. Yes.

Hint: See Fig. A6.4.2, where H and Q are abstract duals to G and $H \ncong Q$. The corresponding edges are labelled by the same letters.

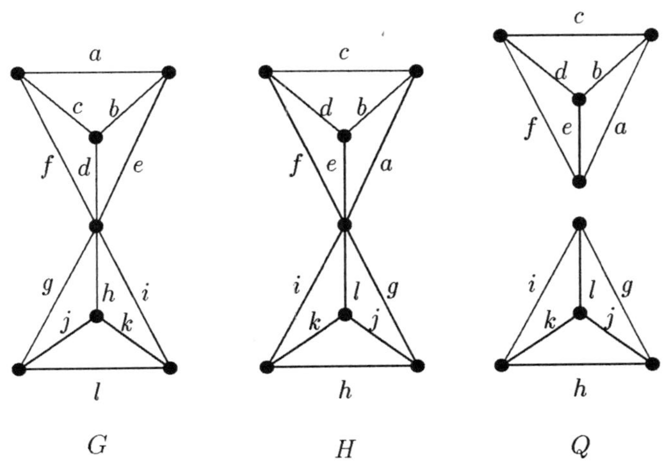

Figure A6.4.2: To Exercise 6.4.18

6.4.19. Hint: Since $n(G) - k(G)$ is the number of edges of the skeleton of the graph G (see Exercise 2.2.11) and $\nu(G)$ is the number of edges to be deleted from G in order to obtain a skeleton of G (see Theorem 2.2.1), the exercise follows from the definition of duality and Theorem 6.4.6.

6.4.20. G, H are dual graphs, H, Q are dual graphs, see Fig. A6.4.3.

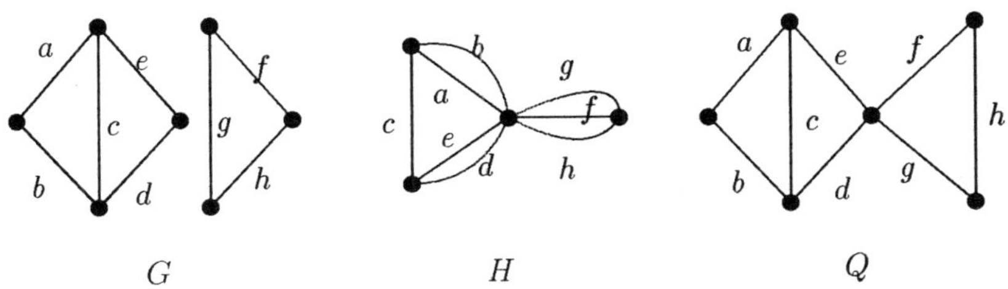

Figure A6.4.3: To Exercise 6.4.20

6.4.21. Not always.

Hint: See the previous exercise: $H = G^+$, $Q = H^+ = G^{++}$, but $Q \ncong G$.

6.4.22. a: Hint: This follows from the Whitney criterion, the Proposition 6.4.5, and the Theorem 6.4.6.

b: See the Exercise 5.2.24.

6.4.23. Solution: a: The equality immediately follows from the definition of the generalized dual graph, if we substitute $H = G$ and notice that $\nu^*(\tilde{G^*}) = 0$.

b: This equality follows from the equality a) and the equality

$$\nu^*(G^*) + \nu(G^*) = m(G^*) = m(G) = \nu^*(G) + \nu(G).$$

c:

$$\nu^*(H) + \nu(\tilde{H^*}) = m(H) - \nu(H) + m(\tilde{H^*}) - \nu^*(\tilde{H^*}) =$$

$$m(H) + m(\tilde{H}) - \nu^*(G^*) = m(G) - \nu(G) = \nu^*(G),$$

since $m(\tilde{H^*}) = m(\tilde{H})$.

6.4.24. Hint: This follows from the equality c) of the previous exercise.

6.4.25. Hint: A possible proof is by induction over the number of edges of a subgraph H considering two cases: 1) after the addition of an edge from G to H the number $\nu(H)$ increases by one; 2) $\nu(H)$ remains unchanged after such operation. In both cases it is easily seen that the equality specifying the generalized duality does not fail. At the same time, this equality is obviously true for $H = O_{|G|}$.

6.5 Measures of Displanarity

6.5.1. Solution: The graphs K_5, K_7, and $K_{4,4}$ are nonplanar but may be embedded onto the torus, see Fig. A6.5.1.

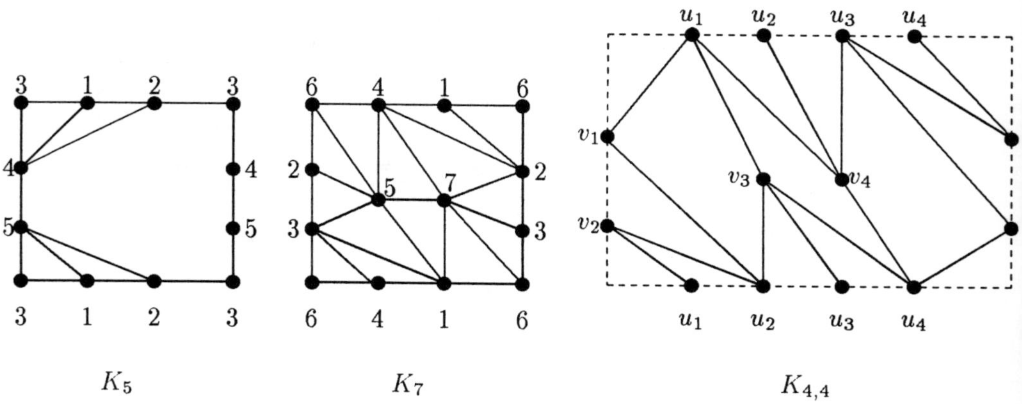

Figure A6.5.1: To Exercise 6.5.1

6.5.2. The genus of the Petersen graph is 1. Hint: It is nonplanar; at the same time, it may be embedded onto the torus, see Fig. A6.5.2.

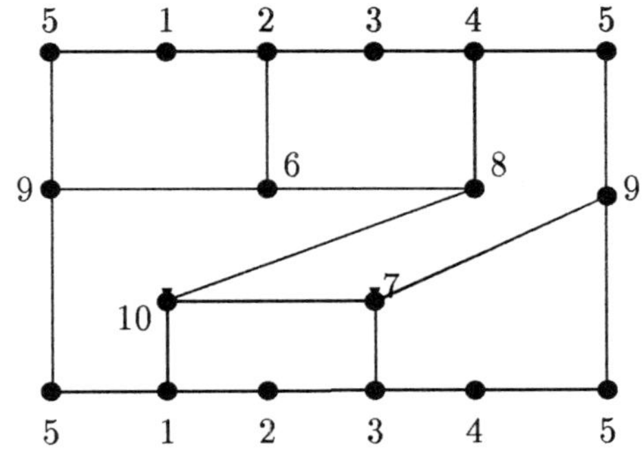

Figure A6.5.2: To Exercise 6.5.2

6.5.3. See Fig. A6.5.3.

6.5.4. Solution: From the formula (6.5.1) for the genus of the complete graph it follows that $\gamma(K_8) = 2$. Therefore all graphs homeomorphic to K_8 are of genus 2.

6.5.5. Hint: Analyze the formula (6.5.1). The next missing value for the genus of K_n is 9.

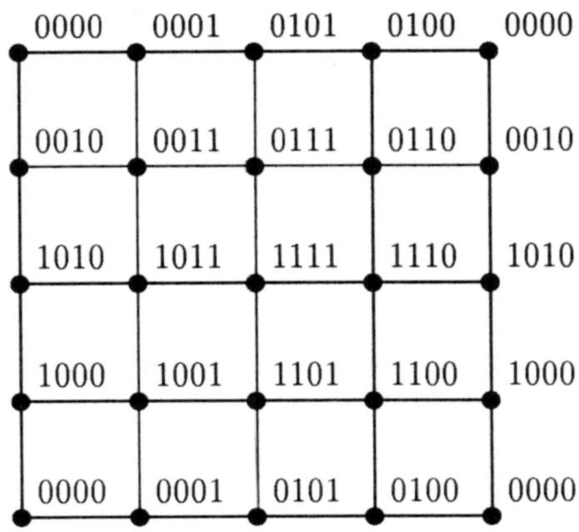

Figure A6.5.3: To Exercise 6.5.3

6.5.6. Hint:

a: With the exception of trivial cases K_1, K_2, P_3, every face of a connected (n, m)-graph is bounded by at least three edges, hence $3f \leq 2m$. Substituting this inequality into the generalized Euler formula (Theorem 6.5.1) and taking into account that the genus is an integer number, we obtain the required inequality.

b: For a triangle-free graph an additional condition is $4f \leq 2m$.

c: This follows from item a.

d,e: Notice that Q_n and $K_{p,q}$ are triangle-free (see Exercise 1.4.28 and König's theorem from Sect. 1.2).

6.5.7. $\gamma(G) \geq \lceil m(1/2 - 1/g) - (n-2)/2 \rceil$.

Hint: See the hint to the previous exercise.

6.5.8. $cr(K_5) = cr(K_{3,3}) = 1$, $cr(\mathcal{P}) = 2$, where $\mathcal{P}$ denotes the Petersen graph.

Hint: Since $\mathcal{P}$ is nonplanar, $cr(\mathcal{P}) > 0$. If it were $cr(\mathcal{P}) = 1$, then the deletion of some edge from $\mathcal{P}$ would produce a planar graph. But one may show that even the deletion of a vertex from the Petersen graph leaves the graph contractible to $K_{3,3}$. On the other hand, it is easy to draw the Petersen graph on the plane with two crossings.

6.5.10.

$$cr(K_{p,q}) \leq \left\lfloor \frac{p}{2} \right\rfloor \left\lfloor \frac{p-1}{2} \right\rfloor \left\lfloor \frac{q}{2} \right\rfloor \left\lfloor \frac{q-1}{2} \right\rfloor$$

for any $p, q \geq 1$.

Hint: Take the vertices a_i, $i = 1, \ldots, p$, at the plane points with coordinates $((-1)^i i, 0)$ and the vertices b_j, $j = 1, \ldots, q$, at the plane points with coordinates $(0, (-1)^j j)$. Connect every a_i with all b_j by line segments and count the number of crossings in such construction (in each quadrant separately).

6.5.11. The statement is false as a whole.

Solution: Let us prove that the sufficiency part fails. Let G be a maximal planar graph ($cr(G) = 0$) that has two vertices u, v at the distance at least 3. Then one may prove by contradiction that $cr(G+e) > 1$, $e = uv$. In fact, suppose that $cr(G+e) = 1$. Then there exists a plane embedding of G such that one may add the edge e to it that crosses only a single edge e' of G. Clearly, $G' = G - e' + e$ is also planar. Therefore G' has a triangular face uvw, see Fig. A6.5.4. But this means that $d_G(u,v) \leq 2$. The contradiction proves that $cr(G+e) > 1$.

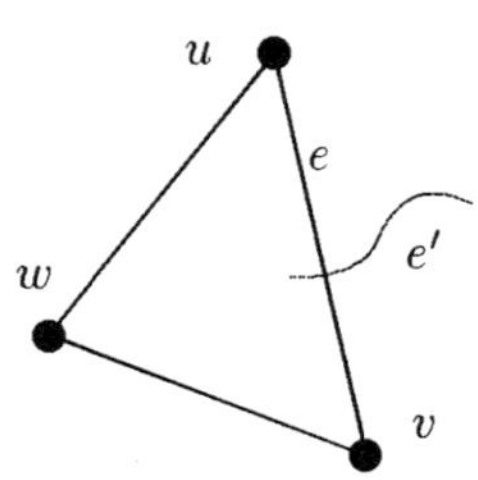

Figure A6.5.4: To Exercise 6.5.11

6.5.13. $t(K_{3,3}) = t(K_5) = t(\mathcal{P}) = 2$, where $\mathcal{P}$ denotes the Petersen graph.

6.5.14. Hint: Consider the graph G' obtained from a nonplanar graph G by subdividing every its edge with two new vertices of degree 2. Its thickness is 2.

6.5.15. Hint: It follows from the Corollary 6.1.5.

6.5.16. Hint: Use the Exercise 6.5.15 and the evident equality $\lceil a/b \rceil = \lfloor (a+b-1)/b \rfloor$ for positive integers a, b.

6.5.17. Hint: Notice that $2n-4 \geq m$ for a connected planar bipartite (n,m)-graph, see Exercise 6.1.13.

6.5.18. Hint: a: Use the previous exercise.

b: For Q_1, Q_2 the statement is obvious. Further, Q_n, $n \geq 3$, has no triangles (see Exercise 1.4.28); it has 2^n vertices and $n2^{n-1}$ edges. Therefore the previous exercise implies $t(Q_n) = \lceil s(n) \rceil$, $n \geq 2$, where $s(n) = n2^{n-1}/2(2^n - 2)$. It is easy to prove that $n/4 < s(n) \leq (n+1)/4$, $n \geq 3$. From the evident inequalities

$$\lceil (n+1)/4 \rceil \geq (n+1)/4; \ \lceil (n+1)/4 \rceil - 1 \leq n/4, \ n \geq 3,$$

we obtain $\lceil s(n) \rceil = \lceil (n+1)/4 \rceil$, $n \geq 3$. This implies $t(Q_n) \geq \lceil (n+1)/4 \rceil = \lfloor n/4 \rfloor + 1$, $n \geq 3$.

6.5.19. Four edges.

Hint: The 4-cube Q_4 contains a maximal (by the number of edges) planar graph without triangles (see Exercise 6.1.27 and Fig. A6.5.5). Since Q_4 has 32 edges, it is necessary to delete 4 edges.

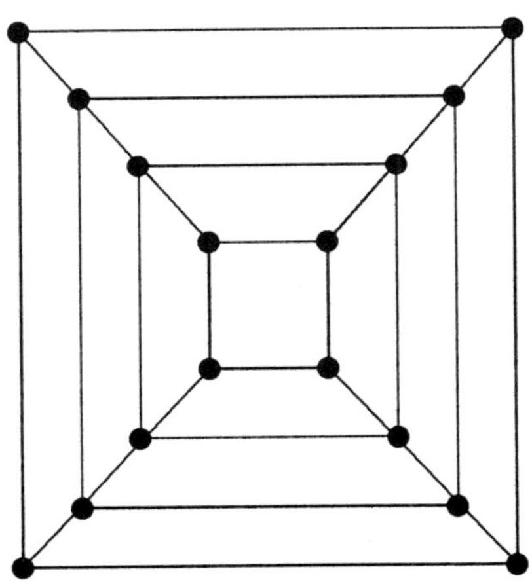

Figure A6.5.5: To Exercise 6.5.19

6.5.20. $sk(K_{3,3}) = sk(K_5) = 1$, $sk(\mathcal{P}) = 2$, where $\mathcal{P}$ denotes the Petersen graph.

Hint: Deleting two nonadjacent edges from the Petersen graph we may obtain a planar graph. Therefore $sk(\mathcal{P}) \leq 2$. On the other hand, the deletion of a vertex from $\mathcal{P}$ produces a graph contractible to $K_{3,3}$ (while proving the latter fact, notice the vertex symmetry of the Petersen graph). Therefore the deletion of any edge from the Petersen graph results in a nonplanar graph; hence $sk(\mathcal{P}) = 2$.

6.5.21. Hint: The formula is true because the maximal number of edges in a planar graph of order n is $3n - 6$ (see Corollary 6.2.3).

6.5.22. $sk(K_{p,q}) = pq - 2(p + q) + 4$, $p, q \geq 2$; $sk(Q_n) = n2^{n-1} - 2^{n+1} + 4$, $n \geq 2$.

Hint: Use the Exercises 6.1.24, 6.1.26.

6.5.23. Hint: Use the Nash-Williams formula (Theorem 6.5.2). Since $m_p \leq 3(p-2)$ for a planar graph G, $\zeta(G) \leq 3$.

6.5.24. A planar triangulation of order 5.

6.5.25. Hint: See the hint to the Exercise 6.5.14.

Answers to Chapter 7:

Graph Traversals

7.1 Eulerian Graphs

7.1.1. Hint: To prove the necessity, show first that if x is a cutpoint of an Eulerian graph that belongs to a pendant block B of the graph then $\deg_B x$ is even. Then apply induction over the number of blocks.

7.1.2. Hint: Consider an algorithm that at every iteration deletes the edges of some simple cycle from the graph G. Then take G_1 to be the graph induced by the set of all deleted edges and take $G_2 = G - EG_1$.

7.1.3. Hint: Use the Exercise 7.1.1.

7.1.4. Hint: Use the fact that the deletion of the edges of some simple cycle from an Eulerian graph produces a graph whose all nontrivial components are Eulerian graphs.

7.1.5. Hint: Use the Exercise 7.1.4.

7.1.6. No.

7.1.7. For every vertex $x \in VG$ the number $\sum_{v \in N(x)} \deg v$ must be even.

7.1.8. The degrees of all vertices are of the same parity.

7.1.10. Hint: Consider the multigraph obtained by the duplication of every edge of the graph.

7.1.11. Hint: Prove that if an Eulerian cycle is not simple then changing the direction of traversal of any of its subcycle produces a new Eulerian cycle (see Exercise 7.1.6).

7.1.12. Hint: Use, e.g., the Exercise 7.1.4.

7.2 Hamiltonian Graphs

7.2.2. No.

7.2.3. No.

7.2.4. Hint: Due to the symmetry of the Petersen graph G, it is sufficient to construct a Hamiltonian cycle in $G - x$ for an arbitrary vertex x.

7.2.5. The required graph may be obtained from $K_{1,3}$ by subdivision of every its edge.

7.2.6. The converse statement is false. Hint: See, e.g., the graph from Fig. A7.2.2.

7.2.7. Yes.

7.2.8. Hint: Use the fact that the deletion of m vertices from C_n produces a graph with at most m components.

7.2.9. Hint: Use the Exercise 7.2.8.

7.2.10. Hint: Use the Exercise 7.2.9.

7.2.11. Hint: Prove that $H_{n,m} = K_m + (O_m \cup K_{n-2m})$ ("+" denotes the operation of join) and use the Exercise 7.2.8.

7.2.12. Hint: Take m to be the minimal index such that $d_m < m$ and $d_{n-m} < n - m$.

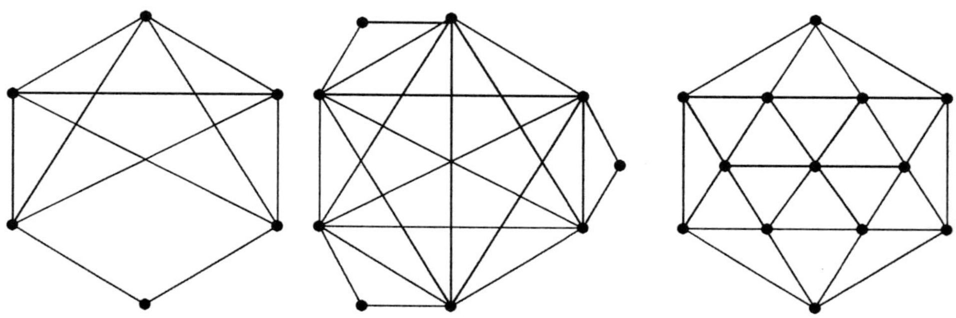

Figure A7.2.1: To Exercise 7.2.23

7.2.13. Hint: Prove that $|EH_{n,m}| \leq \frac{(n-2)^2 + n}{2}$. Further, make use the fact that every non-Hamiltonian graph G does not satisfy the conditions of the Theorem 7.2.1 and by Exercise 7.2.12 it has at most $|EH_{n,m}|$ edges.

7.2.14. Hint: Prove that the line graph does not contain induced subgraphs isomorphic to $K_{1,3}$ and use the Theorem 7.2.2.

7.2.15. Hint: Prove that the line graph is locally connected and use the Theorem 7.2.2.

7.2.16. Hint: Use the Theorem 7.2.2.

7.2.17. Hint: Prove that every edge of G^2 belongs to a triangle and use the Exercise 7.2.15.

7.2.18. Solution: The graph G is Hamiltonian if $m, n \geq 2$ and one of n, m is even. If both n, m are odd then G has an independent set with $(mn+1)/2$ vertices. By Exercise 7.2.10, G is non-Hamiltonian.

7.2.19. Hint: Prove that the graph does not contain induced subgraphs isomorphic to $K_{1,3}$ and use the Theorem 7.2.3.

7.2.20. Hint: Let $d_1 \leq d_2 \leq \ldots \leq d_n$ be the degree sequence of a graph G and let further t be the maximal index such that $1 \leq t < n/2$ and $d_t \leq t$ (the case when $d_t > t$ for all t, $1 \leq t < n/2$, presents no difficulties). Consider the sequence $Q = Q(t) = (q_1, \ldots, q_n)$ with the following elements:

$$\begin{aligned} q_k &= \ 2 \text{ for } k = 1, \ldots, t; \\ q_k &= \ k+1 \text{ for such } k \text{ that } t+1 \leq k < n/2; \\ q_k &= \ n/2 \text{ for such } k \text{ that } n/2 \leq k \leq n-t-1; \\ q_k &= \ n-t \text{ for } k = n-t, n-t+1, \ldots, n. \end{aligned}$$

Prove that $d_k \geq q_k$, $k = 1, 2, \ldots, n$ and $\sum_{k=1}^{n} q_k \geq (n-1)^2/4$.

7.2.21. Hint: Use the Theorem 7.2.1.

7.2.22. Solution: The graph $O_{\lceil \frac{n}{2} \rceil + 1} + K_{\lfloor \frac{n}{2} \rfloor - 1}$ is a $(\lfloor n/2 \rfloor - 1)$-connected graph and by Exercise 7.2.10 it is non-Hamiltonian.

7.2.23. See Fig. A7.2.1.

7.2.24. Petersen graph.

7.2.25. Hint: Suppose the contrary and consider a Hamiltonian path connecting two vertices of the graph whose deletion disconnects the graph.

7.2.26. Hint: Let x be a vertex of maximal degree in G. Prove that if $\deg_G x > (n-1)/2$ then using a Hamiltonian path in $G - x$ and the edges of G incident to x it is possible to construct a Hamiltonian cycle in G.

7.2.27. Hint: Use the fact that $t(X) \leq |VG \backslash X|$ for every subset X of vertices of C_n.

7.2.28. See Fig. A7.2.2.

7.2.29. See Fig. A7.2.3.

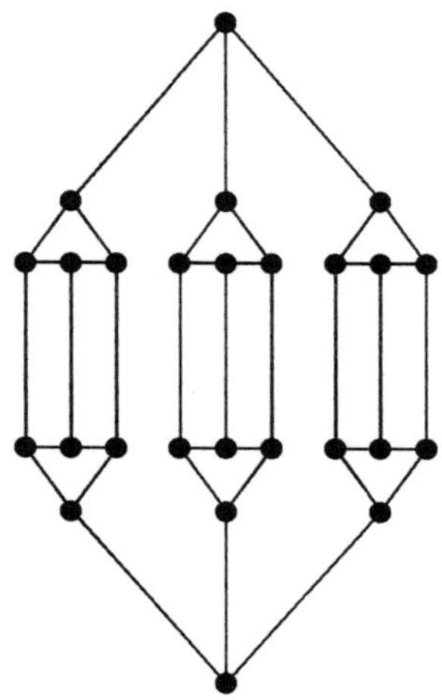

Figure A7.2.2: To Exercise 7.2.28

7.2.30. Solution: A possible solution is by contradiction. Let G contain a vertex v that belongs only to 4-faces, see Fig. A7.2.4a. Deleting this vertex and contracting the resulting cycle, Fig. A7.2.4b, we obtain the graph G' which is also a bipartite 3-polytope. The minimality of G implies that G' is Hamiltonian. But given a Hamiltonian cycle in G', one may easily produce a Hamiltonian cycle in G, see Fig. A7.2.4c,d, hence the contradiction.

7.2.31. Use the Theorem 7.2.5.

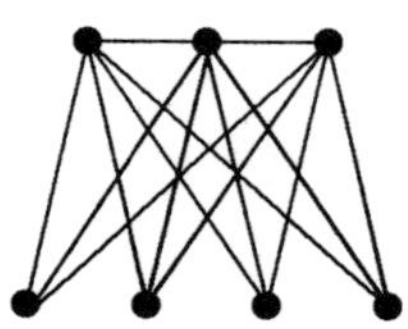

Figure A7.2.3: To Exercise 7.2.29

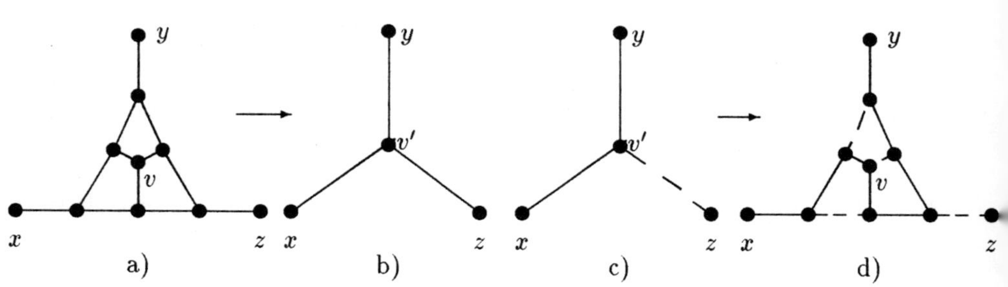

Figure A7.2.4: To Exercise 7.2.30

7.2.32. Hint: Using the induction over the number of vertices prove that for any pendant edge of a caterpillar G belonging to a diametral path of G there exists a Hamiltonian cycle in G^2 containing this edge.

7.2.33. See Fig. A7.2.5.

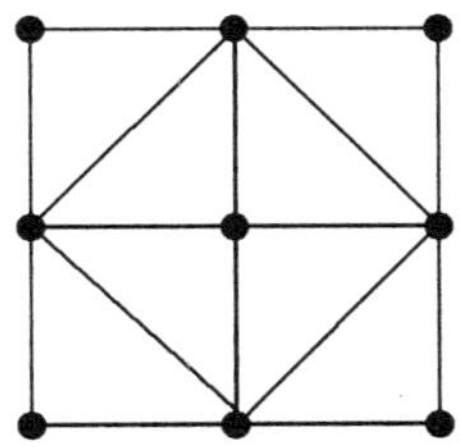

Figure A7.2.5: To Exercise 7.2.33

7.2.34. No. A counterexample is the pair of sequences

$$P = (3, 3, 2, 2, 2),\ Q = (2, 2, 2, 2, 2).$$

7.2.35. Solution: Let us prove that the graph G of order n satisfies a condition of the Theorem 7.2.1, namely, $d_k > k$, for every k, $1 \leq k < n/2$. Suppose the contrary, i.e., $d_t \leq t$ for some t, $1 \leq t < n/2$. Consider a subgraph of G induced by the set of vertices $\{v_1, \ldots, v_t\}$ such that $\deg v_i = d_i$. It is a complete graph since the conditions $l < t$ and $v_l v_t \notin EG$ imply $d_l + d_t \leq n$, contrary to the assumptions. This together with the inequalities $d_i \leq t$, $i = 1, \ldots, t$, imply that every vertex v_i, $i = 1, \ldots, t$, is adjacent to at most one vertex v_j, $j = t+1, \ldots, n$. But $n - t > t$ (since $t < n/2$), hence there exists a vertex v_j, $t + 1 \leq j \leq n$, adjacent to no vertex v_i, $i = 1, \ldots, t$. Therefore $d_j \leq n - t - 1$ and $d_t + d_j \leq n$, which is a contradiction.

7.2.36. Hint: Add a new vertex to the graph, connect it with all other vertices by edges, and use the Exercise 7.2.35.

7.2.37. Solution: The statement is evident if the vertices are adjacent. Without loss of generality assume that $\deg u \geq \deg v$. Let $H = (u, x_1, \ldots, x_{n-2}, v)$ be a Hamiltonian (u, v)-path in G. The inequality $\deg u + \deg v \geq n$ implies that $\{vx_k, ux_{k+1}\} \subseteq EG$ for some k, $1 \leq k \leq n-2$. Therefore G contains a Hamiltonian cycle $C = H' \cup H'' \cup \{vx_k, ux_{k+1}\}$, where H', H'' are the (u, x_k)- and (x_{k+1}, v)-subpaths of H respectively, see Fig. A7.2.6.

7.2.38. $G = G_1 \cup G_2$, where $G_1 \cong K_2$, $G_2 \cong K_{n-1}$, $|VG_1 \cap VG_2| = 1$.

7.2.39. See Fig. A7.2.7.

7.2.40. a: Solution: The necessity follows from the Exercise 7.2.10. The sufficiency will be proved by contradiction. Suppose that the graph G has no Hamiltonian cycles. Let C denote a simple cycle of maximal length in G and

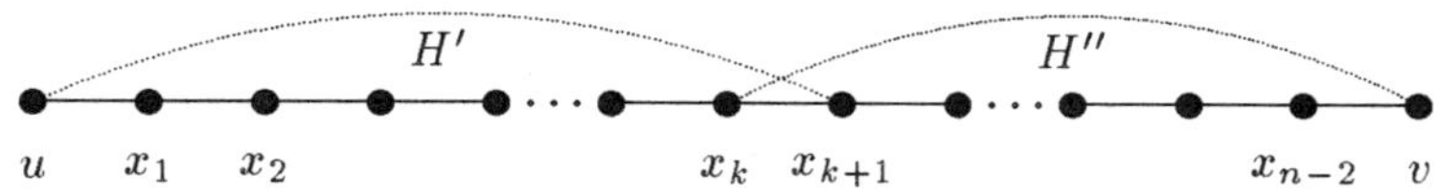

Figure A7.2.6: To Exercise 7.2.37

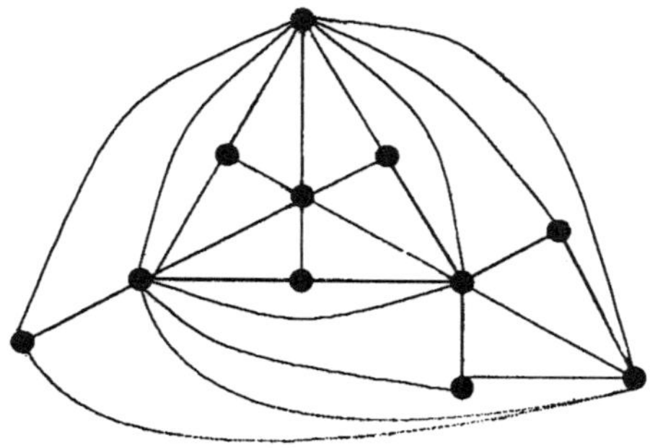

Figure A7.2.7: To Exercise 7.2.39

denote $V' = VG \backslash VC$. It is easily seen that if $x \in V'$ and V_k is the part of G containing x then the vertices of V_k cover all edges of C. Clearly, the graph may have at most two parts with this property. Therefore only two cases are possible.

Case 1: $V' \subseteq V_l$ for a part V_l of G. Then $|VG| = |V_l| + |VC \backslash V_l| < 2|V_l| \leq 2\alpha_0(G)$, contrary to the assumptions of the exercise.

Case 2: $V' \subseteq V_i \cup V_j$ for some parts V_i, V_j of G. Since each of V_i, V_j must cover all edges of C and $\alpha_0(G) \leq |VG|/2$, it follows that G is the complete bipartite graph with parts of equal sizes. This contradicts to the assumption that G is non-Hamiltonian.

b: Solution: Let V_1 denote the largest part of G. Since G is not Hamiltonian, the Exercise 7.2.40a implies $|V_1| = \alpha_0(G) > |G|/2$. Let G' denote the graph obtained from G by the deletion of any $2|V_1| - |G|$ vertices from the part V_1. Clearly, $|G'| = 2\alpha_0(G')$ and hence the Exercise 7.2.40a implies that G' is Hamiltonian, i.e., G contains a simple cycle of length

$$|G'| = |G| - (2|V_1| - |G|) = 2(|G| - \alpha_0(G)) = 2\beta_0(G).$$

7.2.42. Hint: Use the Corollary 5.3.5.

7.2.44. Use the Exercise 7.2.43.

7.2.45. No. See, e.g., $G = P_4$.

Answers to Chapter 8:

Degree Sequences

8.1 Graphical Sequences

8.1.1. a: Three realizations. b: Five realizations. c: The sequence is not graphical.

8.1.3. Hint: Use the fact that the described graph has either a dominating or an isolated vertex.

8.1.7. No.

8.1.8. a: See Fig. A8.1.1.

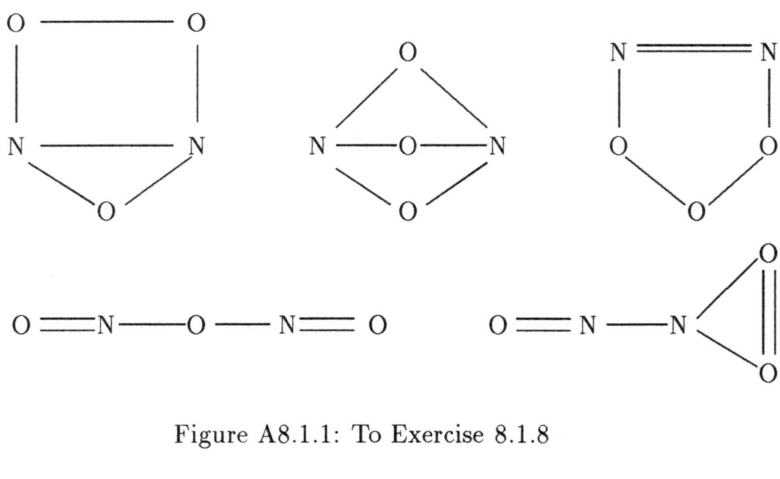

Figure A8.1.1: To Exercise 8.1.8

b: Solution: Let a tree T be the graph of a hydrocarbonate. Let Q denote the subtree of T induced by the vertices of degree four. If $|Q| = n$ then $\sum_{v \in VQ} \deg_Q v = 2(n-1)$. Now it is clear that the number of the pendant vertices (which correspond to the hydrogen atoms) must be $4n - 2(n-1) = 2n+2$.

8.1.9. No. For example, $d = (3, 1^3)$, $d' = (1^4)$.

8.1.10. $s = (15, 62)$, $s = (83, 65)$, $s = (12, 74)$.

8.1.11. No. Hint: Consider the graph shown at Fig. A8.1.2.

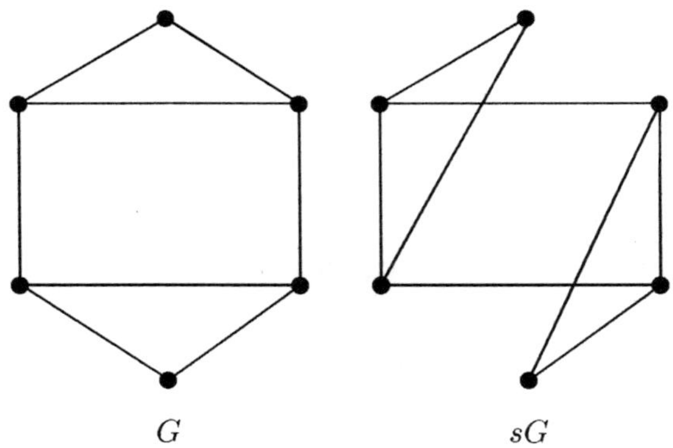

Figure A8.1.2: To Exercise 8.1.11

8.1.12. Solution: Consider a realization G of an n-sequence d in which the vertex v_i is adjacent to the maximal number of vertices of first maximal degrees. Suppose that the claim of the Exercise is false. Then among the first d_i vertices (excluding v_i) there exists a vertex v_j which is not adjacent to v_i. Let v_k be a vertex of maximal degree from $N(v_i)$. Clearly, $\deg v_j \geq \deg v_k$. Therefore there exists a vertex $v_l \in N(v_j)$ not adjacent to v_j. The graph sG for the switching $s = (v_i v_k, v_j v_l)$ contradicts to the choice of G.

8.1.13. One or two.

8.1.14. Hint: For multigraphs the switching is admissible even in the case when a, b and/or c, d are adjacent pairs of vertices. For pseudographs, additionally, any vertices from a, b, c, d may coincide.

8.1.15. May change at most by 1.

8.1.16. a: $G_d = P_3$ (the graph of realizations $G_{d'}$ and the realizations of the sequence d' are shown at Fig. A8.1.3).

c: Hint: Use the Switching Theorem.

d: 0 for $n \leq 5$, $\lfloor n - 4/2 \rfloor$ for $n \geq 6$. Hint: Use the Exercise 8.1.15.

8.1.17. a: Hint: Use the Switching Theorem.

b: Hint: Let realizations G, G' be constructed over the same vertex set V, where $\deg_G v = \deg_{G'} v$ for all $v \in V$. Consider the auxiliary graph $H = (V, E')$, where $E' = (EG \backslash EG') \cup (EG' \backslash EG)$, and prove that every nontrivial component of H may be considered as an alternating cycle of G.

Clearly, a nontrivial component of H contains at least four vertices, therefore $s(d) \leq \lfloor |d|/4 \rfloor$.

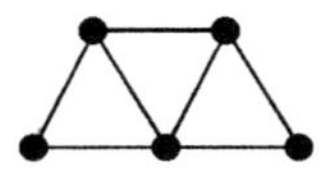

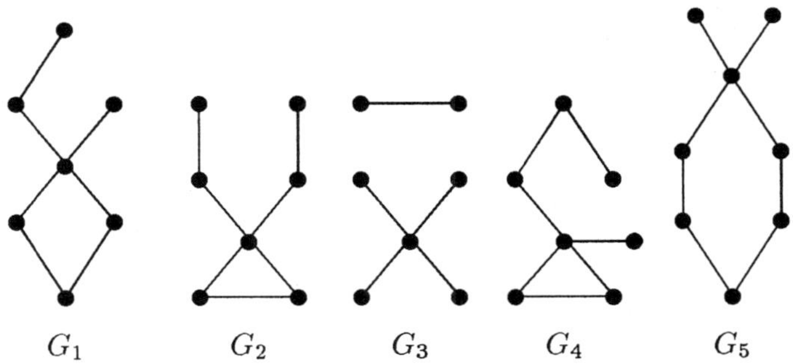

Figure A8.1.3: To Exercise 8.1.16

8.1.18. b: Hint: The sequence $(2^3, 1^2)$ has 7 labelled realizations.

c: $G_d^l \cong K_2$.

d: Assume that the sets $V_1, \ldots, V_k$ are the vertices of G_d. The vertices $V_i, V_j \in VG_d$, $i \neq j$, are adjacent if and only if G_d^l has adjacent vertices $v \in V_i, w \in V_j$.

e: Hint: Use the previous exercise.

f: Hint: The proof of the Switching Theorem implies that the function $s_l(n)$ grows not faster than $O(n^2)$. On the other hand, consider the sequence $d = ((2k)^{4k+1})$ and two its disjoint realizations G, G'. To transform G into G', at least $k(4k+1)/2$ switchings are necessary. Therefore $s(n) \geq s(d) = O(n^2)$, where $n = 4k + 1$.

8.1.19. b: Let G_d be the graph of realizations and let the set $V_\mathcal{P} \subseteq VG_d$ consist of the vertices that correspond to the realizations with the property $\mathcal{P}$. If $\mathcal{P}$ is s-complete then for any graphical sequence d either $V_\mathcal{P} = \emptyset$ or the induced subgraph $G_d(V_\mathcal{P})$ is connected, and vice versa.

c: Hint: Consider two cacti shown at Fig. A8.1.4.

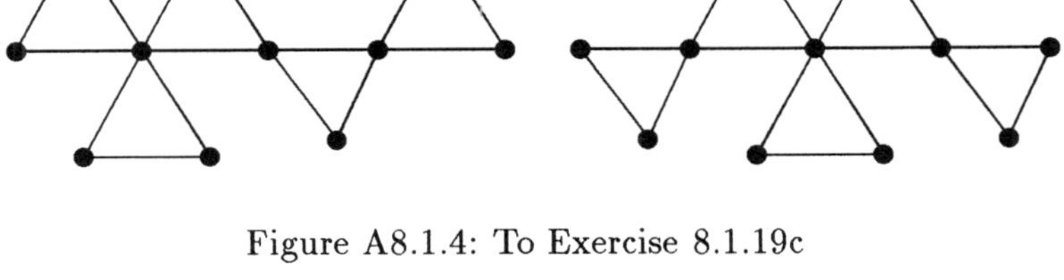

Figure A8.1.4: To Exercise 8.1.19c

8.1.20. No.

8.1.21. Yes. Hint: Consider the Cartesian product of realizations of the sequences d and c.

8.1.23. a: Yes. b: No. c: No.

8.1.24. Hint: Use the Exercise 8.1.12.

8.1.25. Assume that $k = k(d)$ and $m > k$. Then using the inequalities $k + 1 \geq d_{k+1} \geq \ldots \geq d_n$ and $2mk - k^2 - k \leq m^2 - m$ we may compute

$$\sum_{i=1}^{m} d_i = \sum_{i=1}^{k} d_i + \sum_{i=k+1}^{n} d_i \leq k(k-1) + \sum_{i=k+1}^{n} \min\{i, d_i\} + \sum_{i=k+1}^{m} d_i =$$

$$k(k-1) + \sum_{i=k+1}^{n} \min\{i, d_i\} + 2\sum_{i=k+1}^{m} d_i \leq k(k-1) + 2(m-k) + \sum_{i=k+1}^{n} \min\{i, d_i\} \leq$$

$$2mk - k^2 - k + \sum_{i=k+1}^{n} \min\{i, d_i\} \leq m(m-1) + \sum_{i=1}^{n} \min\{i, d_i\}.$$

8.1.26. a: Yes. b: No.

8.1.27. a: nm is even and $m < n$. b: $m + n$ is even and $0 < m \leq n$.
c: For $m^2 + n^2 > 0$. d: For no m, n.

8.1.28. For $n = 4k$ or $n = 4k - 1$, $k \geq 1$.

8.1.29. Hint: Prove that this sequence is realized by the complete k-partite graph.

8.1.31. The necessity is evident. The sufficiency will be proved by induction over $\sigma = \sum_{i=1}^{n} d_i$. The base of the induction ($\sigma = 0$) is evident. Assuming the inductive hypothesis, consider a proper n-sequence d subject to the condition $\sigma \geq 2d_1 > 0$ (the latter implies $d_2 > 0$). Let us construct the sequence $d' = (d_1 - 1, d_2 - 1, d_3, \ldots, d_n)$ and let us order it: $d' = (d'_1, d'_2, \ldots, d'_n)$, $d'_1 > d'_2 > \ldots > d'_n$.

Let us prove that $\sigma' = 2d'$. If $d'_1 = d_1 - 1$ then $\sigma' = \sigma - 2 \geq 2d_1 - 2 = 2d'_1$. Else if $d'_1 = d_1$ then $d_3 = d_1$ and then for $d_1 > 1$ we have $\sigma > 3d_1$. Therefore $\sigma' = \sigma - 2 \geq 3d_1 - 2 \geq 2d_1 = 2d'_1$. Finally, if $d'_1 = d_1 = 1$ then the evenness of σ and the equality $d'_1 = d_1$ imply $d_3 = 1$. Therefore $d_4 = 1$, and hence $\sigma' = \sigma - 2 \geq 4 - 2 = 2 = 2d'_1$.

Thus, by the induction hypothesis, the sequence d' is realizable by a multigraph G'. It remains to connect the vertices u, v of G' with degrees $d_1 - 1$ and $d_2 - 1$ by the edge to obtain a realization G of the sequence d.

8.1.32. Hint: Use the previous exercise.

8.1.33. a:

	5	4	2	3	2	0
4	0	1	1	1	1	0
4	1	0	1	1	1	0
3	1	1	0	1	0	0
2	1	1	0	0	0	0
2	1	1	0	0	0	0
1	1	0	0	0	0	0

c: Hint: Prove that the Fulkerson theorem is equivalent to the Erdös-Gallai one.

8.1.34. Hint: Verify that the sequence $d \cup (d_{n_1})$ satisfies the Erdös-Gallai inequalities.

8.1.35. Hint: Consider the sequence $((8n-3)^{4n}, 2^{8n^2-4n-1})$, $n \geq 1$.

8.1.36. Hint: Use the Theorem 8.1.5.

8.1.37. Hint: Use the Theorem 8.1.5.

8.1.38. a: Yes. b: No.

8.1.39. Hint: Let $d = (d_1, \dots, d_n)$, $d' = (d'_1, \dots, d'_m)$ be two integer sequences. Prove that the pair (d, d') is graphical if and only if the $(m+n)$-sequence $(d_1 + n - 1, \dots, d_n + n - 1, d'_1, \dots, d'_m)$ is graphical.

8.1.40. $a \leq m$, $b \leq n$, $na = mb$.

8.1.41. Hint: Use the König representation of the multigraph, which is defined similarly to the König representation of the graph, see Exercise 1.2.36.

8.1.42. Hint: The Switching Theorem for pairs is stated as follows. Any realization of a graphical pair of sequences may be obtained from any other one by a sequence of bipartite switchings.

8.1.43. a: $n \geq 1$. b: $n \leq 4$. c: $n \leq 5$. d: $n \leq 1$ or $m \leq 1$ or $m = n = 2$.

8.1.44. Yes.

8.1.45. No. Hint: Consider, for example, C_4.

8.1.46. Hint: Use the following statement: graphs G and $\overline{G}$ are unigraphs or not simultaneously.

8.1.47. C_3, C_4, C_5.

8.1.48. K_1, P_4, C_5.

8.1.49. No.

8.1.51. Hint: Prove that the only realization of the sequence is the graph in which the vertices of degree s induce the complete graph and the remaining vertices induce the empty graph, and the neighborhoods of the vertices of degrees $k_1, \dots, k_l$ do not intersect.

8.1.53. Yes.

8.1.54. Yes.

8.1.55. The components are stars and graphs that may be produced from two vertex-disjoint stars by connecting their centers by an edge.

8.1.56. a: Yes. b: No.

8.1.57. b: A minimal FIS set is $\{2K_2, P_4, C_4\}$.

8.1.59. a: Yes. b: Hint: Notice that every vertex v is contained in $\deg v$ edges of the graph. d: The simple paths P_n, $n \geq 5$.

8.1.60. Hint: One may easily produce the degrees of all edges incident to a vertex v from the list $\deg^1 v$.

8.2 $\mathcal{P}$-graphical Sequences

8.2.2. a: The sequence d is forcibly $\mathcal{P}_i$-graphical for $i = 1, 2$.

b,c: These properties cannot be expressed in terms of potentially and forcibly $\mathcal{P}_i$-graphical sequences, $i = 1, 2$.

d: The sequence d is either potentially $\mathcal{P}_1$-graphical or potentially $\mathcal{P}_1$-graphical.

8.2.3. a,b: The first class is contained in the second one.

8.2.4. a: Hint: Use the Exercise 8.2.2a.

b: The inclusion $P_T \subseteq P_A \cap P_C$ is evident. Let us prove the reverse one. Let d be a graphical sequence such that $d \in P_A \cap P_C$. Let $m = \sum_{i=1}^{n} d_i/2$ denote the number of edges in every realization of d; $m \leq n-1$, since $d \in P_A$. On the other hand, $m \geq n-1$, since $d \in P_C$. Thus every acyclic (or connected) realization of d is a tree, i.e., $d \in P_T$.

8.2.5. For those and only for those properties which may be defined in terms of degree sequences.

8.2.7. c: Hint: Let us select any element d_i of the tested n-sequence d. If d is realized by a complete multipartite graph then the size of the part containing a vertex of degree d_i is $n_i = n - d_i$. Let l_i denote the arity (multiplicity) of d_i in d. If $n_i > l_i$, then d cannot be realized by a complete multipartite graph. Otherwise we construct the sequence $d' = d \backslash (d_i^{n_i})$ and apply the above reasoning to it.

8.2.8. Hint: The sufficiency may be proved by induction over n. The inductive step may be based on the Havel-Hakimi theorem.

8.2.9. Hint: a: The sufficiency may be proved by demonstrating that $d_n = 1$ and the sequence $d' = (d_1 - 1, d_2, \ldots, d_{n-1})$ also satisfies the assumptions of the exercise.

8.2.10. $n \geq 3$ and $d_3 \geq 2$.

8.2.12. Solution: The necessity is easily verified. Sufficiency: Let G be a labelled realization of a sequence $d = (d_1, \ldots, d_n)$ with $n \geq 3$, $d_1, \geq d_2 \geq d_3 > 1$, $\sum d_i = 2n$. It is sufficient to prove that d is realizable by a connected graph. Suppose that G is not connected. Then at least one its component H_1 contains a cycle C. Let us switch an edge of C with an edge of a component $H_2 \neq H_1$. Clearly, this produces a realization of d with less components than G has. Repeating such switchings, we ultimately arrive at a connected realization.

8.2.13. $((n-1)^n)$ or $((n-1)^2, (n-1)^{n-3}, 2)$. Hint: Prove that if the line graph $L(G)$ has exactly two dominating vertices then $G = K_1 + (K_2 \cup O_{n-3})$, and then $L(G) = K_2 + (K_2 \cup K_{n-3})$.

8.2.14. Solution: a: Proceed by induction over n. For $n = 1$ and for the empty sequence the statement is evident. Suppose that the inequality holds for any $(n-1)$-sequence. Let $d = (d_1, \ldots, d_n)$ be some nonzero potentially acyclic sequence and let G be one of its realizations. G has a pendant vertex v. We delete it and obtain a graph G' with the degree sequence $d' = (d \backslash (d_j, 1)) \cup (d_j - 1)$, where d_j is the degree of the vertex adjacent to v. By the induction hypothesis, $\sum_{i=1}^{n-1} d'^2_i \leq (n-1)^2 - (n-1)$. Therefore

$\sum_{i=1}^{n} d_i^2 = \sum_{i=1}^{n-1} d'^2_i + 2d'_j + 2 \leq (n-1)^2 - (n-1) + 2(n-2) + 2 = n^2 - n.$

b: The stars.

8.2.15. Let G be a labelled realization of a potentially planar sequence d. Assume that the vertices are indexed in such a way that $\deg v_i = d_i$, $i = 1, 2, \ldots, n$. Denote $X_i = N(v_i) \backslash \{v_1, v_2, v_3\}$, $i = 1, 2, 3$. Then

$$|X_1 \cap X_2| = |X_1| + |X_2| - |X_1 \cup X_2| \geq (d_1 - 2) + (d_2 - 2) - (n - 3).$$

Further,

$$|X_1 \cap X_2 \cap X_3| = |X_1 \cap X_2| + |X_3| - |(X_1 \cap X_2) \cup X_3| \geq$$

$$((d_1 - 2) + (d_2 - 2) - (n-3)) + (d_3 - 2) - (n - 3) = d_1 + d_2 + d_3 - 2n.$$

On the other hand, $|X_1 \cap X_2 \cap X_3| \leq 2$, since a planar graph does not contain $K_{3,3}$.

8.2.16. Any graphical sequence d, $d_1 \geq d_2 \geq \ldots \geq d_n$, with $d_3 \leq 1$ and the sequences $(2^4, 0^{n-4})$, $n \geq 4$.

Solution: Clearly, these sequences are forcibly bipartite. Let us prove that there are no more forcibly bipartite sequences. Consider a forcibly bipartite n-sequence d with $d_1 \geq d_2 \geq \ldots \geq d_n \geq 0$ and $d_3 > 1$. Using the Havel-Hakimi theorem, it is easy to construct a labelled realization G of this sequence in which the vertex v_1 is adjacent to the vertices v_2, v_3. Since G is bipartite, v_1, v_3 are not adjacent. Since we are given that $\deg v_2 = d_2 \geq 2$

and $\deg v_3 = d_3 \geq 2$, the graph G has vertices x and y adjacent to v_2 and v_3 respectively.

Suppose that $x \neq y$ first. The vertices x, y are not adjacent, otherwise the bipartite graph G would contain an odd-length cycle (v_1, v_2, x, y, v_3). The switching $s = (v_2x, v_3y)$ produces a bipartite graph sG. But this contradicts to the fact that G has the triangle (v_1, v_2, v_3).

It remains to consider the case $x = y$. Clearly, the vertices v_1, v_2, v_3, x induce the cycle C_4. Let us prove that in fact G is this cycle. Assume the contrary: there exists a vertex $z \notin VG \backslash VC_4$. Since the considered sequence has no zeroes, $\deg z \geq 1$. If z is adjacent to a vertex of the induced C_4, then $G(VC_4 \cup \{z\})$ has the degree sequence either $(3, 2^3, 1)$ or $(3^2, 2^3)$. Both these sequences are not forcibly bipartite, which contradicts to the forcible bipartiteness of d.

Therefore G has no vertex $z \in VG \backslash VC_4$ adjacent to a vertex of the induced C_4. But if $G \neq C_4$ then it has the induced subgraph $C_4 \cup K_2$ with the degree sequence $(2^4, 1^2)$, which has a nonbipartite realization $C_3 \cup P_3$, which is impossible.

8.2.17. Hint: Use the previous exercise.

8.2.18. An n-sequence d with $d_1 \geq \ldots \geq d_n \geq 1$ is forcibly unicyclic only in the following cases:

1) $d = (2^3)$, $d = (2^4)$, or $d = (2^5)$,

2) $d = (3, 2^4, 1)$,

3) $d = (n-2, 2^3, 1^{n-4})$, $n \geq 5$,

4) $n \geq 4, d_3 \geq 2, d_4 = 1, d_1 + d_2 + d_3 = n + 3$.

Hint: A unicyclic graph is a simple cycle C_k at every vertex v of which a tree T_v is "planted". Use the following scheme for the solution. Let G be a realization of a forcibly unicyclic sequence d. Prove that

1) the length of the cycle C_k is at most 5;

2) every T_v may be assumed to be a star or K_1;

3) For $k = 5$ there may be only a single star $T_v = K_2$;

4) For $k = 4$ there may be only a single star $T_v = K_{1,t}$.

8.2.19. a: No. b: Yes.

8.2.20. A proper graphical n-sequence d is forcibly outerplanar if and only if the sequence $d' = (d_2 - 1, d_3 - 1, \ldots, d_n - 1)$ is forcibly realizable by a disjoint union of simple paths. Therefore $d_2 - 1 \leq 2$ and $d_4 - 1 \leq 1$.

8.2.21. a: $b \leq c$.

8.2.22. a: For example, the property specified by two forbidden induced subgraphs $K_2 \cup K_3$ and P_6.

8.2.23. a: $d^0_m(\mathcal{P}) = \{(2^3),(2^4),(2^5)\}$.

b: $d^0_m(\mathcal{P}) = \{(2^3),(2^5)\}$.

8.2.24. b: For example, $U_{\mathcal{P}} = \mathcal{P}_U$ holds for the property $\mathcal{P}$ "to be a unigraph" and $U_{\mathcal{B}} = \mathcal{B}_U$ does not hold for the property $\mathcal{B}$ "to be a bipartite graph". In fact, the sequence (2^6) is not unigraphical, therefore $(2^6) \notin U_{\mathcal{B}}$. But this sequence has a single bipartite realization C_6, hence $(2^6) \in \mathcal{B}_U$.

8.2.25. Let a proper n-sequence d have a tree realization (see Exercise 8.2.9). Then $d \in T_U$ only in the following cases: 1) $n \le 2$; 2) $n \ge 3$ and $d_3 = 1$; 3) $d_1 = d_2 = d_3 \ge 3$; 4) $d_1 = 2$.

Hint: Let an n-sequence d with $n \ge 3$ be realizable by a tree T. Let W denote the set of nonpendant vertices of T and let us supply every vertex $w \in W$ with a nonnegative integer weight $\delta(w) = \deg_T w - \deg_{T(W)} w$. Let d' denote the degree sequence of the subtree $T(W)$. Two realizations T' and T'' of d' over the vertex set W are called δ-isomorphic if there exists an isomorphism $\varphi : T' \to T''$ that preserves labels, i.e., $\delta(\varphi(w)) = \delta(w)$ for all $w \in W$. Prove that the sequence d is T-unigraphical if and only if any two realizations of d' are δ-isomorphic. Further, find all sequences d' that have single (up to δ-isomorphism) realizations, as well as the corresponding labelling functions for vertices. You will obtain the following: for $d' = (0)$ or $d' = (1^2)$ the labelling function may be arbitrary. For $d' = (2, 1^2)$ the labelling function is $\delta = \text{const} \ge 2$. For $d' = (2^n, 1^2)$ the labelling function is $\delta = \text{const} = 2$.

8.2.26. The converse statement fails. For example, consider the property "to be isomorphic to one of graphs C_8 or $2C_4$", which is not specified by a degree sequence.

8.2.27. No. Let the property $\mathcal{P}$ be "to be a tree" and let the property $\mathcal{P}'$ be "to be a bipartite graph." Clearly, $\mathcal{P} \subseteq \mathcal{P}'$. But $(2^4, 1^2) \in \mathcal{P}_U$ and $(2^4, 1^2) \notin \mathcal{P}'_U$.

8.3 Split and Threshold Graphs

8.3.2. Yes.

8.3.4. a: Yes. b: No.

8.3.5. n.

8.3.7. At most $n + 1$. The maximum is achieved for K_n.

8.3.8. Hint: Use Theorem 8.3.1.

8.3.9. A minimal FIS set is $2K_2, C_4, C_5$.

Solution: It is immediately seen that the above graphs are not split. Further, let G have no induced subgraphs $2K_2, C_4, C_5$. Let us select a maximum cardinality clique K in G such that $G(VG\backslash K)$ is empty.

Suppose the contrary, i.e., $I = VG\backslash K$ has a pair of adjacent vertices u, v. Since the clique K is of maximal cardinality, there are vertices $u', v' \in K$ that are not adjacent to u and v respectively. We may consider that $u' \neq v'$, for if it were impossible to choose $u' \neq v'$, then K would be not of maximal cardinality: $(K\backslash\{u'\}) \cup \{u, v\}$ would be a larger clique.

Since G has no induced subgraphs $2K_2, C_4$, exactly one of $u'v, v'u$ is in EG. Assume, e.g., that $u'v \in EG$ and $v'u \notin EG$.

Let us prove further that v is adjacent to all vertices from $K\backslash\{v'\}$. Suppose the contrary, i.e., that v is not adjacent to $w \in K\backslash\{v'\}$. Clearly, w is adjacent to u' and v'. Moreover, w and u are adjacent, for otherwise it would be $G(\{u, v, v', w\}) = 2K_2$. But then $G(\{u, v, u', w\}) = C_4$, which is also impossible.

Therefore G contains a clique $K' = (K\backslash\{v'\}) \cup \{v\}$ of cardinality equal to $|K|$. The definition of K implies that $|EG(I)| \leq |EG(I')|$, where $I' = VG\backslash K'$. Since $uv \in EG(I)$ and $uv \notin EG(I')$, it follows that there exists a vertex $x \in I$ adjacent to v' but not adjacent to v, see Fig. A8.3.1.

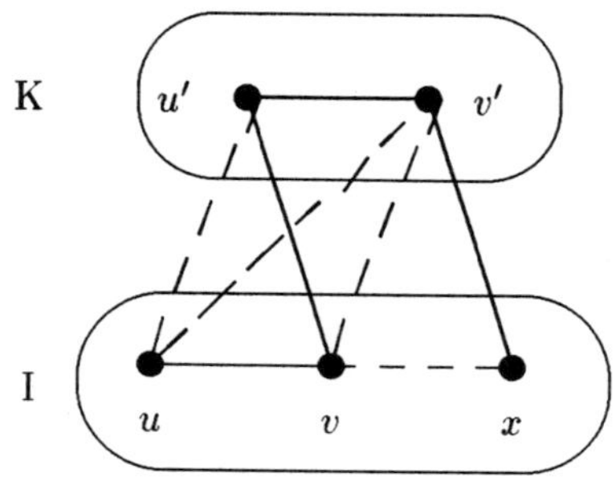

Figure A8.3.1: To Exercise 8.3.9

The vertices u, x are adjacent, for otherwise it would be $G(\{u, v, v', x\}) = 2K_2$. The vertices u', x are not adjacent, for otherwise it would be $G(\{u, v, u', x\}) = C_4$. But then $G(\{u, v, u', v', x\}) = C_5$, which is also impossible.

The resulting contradiction show that $G(I)$ is empty.

8.3.10. a: Hint: Use the Theorem 8.3.1.

b: Hint: Consider the set of vertices with degrees $d_i = i - 1$.

8.3.11. Hint: Use the following mapping of the set of graphs into the set of the split graphs. For a given graph G we construct its König representation $K(G)$ (see Exercise 1.2.36). Then we construct the complete graph over the vertices of $K(G)$ that correspond to the set VG.

8.3.12. No. See the example at Fig. A8.3.2.

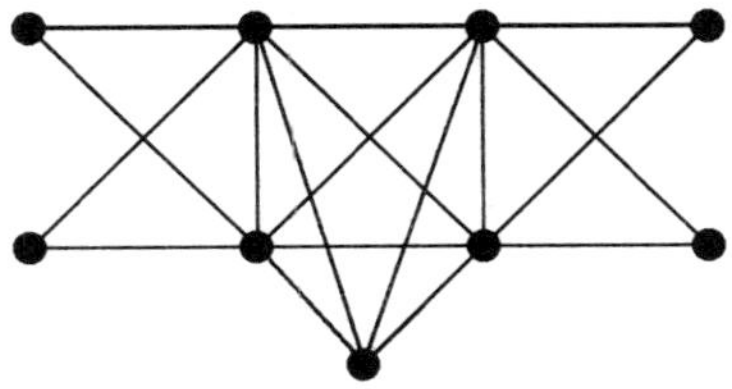

Figure A8.3.2: To Exercise 8.3.12

8.3.13. Trees with diameters at most three.

8.3.14. The complete and the empty graphs.

8.3.15. Hint: Prove that $2K_2, C_4, C_5$ are not threshold and use the characterization of split graphs in terms of forbidden induced subgraphs, see Exercise 8.3.9.

8.3.16. Let us assume that the vertices are numbered in the nondecreasing order of their degrees. Then the table below shows all graphs of order four and the corresponding separating inequalities.

O_4	$x_1 + x_2 + x_3 + x_4 \leq 4$
$K_2 \cup O_2$	$x_1 + x_2 \leq 2$
$P_3 \cup K_1$	$2x_1 + x_2 + x_3 \leq 2$
$K_3 \cup K_1$	$x_1 + x_2 + x_3 \leq 1$
$K_{1,3}$	$3x_1 + x_2 + x_3 + x_4 \leq 3$
$\overline{P_3 \cup K_1}$	$3x_1 + 2x_2 + 2x_3 + x_4 \leq 3$
$K_4 - e$	$2x_1 + 2x_2 + x_3 + x_4 \leq 2$
K_4	$x_1 + x_2 + x_3 + x_4 \leq 1$

The remaining four-vertex graphs are non-threshold.

Hint: See Exercise 3.1.37i.

8.3.17. The vertices of the n-dimensional cube correspond to the $(0,1)$-vectors of length n. A subset of the vertices of this cube correspond to the characteristic vectors of independent sets; let us denote it by I. To construct a separating inequality means to construct a hyperplane in R^n such that I lie in the negative (i.e., containing the origin) closed halfspace and the remaining vertices are in the positive open halfspace. An example of the construction of the separating inequality $2x_1 + x_2 + x_3 = 2$ for $G = P_3$, $VP_3 = \{1,2,3\}, EP_3 = \{12, 13\}$ is shown at Fig. A8.3.3.

8.3.18. Solution: Let $a_1x_1 + \ldots a_nx_n \leq b$ be a separating inequality for a threshold graph G. Clearly, it will also be separating for $G \cup K_1$.

Further, by the previous exercise we may assume that the coefficients a_i are positive. Then the inequality $a_1x_1 + \ldots a_nx_n + bx_{n+1} \leq b$ is separating for $G + K_1$.

8.3.19. Solution: Let us renumber the vertices of a threshold graph G in the nondecreasing order of their degrees. Let u have a larger index than v and

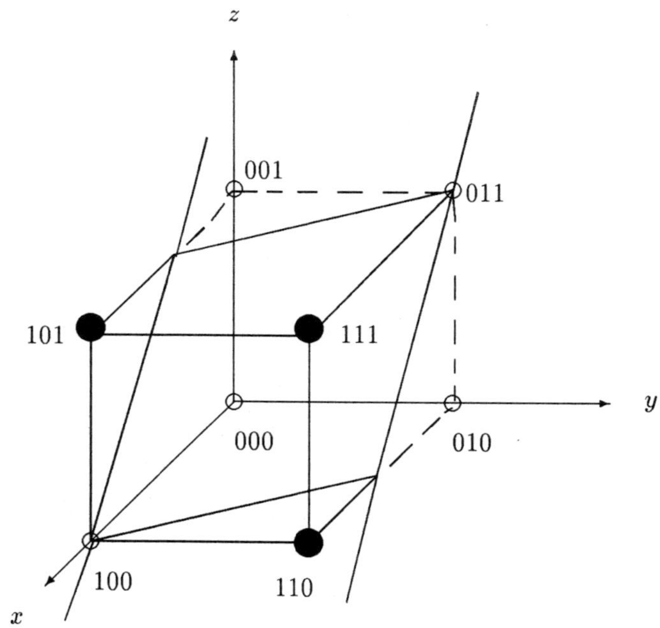

Figure A8.3.3: To Exercise 8.3.17

$N(v) \not\subseteq N(u) \cup \{u\}$, i.e., there exists a vertex $w \in N(v)$ which is distinct from v and not adjacent to it. Since $\deg u \geq \deg v$, there exists a vertex $x \in N(u)$ distinct from u and not adjacent to it. Clearly, $G(\{u, v, w, x\})$ is isomorphic to one of graphs $2K_2$, P_4, C_4. This is impossible, since these graphs are not threshold graphs, and the thresholdness is a hereditary property.

8.3.20. Hint: Use the Theorem 8.3.2.

8.3.21. A minimal FIS set is $2K_2$, P_4, C_4. Hint: Use the Theorem 8.3.2 and the fact that if a graph G does not contain the induced subgraphs $2K_2$, P_4, C_4, then $G \cup K_1$ and $G + K_1$ does not contain such induced subgraphs.

8.3.22. Hint: See Exercise 3.1.37.

8.3.23. Hint: Use the Theorem 8.3.2.

8.3.24. O_m, $K_{1,l}$, $K_{1,l} \cup O_m$, $l \geq 1, m \geq 1$.

8.3.25. Hint: Prove that the threshold graphs and only the ones do not admit switchings.

8.3.26. K_1.

Hint: Use the Exercise 3.1.37j.

8.3.27. Hint: Use the Theorem 8.3.2 and the fact that the thresholdness property is hereditary.

8.3.28. Hint: A possible labelling is the coefficients of the separating inequality constructed for the solution of the Exercise 8.3.22; $t = \beta$.

8.3.29. Hint: Compare the characterizations of split and threshold graphs in terms of forbidden induced subgraphs (see Exercises 8.3.9, 8.3.21).

8.3.30. Hint: Renumber the vertices of each part in the nondecreasing order of their degrees and use the previous exercise.

8.3.31. Hint: Use the fact that the stars are threshold graphs.

8.3.32. a:Hint: Use the following property: if $G = \bigcup_{i=1}^{n} G_i$, $V = VG = VG_i$, $i = 1, 2, \ldots, n$, then the set $I \subseteq V$ is independent in G if and only if it is independent in all G_i, $i = 1, 2, \ldots, n$.

8.3.33. a: $\lfloor n/2 \rfloor$. b: $\min\{m, n\}$. c: $\lfloor n/2 \rfloor$.

8.3.34. a: Hint: Let I be a maximum cardinality independent set of a graph G and $VG \backslash I = \{1, 2, \ldots, t\}$. Construct the spanning subgraphs $G_1, G_2, \ldots, G_t$ in such a way that G_i contains all edges incident to i, $i = 1, 2, \ldots, t$.

b: Hint: : Consider the fact that in a triangle-free graph the edges of every threshold subgraph without isolated vertices constitute a star. Use the equality $\alpha_0(G) = |G| - \beta_0(G)$.

c: Hint: Use the equality $\alpha_1(G) = \beta_0(G)$ for bipartite graphs.

8.3.35. The threshold numbers are 3 and 2 respectively.

8.4 Degree Sets and Arity Partitions

8.4.2. b: Hint: To construct a realization of order $k_1 + 1$, one may apply the induction over even and odd values of $|S|$ separately.

Consider the case of odd $|S|$. Assume that $S = \{k_1, k_2, \ldots, k_n\}$, $k_1 \geq \cdots \geq k_n$. For $n = 1$ the required realization is the complete graph of order $k_1 + 1$. Assume that $n \geq 3$. By the induction hypothesis the set $S' = \{k_2 - k_n, k_3 - k_n, \ldots, k_{n-1} - k_n\}$ is realizable by a graph G' of order $n' = k_2 - k_n + 1$. Consider the complete split graph $H = H(A, B)$ with the complete part A of size k_n and the empty part B of size $k_1 - k_n$. Finally, construct a graph realizing the set S by connecting every vertex of G' with every vertex of the part A of H by an edge.

8.4.3. Hint: See the hint to Exercise 8.4.2b.

8.4.4. b: $\sum_{i=1}^{n}(k_i - 1) + 2$.

8.4.5. Hint: Use the Exercise 3.1.37i.

8.4.6. No.

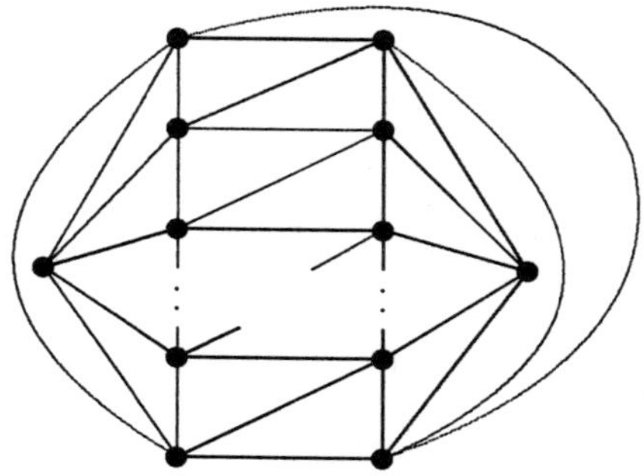

Figure A8.4.1: To Exercise 8.4.9

8.4.7. Hint: Use the induction over n. The induction step for $n \geq 3$ is to be performed as follows: $H = \overline{((H')^o \cup O_{k_1-k_2})} \cup O_{k_n-2}$, where H' is a Hamiltonian realization of the set $S' = \{k_2 - k_n + 2, k_3 - k_n + 2, \ldots, k_{n-1} - k_n + 2\}$, which exists by the induction hypothesis, and $(H')^o$ is the graph obtained from H' by the deletion of the edges of some Hamiltonian cycle.

8.4.8. $1 \leq k_n \leq 5$.

8.4.9. c:Hint: Use the graph shown at Fig. A8.4.1.

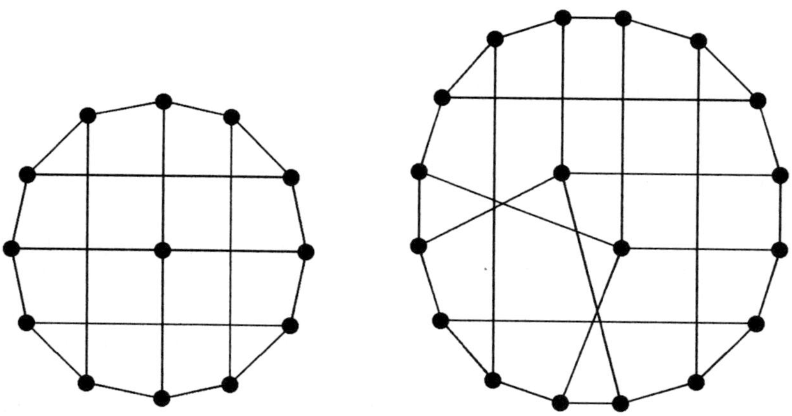

a *b*

Figure A8.4.2: To Exercise 8.4.13

8.4.10. For $n = 2k + 1$ and $n \geq 2k + 3$.

8.4.11. Yes.

8.4.12. a:Hint: First, reduce the problem to the case $|S_1| = |S_2| = 1$ and then construct an appropriate complete bipartite graph.

b: Hint: Construct a bipartite graph realizing the pair of sequences $(k_1^l, k_2^l, \ldots, k_n^l)$ and (l^k), where $k = k_1 + k_2 + \ldots + k_n$.

8.4.13. See Fig. A8.4.2.

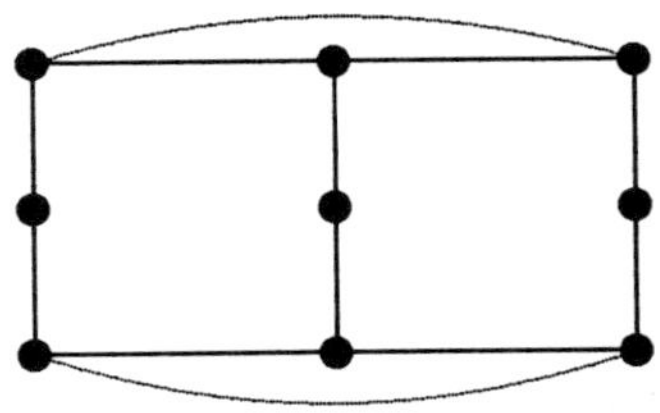

Figure A8.4.3: To Exercise 8.4.14

8.4.14. a: A graph shown at Fig. A8.4.3. b: Such graph does not exist. c: Petersen graph.

8.4.15. a: $(n-2, 2)$. b: (n). c: (m, n).

8.4.16. No.

8.4.18. For $n \geq 2$ and $(p_1, p_2, \ldots, p_n) \neq (1, 1, \ldots, 1)$.

Hint: Use the induction over n. The induction step is performed as follows: add p_n isolated vertices to a realization of the partition $(p_1, p_2, \ldots, p_{n-1})$ and take the complementary graph.

8.4.19. No. Hint: Consider the case $n = 2$.

8.4.20. Yes. Hint: Use the method of the solution of Exercise 8.4.18.

Answers to Chapter 9: Graph Colorings

9.1 Vertex Coloring

9.1.1. c: Hint: Consider the bipartite graph $K_{n,n} - M$, where M is a perfect matching and $n \geq 3$.

9.1.2. b: Set $H_1 = K_1$. If H_{j-1} is already constructed with $VH_{j-1} = \{x_1, \ldots, x_n\}$, H_j is constructed as follows:

$$VH_j = VH_{j-1} \cup \{u_1, \ldots, u_n\} \cup \{v_1, \ldots, v_n\},$$

$$EH_j = EH_{j-1} \cup \{x_i u_i, x_i v_i : i = 1, 2, \ldots, n\},$$

see Fig. A9.1.1.

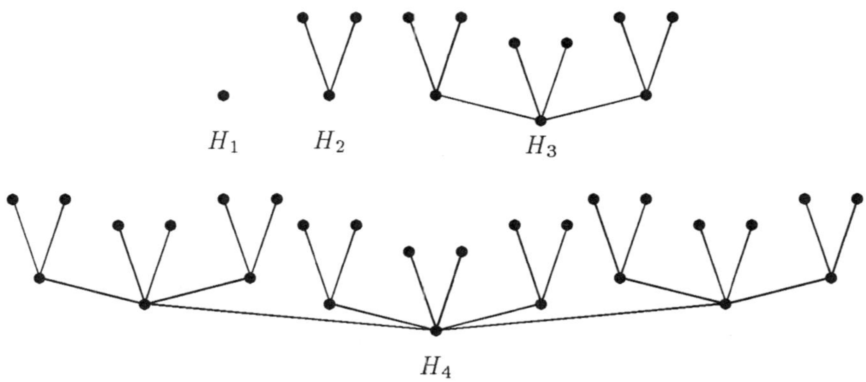

Figure A9.1.1: To Exercise 9.1.2

Clearly the single maximum cardinality independent set of vertices of H_j is $I = VH_j \backslash VH_{j-1}$, where $H_j - I = H_{j-1}$. Therefore the algorithm uses exactly j colors. Since H_j is a tree, $\chi(H_j) = 2$.

9.1.3. Hint: To prove the necessity, show that if a graph G has a cycle of odd length then its vertices must be properly colored in two colors, which is impossible.

9.1.4. a: n. b: 2. c: 2. d: 2 if n is even and 3 otherwise. e: 3.

9.1.5. Hint: Use the induction over the number of blocks.

9.1.6. Hint: Consider the graph $K_{1,n}$.

9.1.7. K_n and C_{2n+1}, $n \geq 1$. Hint: Use the Brooks theorem.

9.1.8. Hint: For $\Delta \geq 3$, consider the neighborhood of some vertex of maximal degree and apply the Brooks theorem.

9.1.9. Hint: Otherwise it would be possible to recolor the vertices of color i in color j.

9.1.10. a: 8, 2, 8, 4. Hint: The castle and bishop graphs contain K_8 as a subgraph. Since the knight changes the color of the square at each move, its graph is bipartite. For the case of the king, use induction over the size of the chessboard.

c: n, 2, n, 4.

9.1.11. Hint: Use the fact that a line graph has no induced subgraphs $K_{1,3}$.

9.1.12. Hint: If two nonadjacent vertices of the octahedron graph are colored in colors 1 and 2 respectively then the remaining four vertices (constituting a 4-cycle) must be colored in two colors 3,4. Otherwise, if every two nonadjacent vertices of the octahedron graph are colored in the same color, then only three colors are used, and some 4-cycle is bicolored.

9.1.13. Solution: Consider the coloring φ in which every color is used at least twice. Let ξ be a minimal coloring which has a maximal number of color classes common with φ. Suppose the contrary, i.e., color 1 is used in ξ only once, say, at vertex x. Let C denote the color class of φ containing x. Convert ξ into a coloring ξ' by coloring the vertices of C in color 1. Clearly, ξ' is a minimal coloring. All vertices of the set $C \backslash \{x\}$ have different colors in the colorings φ and ξ, therefore ξ' has more color classes in common with φ, contrary to the choice of ξ.

9.1.14. Solution: Set $\chi = \chi(G)$. Let the values of n, ξ be fixed. Find the maximal possible value of m. Let $k_1, \ldots, k_\chi$ denote the cardinalities of color classes of some minimal coloring. Then

$$2m \leq k_1(n-k_1)+\ldots+k_\chi(n-k_\chi) = (k_1+\ldots+k_\chi)n - \sum_{i=1}^{\chi} k_i^2 = n^2 - \sum_{i=1}^{\chi} k_i^2.$$

Consider the optimization problem:
minimize $g = \sum_{i=1}^{\chi} k_i^2$ subject to $\sum_{i=1}^{\chi} k_i = n$. It is well known that its

solution is $k_i = n/\chi$, $i = 1, \ldots, \chi$. Therefore $g \geq n^2/\chi$. Then

$$2m \leq n^2 - \sum_{i=1}^{\chi} k_i^2 \leq n^2 - n^2/\chi,$$

and hence $\chi \geq n^2/(n^2 - 2m)$.

9.1.15. Hint: Use the previous Exercise.

9.1.16. Hint: In a minimal coloring, for any two colors i, j there exists an edge whose one end has color i and the other one has color j.

9.1.17. Solution: Suppose that $i_0 = \max\{i : d_i + 1 \geq i\}$, i.e, $d_{i_0} + 1 \geq i_0$, but $d_{i_0+1} + 1 < i_0 + 1$ or $d_{i_0+1} + 1 \leq i_0$. Further,

$$\min\{i, d_i + 1\} = \begin{cases} i, & i \leq i_0; \\ d_i + 1, & i > i_0. \end{cases}$$

Therefore

$$\max_{1 \leq i \leq n} \min\{i, d_i + 1\} = \max\{i_0, d_{i_0+1} + 1\} = i_0.$$

Thus it is necessary to prove that $\chi(G) \leq i_0$.

Let us color the vertices $v_1, \ldots, v_{i_0}$ with degrees $d_1, \ldots, d_{i_0}$ in colors $1, \ldots, i_0$ respectively. The neighborhood of every remaining vertex contains at most $d_{i_0+1} \leq i_0 - 1$ vertices. Therefore the consecutive coloring of the remaining vertices will not use new colors.

9.1.18. a: Hint: Use the fact that the size of every color class does not exceed the independence number.

b: $K_{n-\alpha_0} + O_{\alpha_0}$.

9.1.19. Yes.

9.1.20. Solution: Clearly, $\alpha_0(G \times K_m) \leq |G|$ for any $m \geq 1$, since the vertices of $G \times K_m$ may be partitioned into cliques of size m.

Let us prove that $m \geq \chi(G)$ if and only if $\alpha_0(G \times K_m) = |G|$.

Necessity: Suppose that $\chi(G) = \chi$.

$$V(G \times K_m) = \{(v, i) : v \in VG, i = 1, \ldots, m\}.$$

Let $C_1, \ldots C_\chi$ denote the color classes of G. Consider the sets $R_i = C_i \times i$, $i = 1, \ldots, m$. Clearly, the sets R_i are independent, and moreover, their union R is also independent. Since $|R| = |G|$, the necessity is proved.

Sufficiency: Let I be a maximum cardinality independent set of vertices of a graph $G \times K_m$. Then I contains at most one vertex of type (v, i) for every $v \in VG$. Since $|I| = |G|$, exactly one vertex of kind (v, i) belongs to I for every $v \in VG$. Let us define an m-coloring of G as follows. If $(v, i) \in I$ then v obtains color i. It is easily seen that this coloring is proper, hence $\chi(G) \leq m$.

9.1.21. Yes.

9.1.23. a: Solution: Assume that $VG_1 = VG_2$, for otherwise one could add isolated vertices to G_1 or G_2 without changing the chromatic numbers. Further, every vertex u of $G_1 \cup G_2$ is colored in color (α_1, α_2), where α_i is the color of the vertex u in the minimal coloring of G_i, $i = 1, 2$.

c: $\chi(G_1 \cup G_2) = \max(\chi(G_1), \chi(G_2))$.

9.1.24. Let $C_1, \ldots, C_\chi$ be the color classes in a minimal coloring of G. The sets C_i induce complete graphs in $\overline{G}$. Therefore $\overline{\chi} \geq \max_i |C_i| \geq n/\chi$, i.e., $\chi\overline{\chi} \geq n$. It is known that $(a+b)/2 \geq \sqrt{ab}$ for any positive a, b, hence $\chi + \overline{\chi} \geq 2\sqrt{n}$.

The inequality $\chi + \overline{\chi} \leq n + 1$ is proved by induction over n. It is valid for $n = 1$. Assume that $\chi(G)+\chi(\overline{G}) \leq n$ for any graph G of order $n-1$. Consider a graph H of order n and a vertex $v \in VH$. The graphs $G = H - v$ and $\overline{G} = \overline{H} - v$ have $n-1$ vertices. Clearly, $\chi(H) \leq \chi(G)+1$ and $\chi(\overline{H}) \leq \chi(\overline{G})+1$. If at least one if these inequalities is strict then $\chi(\overline{H})+\chi(H) \leq \chi(\overline{G})+\chi(G)+1$, and we are done.

Suppose that $\chi(H) = \chi(G) + 1$ and $\chi(\overline{H}) = \chi(\overline{G}) + 1$. This means that the deletion of the vertex v from G decreases its chromatic number. From the Exercise 9.1.21 it follows that $\deg_G v \geq \chi(H)$. Similarly, $\deg_{\overline{G}} v = n - \deg_G v - 1 \geq \chi(\overline{H})$. But then $\chi(H) + \chi(\overline{H}) \leq n - 1$.

Finally, the evident inequality $\chi\overline{\chi} \leq (\chi(H) + \chi(\overline{H}))^2/4$ implies $\chi\overline{\chi} \leq (n+1)^2/4$.

9.1.25. a: K_n, $n > 1$. b: $K_q \cup \ldots \cup K_q$ (q terms) for $n = q^2$. c: $K_t \cup O_{t-1}$ for $n = 2t - 1$.

9.1.26. Solution: Suppose that there exists a graph G such that $\chi + \overline{\chi} = 2\sqrt{n}$ and $\chi + \overline{\chi} = (n+1)^2/4$. Then $\overline{\chi} = 2\sqrt{n} - \chi$ implies $\chi\overline{\chi} = \chi(2\sqrt{n} - \chi) = (n+1)^2/4$. This gives

$$\chi^2 - 2\sqrt{n}\chi + (n+1)^2/4 = 0.$$

Since this equation must have real roots, its discriminant is nonnegative: $4n - 4(n+1)^2/4 \geq 0$, i.e., $-(n-1)^2 \geq 0$. Hence $n = 1$, contrary to the assumption.

9.1.29. a: Hint: Take V_1 to be a color class of a minimal coloring of the graph G.

b: Solution: Consider a maximal complete subgraph H_1 of G. Denote $V_1 = VH_1$, $V_2 = VG \backslash VH_1$, $H_2 = G(V_2)$, $|V_1| = k$. We must prove that $\chi(H_2) > \chi(G) - k$. Suppose the contrary, i.e., H_2 admits a proper $(\chi(G) - k)$-coloring in colors $1, 2, \ldots, \chi(G) - k$. Let us color H_1 in colors $\chi(G) - k + 1, \ldots, \chi(G)$. Let T denote the set of all vertices of color 1 adjacent to a vertex u of color $\chi(G)$. Since H_1 is a maximal complete subgraph, every vertex $v \in T$ is not adjacent to some $v' \in H_1$. If we color every $v \in T$ in the color of the corresponding vertex v', the resulting coloring will be proper. Now we color u in color 1 and obtain a proper coloring of G with $\chi(G) - 1$ colors, which is a contradiction.

9.1.30. Solution: For $k = 1$ the statement is trivial. Let us apply induction over k. Assuming the induction hypothesis, consider the graph $G\backslash V_k$. We have

$$\chi(G\backslash V_k) \leq n - |V_k| - (k-1) + 1 = n - |V_k| - k + 2.$$

Let us color $G\backslash V_k$ in colors $1, 2, \ldots, n - |V_k| - k + 2$ and the vertices of V_k in colors $n - |V_k| - k + 3, \ldots, n - k + 2$. Now let us recolor the graph to get rid of one color. By the assumption, every set V_i, $1 \leq i \leq k-1$, has a vertex y_i which is not adjacent to some vertex $x_i \in V_k$. There exists a color, say, color 1, which is used only for the vertices of the set $Y = \{y_1, \ldots, y_{k-1}\}$, since $|VG\backslash Y| = n - k + 1$ is less than the number of colors. Let us recolor every vertex y_i of color 1 in the color of the vertex x_i. This produces a proper coloring with $n-k+1$ colors. (Recall that x_i has a color distinct from colors of all other vertices.)

9.1.31. a: Solution: Let us color G in colors $1, 2, \ldots, \chi(G)$. Then we make the following recoloring. Every vertex of color 1 which is not adjacent to the vertices of color 2 we color in color 2. Every vertex of color 2 which is not adjacent to the vertices of color 3 we color in color 3, and so on up to color $\chi - 1$. This results in a minimal coloring such that every sequence of vertices $v_1, v_2, \ldots, v_\chi$ of colors $1, 2, \ldots, \chi$ is a simple path. Its length is $\chi - 1$, therefore $l \geq \chi - 1$.

9.1.32. Hint: See Fig. A9.1.2.

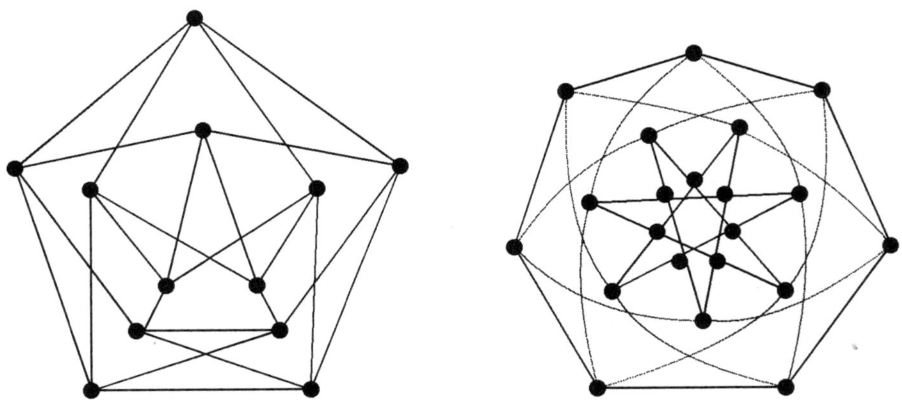

Figure A9.1.2: To Exercise 9.1.32

9.1.33. G_1 and G_3.

9.1.34. Yes.

9.1.35. Hint: See Fig. A9.1.3.

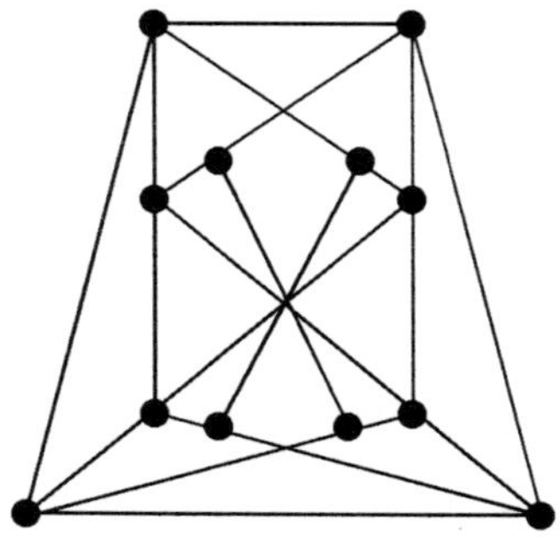

Figure A9.1.3: To Exercise 9.1.35

9.1.36. If the graph G had a vertex v of degree $\deg v \leq k-2$ then for every k-coloring of G it would have been possible to recolor the vertex v preserving the properness of coloring, which is a contradiction.

9.1.37. Solution: Suppose that there exist color classes C, C_1 such that $G(C \cup C_1)$ has at least two components H_1, H_2. Swapping the colors of the vertices from H_1, we obtain a new k-coloring, contrary to its uniqueness for G.

9.1.38. Hint: Let $C_1, \ldots, C_k$ be the color classes of the k-coloring. Verify that $G(C_i \cup C_j)$ has at least $|C_i| + |C_j| - 1$ edges (use the previous exercise). Therefore the total number of edges in the graph is at least

$$\sum_{1 \leq i < j \leq k} (|C_i| + |C_j| - 1) = (k-1)n - C_k^2.$$

9.1.39. Hint: Let φ be a k-coloring of $G * K_k$ (it exists, since $G * K_k$ is a k-partite graph). If for every vertex $v \in VG$ there exists a color which is used twice among the vertices of kind (v, w), $w \in VK_k$, then we may easily construct a proper k-coloring of G by coloring v in this color, contrary to the assumption that $\chi(G) > k$. Therefore there exists a vertex $v_0 \in VG$ such that all vertices (v_0, w), $w \in VK_k$, are colored differently. If $u \in N(v_0)$ then $\varphi(u, w) \neq \varphi(v_0, x)$ for $w \neq x$, since φ is a proper coloring. This implies that $\varphi(u, x) = \varphi(v_0, x)$. By the connectedness of G, the latter equality holds for every vertex v. Therefore the considered k-coloring is induced by a single k-coloring of K_k.

9.1.40. Yes.

9.1.41. Hint: A noncomplete graph with minimal coloring has a pair of vertices of the same color.

9.1.42. Hint: Every k-coloring of the graph $G - v$ is extendible to a $(k+1)$-coloring of G.

9.1.43. The only 2-critical graph is K_2. The 3-critical graphs are the simple cycles of odd length.

9.1.44. Hint: Use the fact that $\chi(G) = \max \chi(B)$, where the maximum is over all blocks B of G.

9.1.45. Hint: Use the fact that if $\chi(G) = k$ and $\deg v < k-1$ then $\chi(G) = \chi(G-v)$.

9.1.46. Solution: The inequality $\chi(G - I) \leq \chi(G) - 1$ holds, since G is critical. The converse inequality is easily proved by contradiction.

9.1.47. Yes. Solution: If a k-chromatic graph G is not k-critical then there exists a vertex $v_1 \in VG$ such that $\chi(G-v_1) = \chi(G)$. The graph $G_1 = G-v_1$ either is k-critical or has a vertex v_2 such that $\chi(G_1 - v_2) = \chi(G_1) = k$. Repeating this reasoning, we ultimately find a k-critical subgraph in G.

9.1.48. Hint: Use the equality $\chi(G_1 + G_2) = \chi(G_1) + \chi(G_2)$.

9.1.49. Hint: a: Delete the vertices u, v from G, color the graph $G-u-v$ in colors $1, \ldots, k-1$, finally color u, v in color k.

b: Color the graph $G - u$ in colors $1, \ldots, k-1$, finally color u in color k.

9.1.50. a: Yes. b: No; Fig. A9.1.4 shows a 4-critical graph with $\chi(G - e) = 4$.

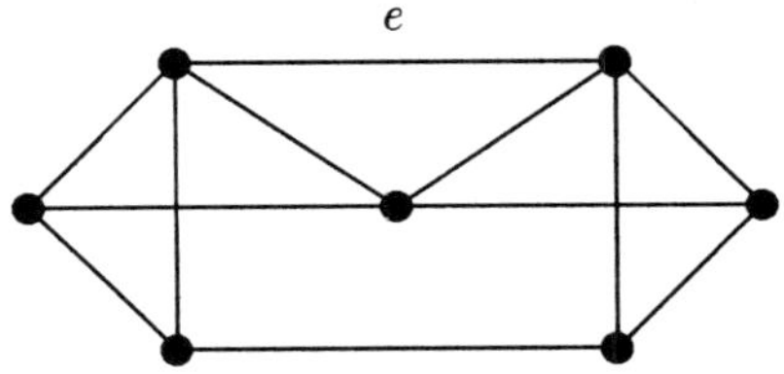

Figure A9.1.4: To Exercise 9.1.50

9.1.51. The simple cycles of odd length.

9.1.52. No. Hint: Consider the graph $(K_1 \cup K_2) + (K_1 \cup K_2)$.

9.1.53. $C_{2n-1} + C_{2n-1}$.

9.1.54. Hint: Prove first that if there exists a proper k-coloring φ of the graph G', then it induces a $(k-1)$-coloring ξ of G as follows. Suppose that $\varphi(w) = 1$. Then $\xi(v) = \varphi(v)$ if $\varphi(v) \neq 1$, otherwise $\xi(v) = \varphi(v')$. But this contradicts to the assumption, therefore $\chi(G') > k$.

Further, prove that $\chi(G'-e) \leq k$ for any edge $e \in EG'$. To this end, consider three cases: $e \in EG$; $e = xy'$, $x \in VG$; $e = wy'$.

9.2 Chromatic Polynomial

9.2.1. a: $t^5 - 7t^4 + 19t^3 - 23t^2 + 10t$. b: $t^5 - 7t^4 + 18t^3 - 20t^2 + 8t$.

9.2.2. a: A vertex v_1 may be colored in any of t colors, v_2 may be colored in any of the remaining $t-1$ colors, etc., v_n may be colored in any of the remaining $t-n+1$ colors.

b: Every vertex may be independently colored in any of t colors.

9.2.3. a: $(t-1)^n + (-1)^n(t-1)$.

Solution: Clearly, $f(C_n, t)$ is equal to the number of proper t-colorings of the path graph P_n such that the terminal vertices are colored differently. Let $f'(P_n, t)$ denote the number of colorings of P_n such that the terminal vertices are colored identically. It is easily seen that $f'(P_n, t) = f(C_{n-1}, t)$. Further,

$$f(C_n, t) = f(P_n, t) - f'(P_n, t) = t(t-1)^{n-1} - f(C_{n-1}, t).$$

Using this recurrence relation, the required formula is easily proved by induction.

b: $t(t-2)^{n-1} + t(-1)^{n-1}(t-2)$.

Hint: $f(W_n, t) = tf(C_n, t-1)$.

9.2.4. a: No. Solution: By Theorem 9.2.3 this graph must have 4 vertices, 3 edges and 2 connected components. The only such graph is $K_3 \cup K_1$, but its chromatic polynomial is given in the item b.

b: Yes.

9.2.5. Hint: Use the fact that every component is colored independently of other components.

9.2.6.

$$f(G, t) = \frac{f(G_1, t) f(G_2, t)}{t(t-1)\dots(t-n+1)}, \quad t \geq n.$$

Hint: Use the Exercise 9.2.2a.

9.2.7. Solution: The number of proper t-colorings of G under which the colors of the vertices u, v are different does not change if we add the edge uv to G. Hence this number is $f(G_1, t)$. Similarly, the number of proper t-colorings of G under which u, v are of the same color is $f(G_2, t)$. Therefore $f(G, t) = f(G_1, t) + f(G_2, t)$.

9.2.8. Solution: Necessity is proved by induction over n. For $n = 1, 2$ the statement is evident. Assume that the chromatic polynomial of a tree of order $n-1$ is $t(t-1)^{n-2}$. Consider a tree T of order n and any its pendant vertex v. By the induction hypothesis, $f(T - v, t) = t(t-1)^{n-2}$. The vertex v may be colored in $n-1$ ways, since it is adjacent to a single vertex of T. Therefore $f(T,t) = f(T-v,t)(n-1) = t(t-1)^{n-1}$.

Sufficiency: Assume that $f(T,t) = t(t-1)^{n-1}$. By Theorem 9.2.3, G is connected (since the coefficient at t is nonzero). The coefficient at t^{n-1} is $C^1_{n-1} = n-1$. Therefore by Theorem 9.2.3 $|EG| = n-1$. Thus G is a tree.

9.2.9. Hint: Use the following evident statement: every skeleton has no less colorings with t colors than the graph itself, and apply the previous exercise.

9.2.10. Solution: Let P be a partition of the vertex set of a graph G into p independent sets; $p \leq n = |G|$. Under this partition, the number of colorings with t colors is $t(t-1)\dots(t-p+1)$. Therefore $f(G,t) = \sum_P t(t-1)\dots(t-p+1)$, where the summation is over all partitioning of VG into independent sets. Since all summands are positive, it follows that for $t > p-1$ we have $f(G,t) > 0$, i.e., no t larger than $p-1$ is a root of $f(G,t)$. It remains to notice that if $t > n-1$ then $t > p-1$.

9.2.11. Hint: Use the Exercise 9.2.6.

9.3 Edge Coloring

9.3.1. 8 for the first graph, see Fig. A9.3.1; 2 for the second graph.

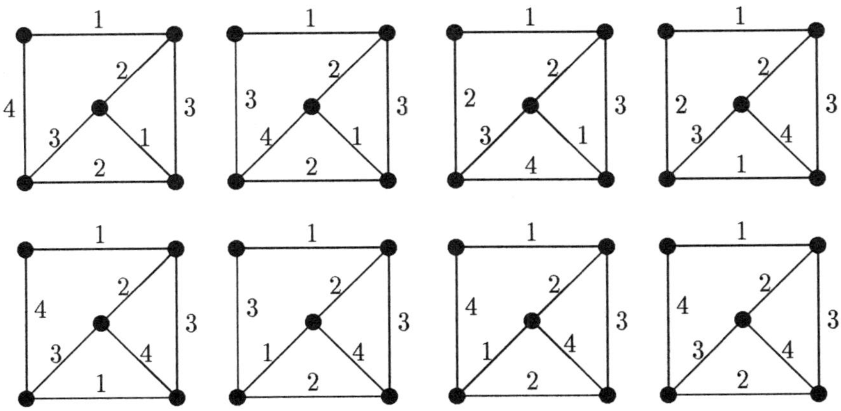

Figure A9.3.1: To Exercise 9.3.1

9.3.2. a: 2, if n is even and 3 otherwise.

b: $n-1$, if n is even and n otherwise; $n \geq 3$.

Solution: For $n = 2t$, a possible partition $E_0, \dots, E_{n-2}$ of the edges of K_n into $n-1$ color classes is as follows. If $VK_n = \{v_0, \dots, v_{n-1}\}$,

$$E_i = \{v_i v_{n-1}\} \cup \{v_{i-j} v_{i+j} : j = 1, 2, \dots, t-1\},$$

where the indices $i - j$ and $i + j$ are calculated modulo n.

The case of odd n is easily reducible to the above.

9.3.4. No.

9.3.5. Disjoint union of simple cycles of even lengths and simple paths.

9.3.6. a: No. b: Yes.

9.3.7. Hint: Start the coloring from the edges incident to a vertex of maximal degree.

9.3.8. Hint: Use the fact that every bipartite graph has a matching that covers all its vertices of maximal degree.

9.3.9. Hint: First analyze the structure of the graph G. If all its cycles are even or there are no cycles then the graph is bipartite and the previous exercise is applicable. If all cycles are odd then it is easily seen that no two cycles have a common edge. Since G is connected and distinct from C_{2n+1}, it follows that $\Delta(G) \geq 3$. Using all the above, it is easy to construct an algorithm of edge coloring of G in $\Delta(G)$ colors.

9.3.10. Hint: If $\chi'(G) \neq 3$ then by the Vizing theorem, $\chi'(G) = 4$. Consider the subgraph H of G induced by the edges of colors 1 and 2. Since G is a cubic graph and $\chi'(G) = 4$, H is a spanning subgraph. If H were not connected, swapping the colors 1, 2 in one of its components would produce a new coloring of G, contrary to the uniqueness of the coloring of G. Therefore H is connected, i.e., its edges constitute a Hamiltonian path or cycle in G. A similar reasoning is valid for colors 3 and 4. Hence G is a cubic graph with two edge-disjoint Hamiltonian paths. However such graph is unique (K_4) and it is 3-edge-colorable. This contradiction implies that $\chi'(G) = 3$, and G has an edge coloring with colors 1, 2, 3. Further, prove that Hamiltonian cycles are constituted only of the edges colored in only two colors.

9.3.11. Solution: If the statement is false then by the Vizing theorem $\chi'(G) = \Delta(G)$. By the assumption, $n = |VG|$ is odd, and hence every color class contains at most $(n-1)/2$ edges. Therefore the total number of edges is at most $\Delta(n-1)/2$. On the other hand, G is regular and by the Handshake Lemma it has $\Delta n/2$ edges, a contradiction.

9.3.12. a: The complete graphs of odd order $n \geq 3$.

b: The nonempty bipartite graphs with maximal vertex degree 2.

9.3.13. Petersen graph.

9.3.14. K_3.

9.3.15. a: The graph shown at Fig. A9.3.2. b: Hint: Use the Exercise 9.3.2b.

9.3.16. a: No. Hint: Consider the graph K_5. b: For odd $n \geq 3$.

c: No. Hint: Consider the graph K_5.

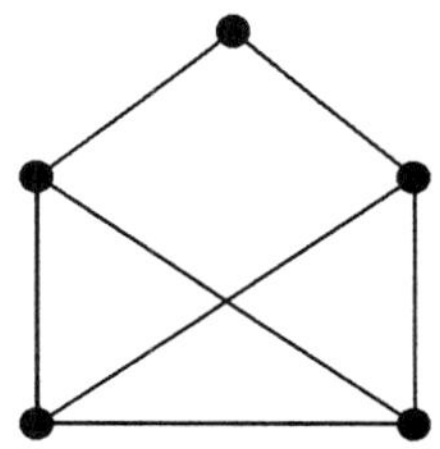

Figure A9.3.2: To Exercise 9.3.15

9.4 Colorings of Planar Graphs

9.4.1. 3.

9.4.2. A plane embedding of K_4.

9.4.3. 4. Hint: Every such arrangement is naturally associated with a plane graph.

9.4.4. Hint: Use the induction over the number of circles. When a new circle is overlaid onto a bicolored map, one must recolor the faces inside of this circle.

9.4.5. Yes.

9.4.6. Hint: Consider the geometric dual graph.

9.4.7. Hint: Use the fact that $\delta(G) \leq 2$ for every outerplanar graph G and apply the induction.

9.4.8. Solution: Let us choose the coordinate system in which the abscissae of the intersection points $(x_1, y_1), \ldots, (x_n, y_n)$ are pairwise different. One may assume that $x_1 < \ldots < x_n$. We shall color the vertices in the indicated order. Each next vertex (x_i, y_i) is adjacent to at most two previous vertices, since exactly two lines pass through every intersection point. Therefore the coloring requires at most three colors.

9.4.9. Solution: Use the induction over $n = |G|$. Consider a plane embedding of G. For $n \leq 5$ there is nothing to prove. Assume that the statement is true for graphs with less than n vertices.

If G has a vertex v with degree at most 3 then a proper 4-coloring of G may be obtained from a proper 4-coloring of $G - v$. Therefore we may assume that $\deg v \geq 4$ for all $v \in VG$. Assume that $\deg v_0 = 4$. The set $N(v_0) = \{v_1, v_2, v_3, v_4\}$ has two nonadjacent vertices, say, v_1 and v_2, for otherwise G would contain K_5 and hence would be nonplanar. The graph G' obtained from $G - v_0$ by merging the vertices v_1, v_2 into $\tilde{v}$ is a plane graph, and by the induction hypothesis it is 4-colorable. Now it is easy to produce a proper coloring of $G : v_1, v_2$ are colored in the color of $\tilde{v}$, all vertices distinct from v_0 inherit their colors from G', finally, v_0 is colored in a color unused in $N(v_0)$.

9.4.10. Hint: Prove that the faces inside (outside) the Hamiltonian cycle may be colored in two colors, and use the colors 1.2 (respectively, 3,4) for the faces inside (respectively, outside) the Hamiltonian cycle.

9.4.11. Hint: Use the inequality $\chi(G) \leq \Delta(G) + 1$.

9.4.12. Hint: The contraction of a bridge does not influence on the adjacency of faces. The vertices of degrees greater than 3 may be transformed as shown at Fig. A9.4.1.

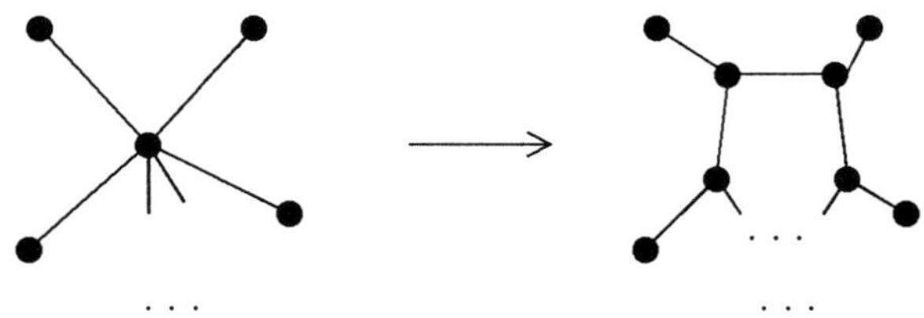

Figure A9.4.1: To Exercise 9.4.12

9.4.13. Hint: Let the faces be colored in colors 1, 2, 3, 4. Every edge borders on two faces. An edge e is said to be of type ij if it borders on faces colored in colors i and j. Prove that the following coloring α of the edges is proper:

$\alpha(e) = 1$, if e is an edge of type 12 or 34;

$\alpha(e) = 2$, if e is an edge of type 13 or 24;

$\alpha(e) = 3$, if e is an edge of type 14 or 23.

Conversely, let there exist a proper coloring of the edges in colors 1, 2, 3. Prove that the maps G_{12} and G_{23} produced respectively by the edges of colors 1,2 and 2,3 are bicolored. Using their bicolorings, construct the 4-coloring of G.

9.4.14. Solution: Let C_1, C_2, C_3, C_4 be the color classes of a 4-coloring of a planar graph G, $|C_i| = n_i$. By Exercise 9.1.37, all induced subgraphs $G(C_i \cup C_j)$, $i \neq j$, are connected. Therefore the number of edges of G is at least

$$\sum_{1 \leq i < j < 4} (n_i + n_j - 1) = 3(n_1 + n_2 + n_3 + n_4) - 6 = 3n - 6.$$

By the corollary 6.1.5, G is a maximal planar graph.

9.4.15. Hint: Notice that a trichromatic graph always has a cycle.

9.4.16. Hint: For $n = 5$ the Hadwiger conjecture states that a 5-chromatic graph is contractible to K_5. But this means that the graph is nonplanar.

9.4.17. Yes.

9.4.18. <u>Hint</u>: Use the stereographic projection.

9.4.19. <u>Hint</u>: Fig. A9.4.2 shows a map on the torus that requires at least 7 colors.

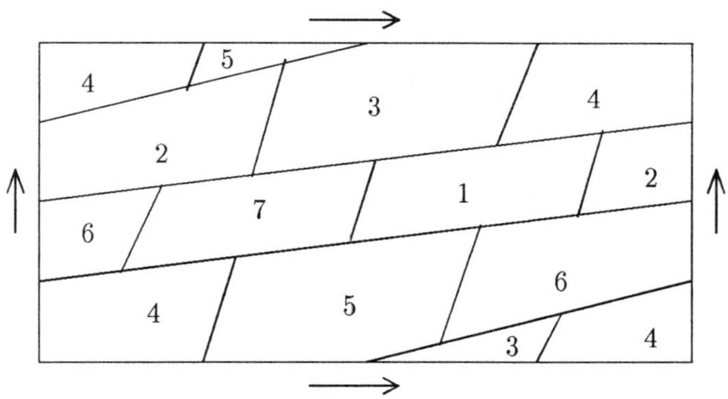

Figure A9.4.2: To Exercise 9.4.19

9.5 Perfect Graphs

9.5.1. No.

9.5.3. <u>Hint</u>: Use the equalities $\chi(G+H) = \chi(G) + \chi(H)$ (Exercise 9.1.22) and $\varphi(G+H) = \varphi(G) + \varphi(H)$.

9.5.4. <u>Hint</u>: Prove that $\chi(G \cup H) = \max\{\chi(G), \chi(H)\}$ and $\varphi(G \cup H) = \max\{\varphi(G), \varphi(H)\}$.

9.5.6. <u>Hint</u>: Use the Exercise 9.5.5a and the Corollary 9.5.2.

9.5.7. Bipartite graphs.

9.5.8. <u>Solution</u>: a: Let us prove that $\chi(G) \leq \varphi(G)$ (and hence $\chi(G) = \varphi(G)$). Let $\xi(x)$ denote the length of the longest path $x = x_1 \leq x_2 \leq \ldots \leq x_\xi$ starting at a vertex x. Clearly, $1 \leq \xi(x) \leq \varphi(G)$, because the vertices of such path induce a clique in G. Let us prove that ξ is a proper coloring of G. Let the vertices $x \geq y$ be adjacent in G. Then the longest path from x is surely shorter than the longest path from y. Hence $\xi(y) \neq \xi(x)$. Therefore $\chi(G) \leq \max_{x \in P} \chi(x) \leq \varphi(G)$, and hence $\chi(G) = \varphi(G)$.

b: Yes.

9.5.9. <u>Solution</u>: Let H_i denote a maximal complete subgraph of G_i, $i = 1, 2$. Then $VH_1 \times VH_2$ induces a complete subgraph of $G_1 \cdot G_2$. Therefore $\varphi(G_1 \cdot G_2) \geq \varphi(G_1)\varphi(G_2)$.

Further, let ξ_i be a minimal coloring of G_i, $i = 1, 2$. Let us define the coloring ξ of $G_1 \cdot G_2 : \xi(x_1, x_2) = (\xi(x_1), \xi(x_2))$. It is easily seen that ξ is a proper coloring of $G_1 \cdot G_2$ using $\chi(G_1)\chi(G_2)$ colors. Therefore

$$\chi(G_1 \cdot G_2) \leq \chi(G_1)\chi(G_2) \leq \varphi(G_1)\varphi(G_2) \leq \varphi(G_1 \cdot G_2).$$

The converse inequality $\chi(G_1 \cdot G_2) \geq \varphi(G_1 \cdot G_2)$ is evident.

b: No. Hint: Consider the graph from Exercise 9.5.1.

9.5.10. Hint: Necessity: A possible required independent set I is a color class in some $\varphi(H)$-coloring.

Sufficiency may be proved by induction over the number of vertices. It is easily seen that

$$\chi(H) \leq \chi(H - I) + 1, \ \varphi(H) \geq \varphi(H - I) + 1.$$

By the induction hypothesis, $\varphi(H - I) = \chi(H - I)$. Therefore $\chi(H) \leq \varphi(H)$, and the converse inequality is evident.

9.5.11. b: Hint: Use the fact that every interval graph is triangulated; see Exercise 9.5.20.

9.5.12. H.

9.5.13. Yes.

9.5.14. a:No. b: No.

9.5.15. If it is a forest.

9.5.17. No.

9.5.19. Hint: Consider two nonadjacent vertices u, v and a minimal set S of vertices separating them. Prove by induction over the number of vertices that the components G_u and G_v of $G \backslash S$ that contain u and v respectively have simplicial vertices. Further, apply the Theorem 9.5.4.

9.5.20. Hint: Fig. A9.5.1 shows a triangulated graph which is not an interval graph.

9.5.21. Solution: Necessity: Let H be an induced subgraph of a triangulated graph G. Let us construct a perfect dismantling of H, i.e., a sequence of graphs

$$H_1 = H, \ H_2 = H_1 - v_1, \ \ldots, \ H_{n'} = H_{n'-1} - v_{n'-1},$$

where $n' = |H|$ and every vertex v_i is simplicial in H_i, $i = 1, 2, \ldots, n' - 1$.

Then $s = \max_i \deg_{H_i} v_i \leq \varphi(H) - 1$. Let H' be an induced subgraph of H. Then

$$\delta(H') = \deg_{H'} u \leq \deg_{H'} v_j \leq \deg_{H_j} v_j \leq s,$$

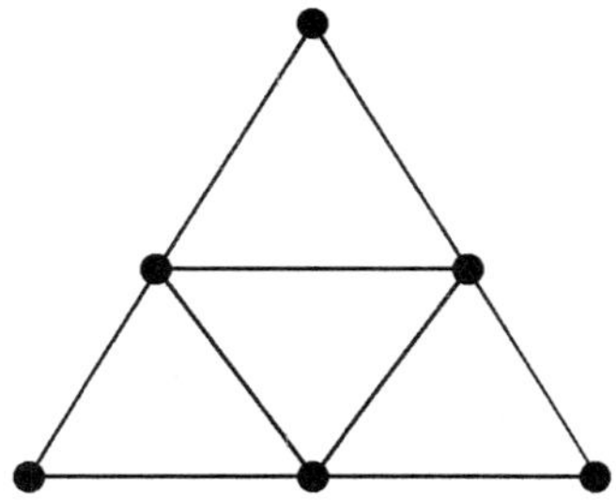

Figure A9.5.1: To Exercise 9.5.20

where $u \in VH'$ is a vertex of maximal degree in H' and v_j is the first one in the dismantling sequence $v_1, \ldots v_{n'-1}$ that belongs to H'. We obtain $w(H) \leq s \leq \varphi(H) - 1$. On the other hand, the inequality $w(H) \geq \varphi(H) - 1$ is evident.

Sufficiency: Suppose that G is not triangulated, i.e., it contains an induced simple cycle C_k, $k \geq 4$. Then $\delta(C_k) = 2 = \varphi(C_k)$, contrary to the assumptions of the exercise.

9.5.22. a: Hint: Proof is by induction over k. For $k = 2$ the statement is trivial. The inductive hypothesis is $\bigcap_{i=1}^{k-1} VT_i = U_{k-1} \neq \emptyset$. Suppose that $VT_k \cap U_{k-1} = \emptyset$. Let us prove that there exists a vertex $v \in VT_k$ (and it is unique) such that every path connecting any pair $x \in U_{k-1}$, $y \in VT_k$ passes through v. Let us take $u \in U_{k-1}$. Then every tree T_i contains the vertex u and some vertex of T_k, and hence the simple path P connecting these vertices. The vertex v belongs to P, therefore $v \in VT_i$, $i = 1, 2, \ldots, k$, which contradicts to the supposition $VT_k \cap U_{k-1} = \emptyset$, $v \notin U_{k-1}$.

9.5.23. Solution: Suppose that the comparability graph G is not triangulated, i.e., G contains an induced simple cycle $C_k = (v_1, v_2, \ldots, v_k, v_1)$, $k \geq 4$. Consider three consecutive vertices v_i, v_{i+1}, v_{i+2} of this cycle. If $v_i \leq v_{i+1}$ and $v_{i+1} \leq v_{i+2}$ then $v_i \leq v_{i+2}$, and hence G has an edge $v_i v_{i+2}$, which is impossible, since C_k is an induced subgraph. The case $v_i \geq v_{i+1}$ and $v_{i+1} \geq v_{i+2}$ is excluded in a similar way. Therefore either $v_i \leq v_{i+1}$ and $v_{i+1} \geq v_{i+2}$ or $v_i \geq v_{i+1}$ and $v_{i+1} \leq v_{i+2}$, i.e. the relation "$\leq$" on the vertex set of C_k alternates along the cycle, hence the length of C_k must be even.

9.5.24. Hint: Use the Theorem 9.5.5 and the hereditarity of the property "to be a triangulated graph".

9.5.25. Hint: Use the hereditarity of the property "to be a triangulated graph".

9.5.26. Hint: Use a normal dismantling of the graph.

9.5.27. Hint: Consider the graph $2C_4$.

9.5.28. Hint: Perform the proof by contradiction considering a minimal non-perfect quasitriangulated graph and applying the Corollary 9.5.2.

Answers to Chapter 10:

Directed Graphs

10.1 Directed Graphs: Basic Notions

10.1.1. n^2.

10.1.3. 2^{n^2}.

10.1.7. a,b: Empty digraphs. c,d: Complete digraphs.

10.1.8. No. Hint: See Fig. A10.1.1

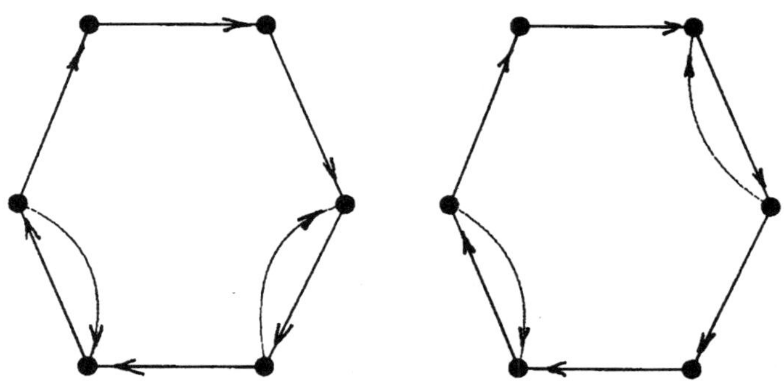

Figure A10.1.1: To Exercise 10.1.8

10.1.9. The isomorphic digraphs are shown at Fig. 10.1.3a,b.

10.1.10. Hint: There are 16 such digraphs.

10.1.11. a: $|VL(G)| = |AG|$, $|AL(G)| = \sum_{v \in VG} d^-(v)d^+(v)$.
b: Hint: See the hint to the Exercise 1.1.21.

10.1.12. Hint: Consider, for example, the sequences $(3, 2, 1)$ and $(3, 3, 0)$.

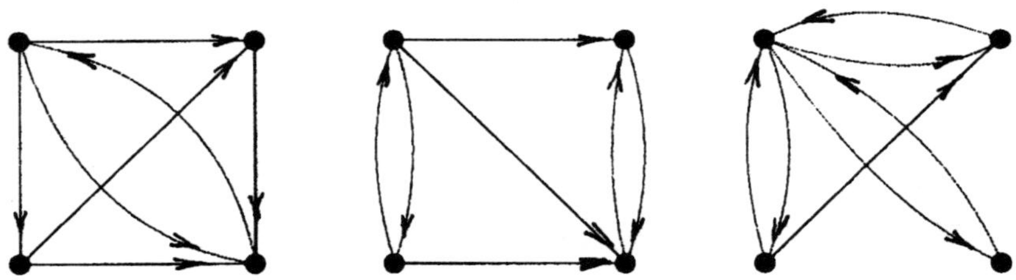

Figure A10.1.2: To Exercise 10.1.13

10.1.13. See Fig. A10.1.2.

10.2 Reachability and Components

10.2.6. No. Hint: See Fig. A10.2.1

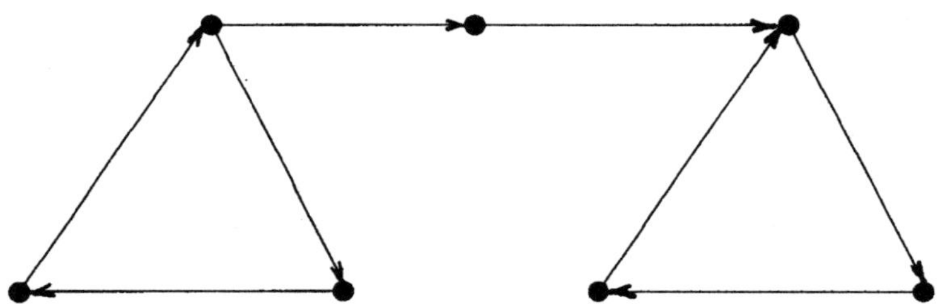

Figure A10.2.1: To Exercise 10.2.6

10.2.9. Hint: To prove the sufficiency, extract a walk (semiwalk) connecting the required vertices from the existing spanning walk (semiwalk).

10.2.10. Solution: Necessity. Suppose that a digraph G is strong. Consider a partition of the vertex set of the digraph into two disjoint sets V_1, V_2 and take $v_1 \in V_1$, $v_2 \in V_2$. Since the digraph is assumed to be strong, v_1, v_2 are reachable from each other. Hence there exists an arc outgoing from V_1 and ingoing into V_2 and an arc outgoing from V_2 and ingoing into V_1. Hence G is irreducible.

Sufficiency. Suppose that a digraph G is irreducible. Consider a vertex $u \in VG$ and the set $V(u)$ of vertices reachable from u. Suppose that $V(u) \neq VG$. Then the irreducibility of G implies that there exists an arc (v, w) such that $v \in V(u)$ and $w \notin V(u)$. This contradicts to the definition of $V(u)$. Hence $V(u) = VG$. This means that every vertex is reachable from every other vertex, or G is strong.

10.2.11. a: <u>Hint</u>: Use the hint for the Exercise 10.2.9. b: <u>Hint</u>: Use the Exercise 10.2.10. c: <u>Hint</u>: Use the Exercise 10.2.8.

10.2.15. <u>Hint</u>: a: To prove the inequality $m \leq (n-1)^2$, consider a digraph with the vertex set $\{v_1, \ldots, v_n\}$ and the arc set

$$\{(v_i, v_j),\ i, j = 1, 2, \ldots, n-1,\ i \neq j\} \cup \{(v_i, v_n),\ i = 1, 2, \ldots, n-1\}.$$

Prove that G has the maximal number of arcs among connected but not strongly connected digraphs of order n.

c: To prove the inequality $m \leq n(n-1)/2$, consider a digraph with the vertex set $\{v_1, \ldots, v_n\}$ such that $(v_i, v_j) \in AG$ if and only if $i < j$. Prove that G has the maximal number of arcs among connected contourless digraphs of order n.

10.2.16. <u>Hint</u>: Consider a vertex from which one can reach a maximal number of vertices of the digraph and prove that this vertex is a source.

10.2.17. <u>Solution</u>: The previous exercise implies that a unilateral digraph G has a vertex u from which all other vertices are reachable. The opposite digraph $\overleftarrow{G}$ is also unilateral, and it has a vertex v with similar properties. In G, the vertex v is reachable from all vertices. Therefore $G + (v, u)$ is strong.

10.2.18. <u>Hint</u>: Consider the following numbering procedure. A vertex of the digraph G with zero indegree obtains the number 1. If the vertices from $S \subseteq VG$ have already obtained the numbers $1, 2, \ldots, k$, a vertex of the digraph $G - S$ with zero indegree obtains the number $k+1$. Proceed in this way until all vertices are numbered.

The existence of vertices with zero indegree in all considered digraphs follows from the Exercise 10.2.4.

10.2.19. See Fig. A10.2.2.

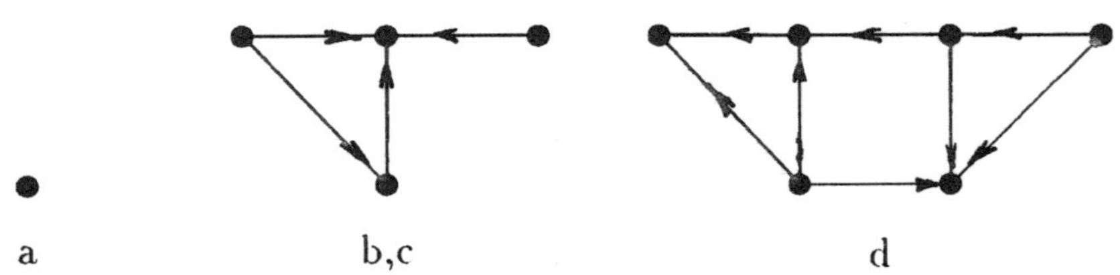

Figure A10.2.2: To Exercise 10.2.19

10.2.20. a $\Rightarrow$ d: <u>Hint</u>: Use the fact that every vertex of a contourless digraph constitutes a strong component. Perform the topological sorting of the digraph (Exercise 10.2.18).

10.2.23. a: $(n-k+1)^2+(k-1)$. Solution: Let us demonstrate that a digraph G containing a maximal number of edges has $k-1$ single-vertex components. Suppose the contrary; let G have components G_1 and G_2, $|G_1|=n_1$, $|G_2|=n_2$, $n_1 \geq n_2 > 1$. Clearly, G_1 and G_2 are complete digraphs with $|EG_1|+|EG_2|=n_1^2+n_2^2$. Consider a digraph $\tilde{G}$ in which the components G_1 and G_2 are replaced by complete graphs $\tilde{G}_1$ and $\tilde{G}_2$ with $|\tilde{G}_1|=n_1+1$ and $|\tilde{G}_2|=n_2-1$. One may easily verify that this increases the number of edges, contrary to the choice of the digraph G.

b: $n(n-k+1)+k(k-1)/2$. Hint: Argue similarly to the item a, in terms of strong components.

10.2.24. b: Hint: To prove the sufficiency, consider a maximal under inclusion directable subgraph $\tilde{G}$ of a bridgeless graph G. If $V\tilde{G} \neq VG$ then prove the existence of a simple (u,v)-path L such that $u,v \in V\tilde{G}$, and the remaining vertices of L do not belong to $\tilde{G}$. Prove that the subgraph $\tilde{G} \cup L$ is directable, contrary to the choice of $\tilde{G}$.

10.2.25. Hint: Consider some numbering of the vertices of the graph and direct every edge in such a way that the number of the tail of the resulting arc is smaller than the number of the head.

10.2.26. Hint: Use the Exercise 10.2.24b.

10.2.29. Hint: Prove that a vertex u_1 of a quasistrong digraph G from which a maximal number k of vertices $v_1,\ldots,v_k$ is reachable is a source.

10.2.30. Solution: a) implies b): If G is a rooted tree then G is connected; otherwise G has either more than one vertex with zero indegree or a contour. Consider a semiwalk in G connecting the root with some other vertex. Since $d^-(u)=1$ for all vertices distinct from the root, it follows that this semiwalk is a walk, and moreover, this walk is unique.

b) implies c): Let v denote a vertex which is connected with every vertex of the digraph by a single walk. Suppose that $d^-(v)>0$ and consider $(u,v) \in AG$. Then the existence of the walk $L=(v,v_1,\ldots,v_l,u)$ would imply the existence of other walks connecting v and u, e.g., $L_1=(v,v_1,\ldots,v_l,u,v,v_1,\ldots,v_l,u)$. Therefore $d^-(v)=0$. Similarly we may observe that $d^-(u)=1$ for all vertices u other than v.

c) implies a): It is necessary to prove that G has no contours. Suppose the contrary, i.e., G has a contour. If a vertex v belongs to it, then $d^-(v)>0$. If v does not belong to the contour then the existence of walks connecting v to the vertices of the contour implies that $d^-(u)>1$ for some vertex of the contour. In both cases we have a contradiction.

a) and b) imply d): Clearly, a rooted tree is a quasistrong digraph. Suppose that it remains quasistrong after the deletion of an arc (u,w). The quasistrongness implies that there exists a vertex z from which u,w are reachable. Hence the original digraph has two walks from z (and hence from the root) to w, which contradicts to item b).

d) implies a): Since G is quasistrong, by Exercise 10.2.29, it has a source s. Therefore $d^-(v) > 0$ for all vertices $v \neq s$. Suppose that there exists a vertex v with $d^-(v) > 1$, i.e., the digraph has two different arcs (u, v), (w, v). After the deletion of one of them the digraph remains quasistrong, contrary to the assumption. Therefore $d^-(v) = 1$. Similarly one may prove that $d^-(s) = 0$.

10.2.33. Hint: The sufficiency follows from Exercise 10.2.30d.

10.2.34. a: n^{n-2}. Solution: A bijection exists between the sets of rooted edge-labelled trees of order n and labelled trees, and the required number follows from the Kelly theorem, Sect. 2.2. A possible bijection is as follows. The root obtains the label n and every remaining vertex obtains the label of the arc on the shortest path from the root to it.

b: n^{n-3}. Hint: Use the item a.

10.2.35. a: See Fig. A10.2.3. b: See Fig. A10.2.4.

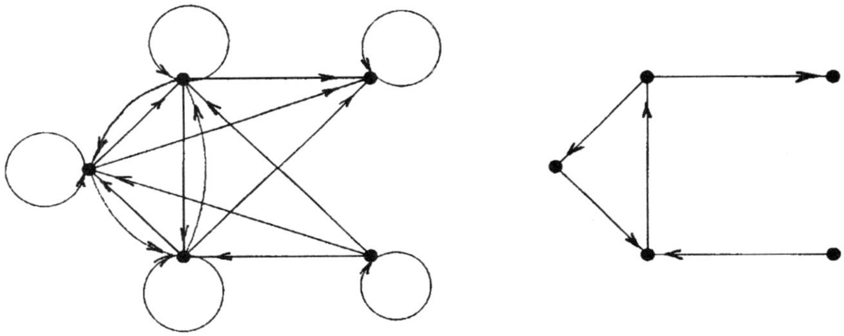

Figure A10.2.3: To Exercise 10.2.35a

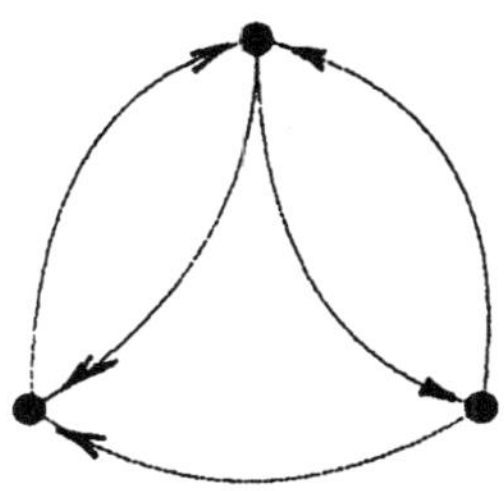

Figure A10.2.4: To Exercise 10.2.35b

c: Hint: Suppose the contrary and consider two transitive reductions R_1, R_2 of a contourless digraph G. Suppose that $(u, v) \in AR_1$, $(u, v) \notin AR_2$. Since

$(u, v) \in AG$, there exists a path $L = (u, w_1, \ldots, w_k, v)$ in R_2. By the definition of transitive reduction, (u, v) and all arcs of L cannot simultaneously belong to R_1. Assume that $(w_l, w_{l+1}) \notin R_1$. Since $(w_l, w_{l+1}) \in AG$, there exists a path $L = (w_l, w_1', \ldots, w_{l+1})$ in R_1. Since G is contourless, L_2 does not pass through the vertices of L_1 other than w_l and w_{l+1}.

Just the same, some arc $(w_p', w_{p+1}') \in L_2$ cannot belong to R_2, and its ends may be connected by a (w_p', w_{p+1}')-path with all arcs from R_2, and so on. But since the digraph has a finite number of arcs, at some step it will become impossible to select a new path.

10.2.39. a: See Fig. A10.2.5.

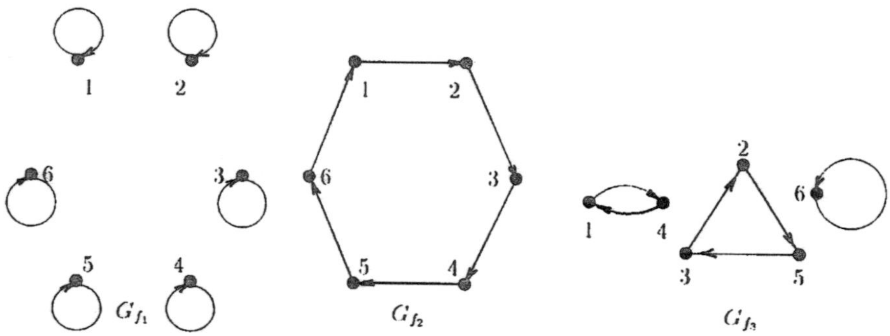

Figure A10.2.5: To Exercise 10.2.39

b: The digraph G_f is a disjoint union of contours.

c: Digraphs G_{f_1}, G_{f_2} are isomorphic if and only if the permutations f_1, f_2 have the same numbers of cycles of the same length (including the cycles of length 1).

10.3 Matrices Associated with Digraph

10.3.1. $A(G) = \begin{bmatrix} 0 & 1 & 0 & 0 & 0 & 0 \\ 0 & 0 & 0 & 1 & 1 & 0 \\ 1 & 0 & 0 & 0 & 0 & 0 \\ 1 & 0 & 1 & 0 & 1 & 0 \\ 0 & 0 & 0 & 0 & 0 & 0 \\ 0 & 0 & 0 & 0 & 0 & 0 \end{bmatrix}$. $R(G) = \begin{bmatrix} 1 & 1 & 1 & 1 & 1 & 0 \\ 1 & 1 & 1 & 1 & 1 & 0 \\ 1 & 1 & 1 & 1 & 1 & 0 \\ 1 & 1 & 1 & 1 & 1 & 0 \\ 0 & 0 & 0 & 0 & 1 & 0 \\ 0 & 0 & 0 & 0 & 0 & 1 \end{bmatrix}$.

$$I(G) = \begin{bmatrix} 1 & 0 & -1 & -1 & 0 & 0 & 0 \\ -1 & 1 & 0 & 0 & 1 & 0 & 0 \\ 0 & 0 & 1 & 0 & 0 & -1 & 0 \\ 0 & 0 & 0 & 1 & -1 & 1 & 1 \\ 0 & -1 & 0 & 0 & 0 & 0 & -1 \\ 0 & 0 & 0 & 0 & 0 & 0 & 0 \end{bmatrix}.$$

10.3.3. The matrices $A(\overleftarrow{G})$ and $R(\overleftarrow{G})$ are the transposed $A(G)$ and $R(G)$ respectively, and $I(\overleftarrow{G})$ is obtained from $I(G)$ by changing the signs of all entries.

10.3.4. Hint: See the hints to Exercise 1.5.4.

10.3.5. Hint: A graph and its transitive closure have the same reachability matrix.

10.3.7. Hint: Use the topological sorting (see Exercise 10.2.18) of the vertices of the condensation of the digraph.

10.3.8. Hint: Use the topological sorting (see Exercise 10.2.18) of the vertices of the digraph.

10.3.9. a: Hint: The sum of all rows of the matrix $I(G)$ is equal to the zero row, therefore rank $I(G) \leq n-1$. Since the digraph G is connected, the deletion of any $k < n$ rows from $I(G)$ results in a matrix which has a column with exactly one nonzero element. The rows of such matrix cannot be linearly dependent.

b:Hint: Use the item a.

10.3.10. Solution: Proof is by induction over the dimensions of the minor. For the dimensions 1×1 the statement is evident. Suppose that it is valid for the dimensions $(n-1) \times (n-1)$. Consider a submatrix A of dimensions $n \times n$. If A has a zero column then $\det A = 0$. If A has a column with exactly one nonzero element then we decompose its determinant over the elements of this column and obtain $\det A = \pm A'$, where A' is the cofactor of the nonzero element of the column, which is either 0 or ± 1, by the induction hypothesis. If every column of A has exactly two nonzero elements then its rows are linearly dependent and $\det A = 0$.

10.3.11. Hint: Use induction over the length of the walk.

10.3.12. a: $A^{(1)} = \begin{bmatrix} 0 & 1 & 0 & 0 & 1 \\ 0 & 0 & 1 & 1 & 0 \\ 0 & 1 & 0 & 0 & 0 \\ 1 & 0 & 0 & 0 & 1 \\ 0 & 0 & 0 & 0 & 1 \end{bmatrix}$; $A^{(2)} = \begin{bmatrix} 0 & 1 & 1 & 1 & 1 \\ 1 & 1 & 1 & 1 & 1 \\ 0 & 1 & 1 & 1 & 0 \\ 1 & 1 & 0 & 0 & 1 \\ 0 & 0 & 0 & 0 & 1 \end{bmatrix}$;

$$A^{(3)} = A^{(4)} = \begin{bmatrix} 1 & 1 & 1 & 1 & 1 \\ 1 & 1 & 1 & 1 & 1 \\ 1 & 1 & 1 & 1 & 1 \\ 1 & 1 & 1 & 1 & 1 \\ 0 & 0 & 0 & 0 & 1 \end{bmatrix}.$$

c: Solution: Consider the proof for the case $k = 2$. Suppose that $A(G) = A^{(2)}(G)$ for a digraph G but G is not transitive. This means that there exists a path $(u_1, \ldots, u_p)$, $p \geq 3$ such that $(u_1, u_p) \notin AG$. Let $L = (u_1, u_2, u_3, \ldots)$ be the shortest among the (u_1, u_p)-paths. Since $A(G) = A^2(G)$, it follows that $(u_1, u_3) \in AG$, and the path $(u_1, u_3, \ldots)$ obtained from L is shorter than L, contrary to the selection of L.

For $k > 2$ the proof may be performed in a similar way.

10.3.13. Hint: Notice that the reachability matrices of all these graphs coincide.

10.3.14. Hint: Use the fact that a term of the sum $R_{ii}^{(2)} = \sum_{k=1}^{n} R_{ik}R_{ki}$ is equal to 1 if and only if the vertices i and k are in the same strong component.

10.3.15. Hint: Necessity: Perform the topological sorting of the vertices of the rooted tree G (Exercise 10.2.18) and show that the matrix $M_1(G)$ under the new numeration of G is obtained from the original matrix $M(G)$ by the same permutations of rows and columns. Next, prove that the determinant of the matrix obtained by the deletion of the first row and the first column from $M_1(G)$ is equal to 1.

Sufficiency: From the hypothesis of the exercise it follows that $d^-(r) = 0$ and $d^-(i) = 1$ for $i \neq r$. Suppose that the digraph has a contour. Determine the structure of the connected component T of the digraph containing this contour. Prove that $\det M(F) \neq 1$ for the digraph F induced by the vertices of the contour. Then using induction over the number of vertices prove that $\det M(T) \neq 1$.

10.4 Tours and Paths

10.4.2. Hint: This digraph cannot be connected.

10.4.4. Hint: To prove the implication $1 \Rightarrow 2$, notice that a cyclic Eulerian chain ingoes into every vertex and outgoes from it the same number of times.

To prove $2 \Rightarrow 3$, select some contour L in the digraph, notice that claim 2 holds for the connected components of the digraph $G \backslash AL$, and use induction over the number of contours.

To prove $3 \Rightarrow 1$, use the fact that the union of two Eulerian digraphs with a single common vertex is an Eulerian digraph.

10.4.5. Hint: Add a path $L = (v_1, w, v_2)$ to the digraph G, where w is a new vertex, and use the Theorem 10.4.1.

10.4.6. Hint: Add k paths $L_1 = (u, w_1, v)$, ..., $L_k = (u, w_k, v)$ to the digraph G, where w_i is new vertices, and use the Theorem 10.4.1.

10.4.7. Hint: Turn the graph into an Eulerian multigraph by augmenting it with edges connecting pairs of odd-degree vertices. Further, direct the edges according to the traversal of an Eulerian cycle of the multigraph, and then delete the arcs corresponding to the added edges.

10.4.8. Hint: Consider a digraph shown at Fig. A10.4.1.

10.4.9. Hint: Use the Theorem 10.4.2.

Figure A10.4.1: To Exercise 10.4.8

10.4.10. Hint: Add a new vertex v to the digraph and connect it by two arcs (v, u) and (u, v) with every vertex u of the digraph. Prove that the resulting digraph has a Hamiltonian contour.

10.4.11. Hint: Use the Theorem 10.4.3.

10.4.12. Hint: a: Prove that $\alpha_0(G) = 5$ and use the Theorem 10.4.3.

b: Prove that $\chi(G) = 4$ and use the Theorem 10.4.4.

10.4.13. Hint: To prove the inequality $l(G) \geq \alpha_0(G)$, notice that every path contains at most one vertex from an independent set of vertices of a digraph.

The opposite inequality follows from the Theorem 10.4.3.

10.4.14. Solution: Consider a coloring $g : VG \to \{1, 2, \ldots, \chi(G)\}$ of the graph G. Direct every edge in such a way that the tail of the resulting arc has color of smaller number than the head. The resulting digraph has no paths longer than $\chi(G)$. Therefore $\tau(G) \leq \chi(G) - 1$. The opposite inequality $\tau(G) + 1 \geq \chi(G)$ follows from Theorem 10.4.4.

10.4.15. Hint: Consider a k-coloring of the arcs of a digraph G corresponding to a minimal k-coloring of vertices of the digraph $L(G)$. Prove that the substrate of the directed subgraph of G induced by the arcs of the same color is a bipartite graph and the substrate G_s of G is the union of k spanning bipartite graphs: $G_s = \sum_{i=1}^{k} G_k$. The vertices of every graph G_i may be colored in two colors $0, 1$. Therefore every vertex v of G may be associated with a sequence $(a_1, \ldots a_k)$, where a_i is the color of v in G_i. Clearly, these $(0,1)$-sequences produce a proper coloring of G; therefore $\chi(G) \leq 2^k$, or $\chi(L(G)) = k \geq \log_2 \chi(G)$.

10.4.16. Solution: a: Suppose that $k_1 = \min_{v \in VG} d^+(v)$. Then at least k_1 arcs outgoes from a vertex v. Taking any of them, say, (v, v_1), we have a path of length 1. Assume that we have already constructed a path $L = (v, v_1, \ldots, v_l)$ of length $l < k_1$. Then there are at least $k_1 - l$ arcs outgoing from v_l that are not incident to the vertices $v, \ldots, v_{l-1}$. Any one of them may be used to augment the path L. This procedure allows to construct a path of length k_1.

Since at the first step we have at least k_1 arcs to select from, at the second step we have $k_1 - 1$ possibilities, etc., there are at least $k_1!$ possible paths of length k_1.

b: The solution is similar to that for item a.

c: Suppose that $\max\{k_1, k_2\} = k_1 = k$. Item a implies that G has a path $L = (v, v_1, \ldots, v_k)$ of length k. If G has an arc (v_k, v), then we have the required contour. Otherwise L may be augmented by an arc (v_k, v_{k+1}), where $v_{k+1} \notin \{v_1, \ldots, v_k\}$. If G has an arc (v_{k+1}, v) or (v_{k+1}, v_1) then we have the required contour. Otherwise the path may be augmented still further, and so on. Clearly, at some moment the path cannot be extended further, and the addition of a new arc produces a contour of length at least $k+1$.

10.4.17. Solution: Sufficiency: Construct a bipartite graph $\tilde{G} = (V, V'; E)$ for a digraph G in the following way. V, V' are two disjoint copies of the vertex set of the digraph G, and $uv' \in E$ if and only if $uv \in AG$, see Fig. A10.4.2. Clearly, $\delta_0(G)$ is equal to the deficiency $\delta(V, V', E)$ of the bipartite graph

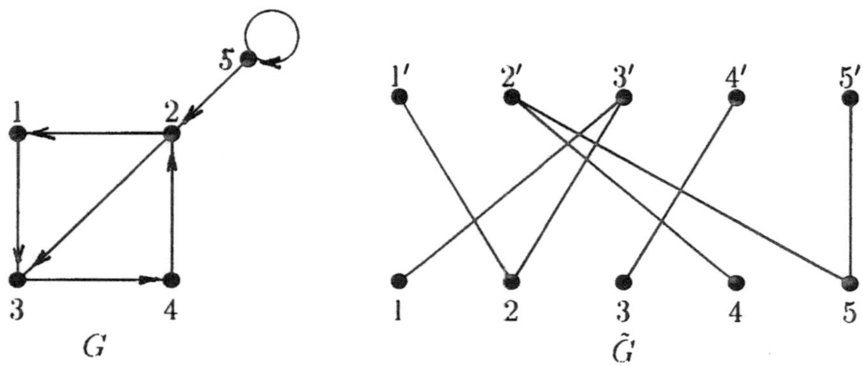

Figure A10.4.2: To Exercise 10.4.17

$\tilde{G}$. Theorem 3.5.4 implies the existence of a perfect matching in $\tilde{G}$. The matching $a_0a'_1, a_1a'_2, a_2a'_3 \ldots$ is associated with the collection of contours and loops $(a_0, a_1, a_2, \ldots, a_{k_1}, a_0)$, $(a_{k_1+1}, a_{k_1+2}, \ldots, a_{k_2}, a_{k_1+1}), \ldots$ in G. It is easily seen that this collection is the required spanning subgraph.

Necessity may be proved using the same construction.

10.5 Tournaments

10.5.1. See Fig. A10.5.1.

10.5.2. Hint: Use the Theorem 10.4.3.

10.5.3. Hint: Use the Exercise 10.5.2.

10.5.4. Hint: Use the Exercise 10.5.2.

10.5.5. Hint: To prove the uniqueness, notice that a transitive tournament has no contours.

10.5.6. Hint: Use the Corollary 10.5.2.

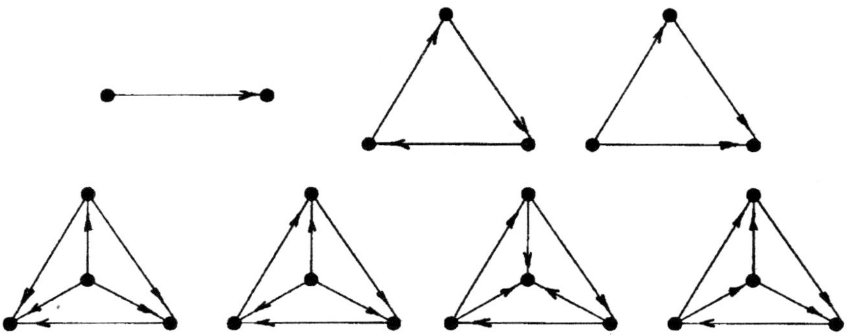

Figure A10.5.1: To Exercise 10.5.1

10.5.7. Solution: Let $L = (v_0, v_1, \ldots, v_{n-1}, v_0)$ be a Hamiltonian contour of a tournament T. If T has an arc of form (v_i, v_{i+2}), $0 \le i \le n-1$, (the addition of indices is modulo n here) then v_i is the required vertex. Suppose now that T has the arcs of form (v_{i+2}, v_i), $0 \le i \le n-1$. If n is odd then the tournament $T - v_0$ has the path $L_1 = (v_{n-1}, v_{n-3}, \ldots, v_3, v_1)$ from v_{n-1} to v_1. If n is even then the tournament $T - v_0$ has the path $L_2 = (v_{n-1}, v_{n-3}, \ldots, v_2, v_3, v_1)$ from v_{n-1} to v_1. Therefore $T - v_0$ is strong, and by Corollary 10.5.2 it is Hamiltonian.

10.5.8. Hint: If the automorphism group of a tournament were of even order then it would have a subgroup of second order, which is impossible, since a tournament cannot have both (u, v) and (v, u) arcs.

10.5.9. a: Solution: Let u be a vertex of a tournament T with maximal outdegree and let v be an arbitrary vertex of T. If $(u, v) \in AT$ then $\rho(u, v) = 1$. If $(v, u) \in AT$ then let us consider the vertices from Γu. Suppose that there exists an arc $(w, v) \in AT$ with $w \in \Gamma u$. In this case $\rho(u, v) = 2$. If T has the arc (v, w) for every vertex $w \in \Gamma u$, then $d^+(v) > d^+(u)$, contrary to the assumption of the exercise.

b: Hint: Consider the opposite digraph $\overleftarrow{T}$ and use the item a.

10.5.10. Solution: a: The equality follows from the Theorem 10.1.1.

b: Consider the tournament T_k induced by the vertices corresponding to the list $(s_1, \ldots, s_k)$. If $(\overline{s}_1, \ldots, \overline{s}_k)$ is the list of outdegrees of T_k, then the equality from item a) of the exercise implies that $\sum_{i=1}^{k} \overline{s}_i = k(k-1)/2$, $k = 1, \ldots, n-1$, which in turn implies the required inequality.

If T is a strong tournament then there exists an arc outgoing from a vertex of the set T_k. Therefore $\sum_{i=1}^{k} s_i > k(k-1)/2$.

Suppose now that $\sum_{i=1}^{k} s_i > k(k-1)/2$ holds for all $k = 1, \ldots, n-1$, and let us prove that T is strong. Suppose the contrary, i.e., T is not strong.

Let $T_1, \ldots T_k$ be its strong components. Assume that T_1 contains a vertex v_1 with minimal outdegree. Consider two components T_1 and T_i, $i \geq 2$. The arcs connecting them must either outgo from every vertex of T_1 and ingo into every vertex of T_i or vice versa. Suppose that T has an arc (v_1, v_p), where $v_p \in T_i$. Should there exist an arc (u, v_1) with $u \in \Gamma v_p$ then v_1, v_p would be in the same strong component. Therefore $(v_1, u) \in AT$ for all $u \in \Gamma v_p$, see A10.5.2 But then $s_1 > s_p$, contrary to the choice of T_1.

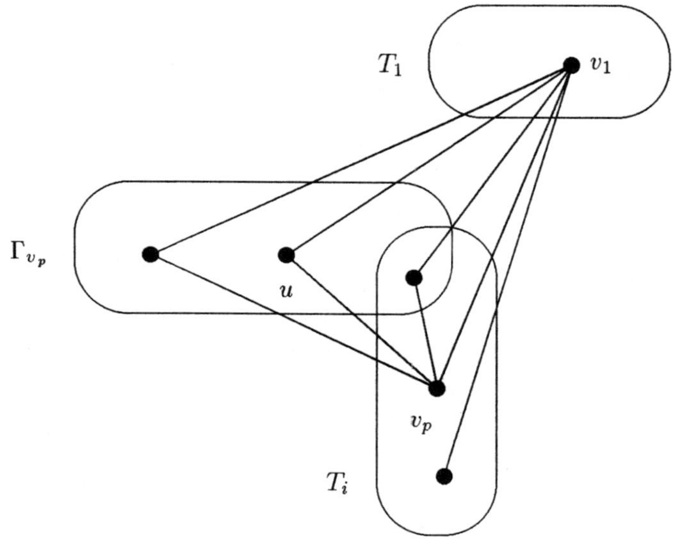

Figure A10.5.2: To Exercise 10.5.10

Thus it is proved that for every vertex v of T_1 there are arcs outgoing from all vertices of the remaining components and ingoing into v. Therefore if s_t is the maximal outdegree for the vertices from T_1 then all vertices with outdegrees at most s_t belong to T_1. But in this case $\sum_{i=1}^{t} s_i = t(t-1)/2$, contrary to the assumptions of the exercise.

c: To prove the left-hand inequalities, we shall use item b.

$$k(k-1)/2 \leq \sum_{i=1}^{k} s_i \leq k s_k, \ k = 1, \ldots, n-1.$$

Hence $s_k \geq (k-1)/2$.

To prove the right-hand inequalities, consider the opposite tournament $\overleftarrow{T}$ and its list of outdegrees $(\overline{s}_1, \ldots, \overline{s}_n)$ arranged in the nondecreasing order. Since the outdegree of a vertex of a digraph is equal to the indegree of the vertex in the opposite digraph, we have $s_k + \overline{s}_{n+1-k} = n-1$. The fact $s_k \geq (k-1)/2$ implies that $\overline{s}_{n+1-k} \geq ((n+1-k)-1)/2 = (n-k)/2$. Hence $s_k \leq (n+k-2)/2$.

d: Let T be a transitive tournament. Let us perform topological sorting of the vertices of T, see Exercise 10.2.18. Let $1, 2, \dots n$ be the numbers assigned to the vertices by this sorting. Then $s_1 = d^+(n) = 0$, $s_2 = d^+(n-1) = 1$, $\dots$, $s_k = d^+(n+1-k) = k-1$, $k = 1, \dots, n$.

Suppose now that $s_k = k-1$ holds for all $k = 1, \dots, n$. Since $s_n = n-1$, the arcs outgoing from the vertex corresponding to the number s_n ingo into all remaining vertices of the tournament. Since $s_{n-1} = n-2$, all arcs outgoing from the vertex corresponding to the number s_{n-1} ingo into all remaining vertices of the tournament except of v_n. Arguing in a similar way for all vertices, we conclude that the tournament is transitive.

10.5.11. Hint: Since $d^+(v) + d^-(v) = n-1$ for every vertex v of a tournament, it is sufficient to prove the equality

$$\sum_{v \in VT} (d^+(v))^2 = \sum_{v \in VT} (n-1-d^+(v))^2,$$

which may be deduced from

$$\sum_{v \in VT} d^+(v) = n(n-1)/2.$$

10.5.12. Solution: Proof is by induction over the number of vertices of a tournament. For a tournament with two vertices the statement is evident. Suppose that it is valid for tournaments T of order n, i.e., $|A\tilde{T}| \geq \lfloor n/2 \rfloor \lfloor (n+1)/2 \rfloor$. Consider a tournament T_1 of order $n+1$. Since $\sum_{v \in VT_1} d^+(v) = n(n+1)/2$, there exists a vertex v_0 with the outdegree at least $\lfloor n/2 \rfloor = \lfloor (n+1)/2 \rfloor$ for even n or at least $\lfloor (n+1)/2 \rfloor$ for odd n. Let T_2 denote a maximal contourless subgraph of $T_1 - v_0$. If we add the arcs outgoing from v_0 into T_2, we obtain a contourless digraph. By the induction hypothesis, we deduce

$$|A\tilde{T_1}| \geq |AT_2| + d^+(v_0) \geq \lfloor n/2 \rfloor \lfloor (n+1)/2 \rfloor + \lfloor (n+1)/2 \rfloor = \lfloor (n+1)/2 \rfloor \lfloor (n+2)/2 \rfloor.$$

10.5.13. No. Hint: See Fig. A10.5.3.

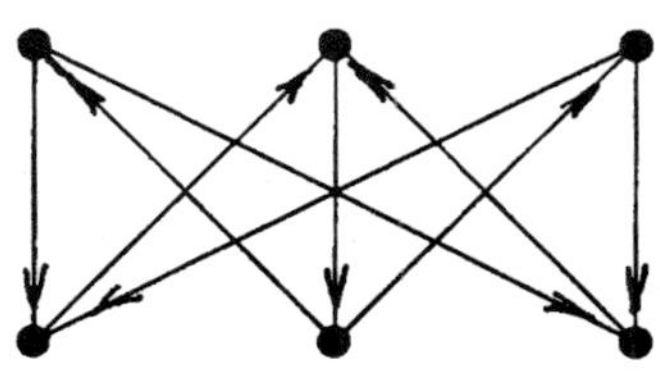

Figure A10.5.3: To Exercise 10.5.13

10.6 Base and Kernel

10.6.2. Hint: Notice that no one of these vertices is reachable from any other one.

10.6.3. Hint: Use the fact that all vertices of a strong component are mutually reachable and every vertex of a non-base component is reachable from every vertex of some base component.

10.6.4. Hint: Use the Exercise 10.6.3.

10.6.5. All vertices with zero indegree. Hint: Use the Exercise 10.6.3.

10.6.6. Hint: Use the Exercise 10.6.3.

10.6.7. Solution: Suppose that v is a vertex from which a maximal number of vertices is reachable and v does not belong to a base B. Then there exists a vertex $u \in B$ such that v is reachable from u. Since all vertices reachable from v are also reachable from u, u is the required vertex for B.

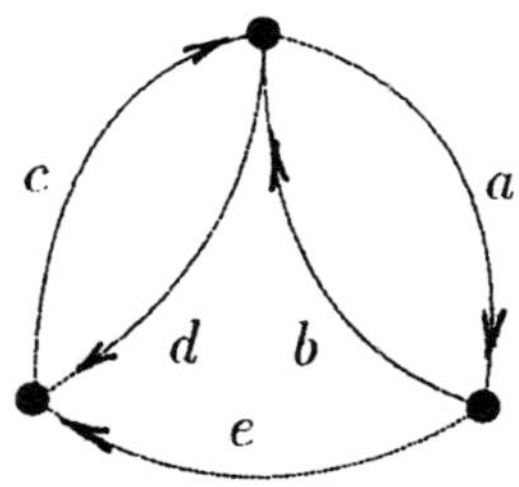

Figure A10.6.1: To Exercise 10.6.13

10.6.9. Hint: Use the Exercise 10.6.8.

10.6.11. Solution: Let A be an arc base of a digraph G. If $(x, y) \in AG$ then $y \in D(x)$, and by the definition $y \in D(x, A)$, i.e., the condition (1) holds. If $(x, y) \in A$ then the existence of a path L that consists of arcs from the set $A' = A \backslash \{(x, y)\}$ would imply $D(u, A) = D(u, A')$ for every vertex u of the digraph, contrary to the definition.

Suppose now that A satisfies both conditions (1), (2). The first one implies that for every (x, y)-path we may replace its arcs which are not in A by paths that consist of arcs from A, i.e., $D(x) = D(x, A)$ for every $x \in VG$. Suppose that $A' \subset A$, and $(x, y) \in A \backslash A'$. Condition (2) implies that $D(x, A) \backslash D(x, A') \neq \emptyset$. Therefore A is an arc base of G.

10.6.12. Hint: Consequently delete arcs from a digraph G according to the following rule: an arc (x, y) is deleted if after its deletion y remains reachable from x. Using the Exercise 10.6.11 prove that this procedure produces an arc base of G.

10.6.13. For a digraph shown at Fig. A10.6.1 the sets $\{a, b, c, d\}$ and $\{a, c, e\}$ are the bases.

10.6.14. Solution: Suppose the contrary, i.e., a digraph G has two arc bases $U_1 \neq U_2$, and suppose that an arc $(x, y) \in U_1 \backslash U_2$. Since U_2 is an arc base, there exists a walk $L = (x, x_1, \ldots, x_p, y)$ with all arcs in U_2. Since U_1 is a base, there exist walks with all arcs in U_1 connecting x with x_1, x_1 with $x_2, \ldots, x_p$ with y. Some of these walks contain the arc (x, y), for otherwise there would exist an (x, y)-walk with all arcs in $U_1 \backslash \{(x, y)\}$, which is impossible, see Exercise 10.6.11. Suppose that a walk $L_i = (x_{i-1}, v_1, v_2, \ldots, x, y, u_1, \ldots, u_s, x_i)$ contains the arc (x, y). Consider two cyclic walks:

$$S_1 = (x, x_1, x_2, \ldots, x_{i-1}, v_1, v_2, \ldots, x),$$

$$S_2 = (y_1, u_1, \ldots, u_s, x_i, x_{i+1}, \ldots, x_p, y).$$

At least one of them contains two different vertices (otherwise G would contain parallel arc connecting x and y). One may select a contour from this walk (see Exercise 10.2.1), contrary to the assumption of the exercise.

10.6.16. Hint: Use the Exercise 10.6.15.

10.6.17. Hint: Consider a subgraph D of a strong digraph induced by the arcs of an arc base. By Exercise 10.6.15 D is strong. Further, use the induction over the number of contours whose union is D (see Exercise 10.2.11).

10.6.18. See Fig. A10.6.2.

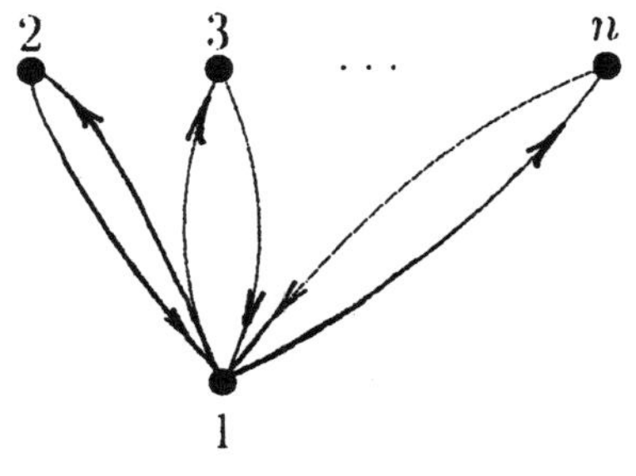

Figure A10.6.2: To Exercise 10.6.18

10.6.19. Hint: Consider a digraph induced by the arcs of an arc base.

10.6.21. See Fig. A10.6.3.

10.6.23. See, e.g., Fig. A10.6.4, where the required vertex sets are marked.

10.6.25. Hint: Prove that the set $N \subseteq VG$ is dominating if and only if $VG \backslash N \subseteq \Gamma^{-1}N$, and it is independent if and only if $\Gamma^{-1}N \subseteq VG \backslash N$.

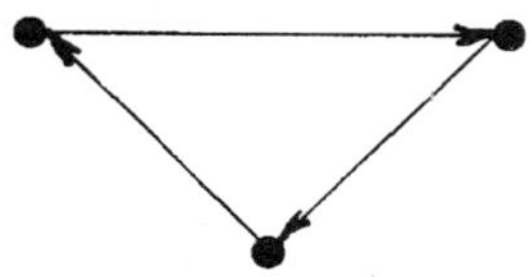

Figure A10.6.3: To Exercise 10.6.21

Figure A10.6.4: To Exercise 10.6.23

10.6.26. See Fig. A10.6.5. Here the required inkernels are marked by "–" and the outkernels are marked by "+". At Fig. A10.6.5e, $\{2, 4\}$ and $\{1, 3, 5\}$ are kernels.

10.6.28. A complete loopless digraph.

10.6.29. Solution: Suppose that $N \subseteq VG$ satisfies (a), (b). Then $\Gamma N \subseteq VG \backslash N$.

Suppose that $V' = VG \backslash (V^+ \cup \Gamma V^+) \neq \emptyset$ and $N' = N \backslash V^+$ is the outkernel of the subgraph G', i.e., $\Gamma N' = V' \backslash N'$. Then

$$\Gamma N = \Gamma N' \cup \Gamma V^+ = (V' \backslash N') \cup \Gamma V^+ =$$

$$((VG \backslash (V^+ \cup \Gamma V^+)) \backslash N') \cup \Gamma V^+ = (VG \backslash (N \cup \Gamma V^+)) \cup \Gamma V^+.$$

Since $N \cap \Gamma V^+ = \emptyset$, it follows that $\Gamma N = VG \backslash N$, i.e., N is an outkernel.

Suppose now that N is an outkernel. Its definition implies conditions (a) and (b). If the set N' is nonempty then for the set $\Gamma' N'$ of vertices of G' adjacent to N',

$$\Gamma' N' = \Gamma N \backslash \Gamma V^+ = (VG \backslash N) \backslash \Gamma V^+ = V' \backslash N'.$$

10.6.30. Hint: Replace Γ by Γ^{-1} and V^+ by V^-, where V^- is the set of vertices of the digraph with zero outdegree.

10.6.31. Hint: Use the fact that the substrate of the digraph is a bipartite graph and prove that its parts are the required in- or outkernels.

To demonstrate that the statements cease to be valid, consider the digraph shown at Fig. A10.6.6.

10.6.32. Solution: a: Let N_1, N_2 be different inkernels of G. From Exercises 10.6.25, 10.6.22 in follows that neither $N_1 \subset N_2$ nor $N_2 \subset N_1$.

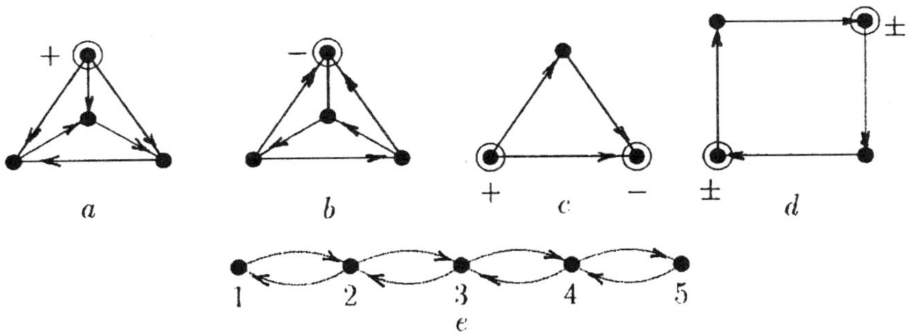

Figure A10.6.5: To Exercise 10.6.26

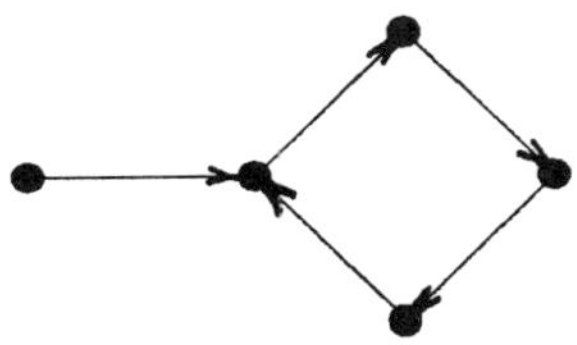

Figure A10.6.6: To Exercise 10.6.31

Let us pick a vertex $u_1 \in N_1 \backslash N_2$. Since N_2 is an inkernel, there exists a vertex $v_1 \in N_2$ such that $u_1 \in \Gamma^{-1} v_1$ and $v_1 \notin N_1$, since N_1 is independent. Therefore $v_1 \in N_2 \backslash N_1$. By the same reason there exists a vertex $u_2 \in N_2 \backslash N_1$ such that $v_1 \in \Gamma^{-1} u_2$, and so on. Since G is finite, the vertices $u_1, v_1, u_2, v_2, \ldots$ cannot be all different, hence G contains a contour, contrary to the statement of the exercise.

b: The solution is similar to that for item a.

10.6.34. Hint: Use the Theorem 10.6.1 and the Exercise 10.6.24.

10.6.35. Hint: Select a maximal independent set in the graph and direct the edges in an appropriate way.

10.6.36. a: See Fig. A10.6.7.

b: Hint: Select a maximal independent set $S_1 \subseteq VG$ and set $g(v) = 0$ for all $v \in S_1$. Then select a maximal independent set $S_2 \subseteq VG \backslash S_1$ and set $g(v) = 1$ for all $v \in S_2$, and so on. Prove that this process produces a Grundy function.

c: See Fig. A10.6.8. d: See Fig. A10.6.9. e: See Fig. A10.6.10.

f: Hint: Prove that the vertices of a digraph for which a Grundy function is equal to zero constitute a kernel of the digraph.

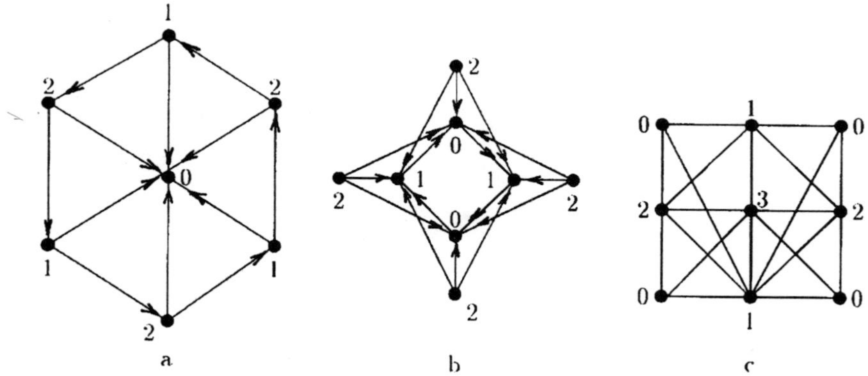

Figure A10.6.7: To Exercise 10.6.36a

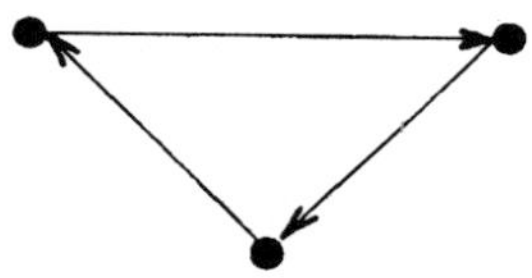

Figure A10.6.8: To Exercise 10.6.36c

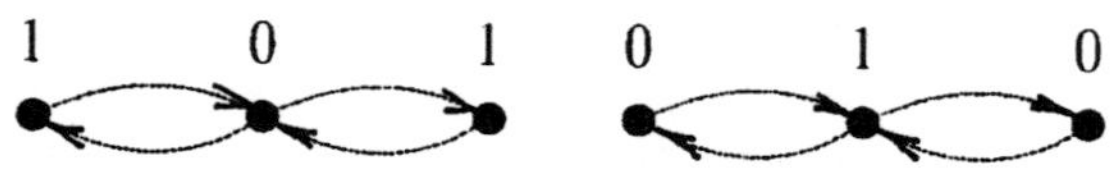

Figure A10.6.9: To Exercise 10.6.36d

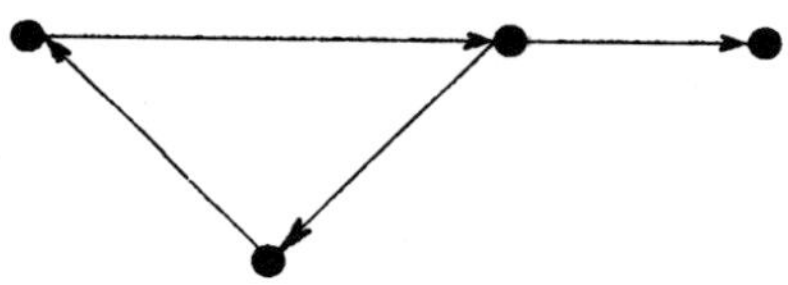

Figure A10.6.10: To Exercise 10.6.36e

g: Hint: Let S_1 be a kernel of a digraph G. Set $g(v) = 0$ for all $v \in S_1$. Let further S_2 be a kernel of $G - S_1$. Set $g(v) = 1$ for all $v \in S_2$, and so on. Prove that this process produces a Grundy function.

h: Hint: Use the Theorem 10.6.1 and item g of the exercise.

i: Hint: Use, e.g., item h.

j: Hint: If such a Grundy function exists then we may assume its values to be colors of the vertices. Prove that such coloring is proper.

k: If G is a p-chromatic graph and $VG = \bigcup_{i=0}^{p-1}$ is a partition of its vertex set into color classes then one may construct another partition as follows. V_0' is V_0 plus all vertices from $VG \backslash V_0$ adjacent to no vertices of V_0. V_1' is $V_1 \backslash V_0'$ plus all vertices from $VG \backslash (V_1 \cup V_0')$ adjacent to no vertices of $V_1 \backslash V_0'$, and so on. Prove that the function $g(v) = k$, $v \in V_k'$ is a Grundy function.

Answers to Chapter 11:

Hypergraphs

11.1 Hypergraphs: Basic Notions

11.1.1. $2^{C_{10}^3} - 1$.

11.1.2. a: C_{63}^3.

b: C_{189}^3.

Hint: a: The number of different nonempty subsets of a six-element set is $2^6 - 1 = 63$. Therefore the maximal possible number of edges in a hypergraph of order six without multiple edges is 63. Therefore the number of labelled $(6,3)$-hypergraphs without multiple edges is C_{63}^3.

b: Continuing the reasoning of item a, the maximal possible number of edges in a hypergraph of order six with edge multiplicity at most 3 is $63 \cdot 3 = 189$, hence the given answer.

11.1.3. a: $\sum_{v \in VH} \deg v = \sum_{e \in \mathcal{E}H} |e|$.

b: $\sum_{v \in VH} \deg v = k|\mathcal{E}H|$.

11.1.4. a: The independent sets of the matroid M are the independent sets of vertices of the hypergraph $\mathcal{C}(M)$.

b: $\alpha_0(\mathcal{C}(M)) = \rho(M)$.

c: If M is the partition matroid for $V = V_1 \cup \ldots \cup V_k$ specified by the vector $m = (m_1, \ldots, m_k)$, $1 \leq m_i < |V_i|$, $i = 1, 2, \ldots, k$, then the hypergraph $\mathcal{C}(M)$ has k components $H_i = (V_i, \mathcal{E}_i)$, $i = 1, 2, \ldots, k$, where H_i is the complete $(m_i + 1)$-uniform hypergraph.

11.1.6. The cycle C_{2n}.

11.1.7. Yes.

11.1.11. Hint: Consider the König representation of such hypergraph. What property must it possess?

11.1.12. Yes.

11.1.13. b: <u>Hint</u>: Use the inequality $m \le 3n - 6$ valid for planar (n, m)-graphs.

11.1.14. a: Possible examples are $\{\{1,2\}\}$ for hypergraph H_1, $\{\{1,2\},\{4,5\}\}$ for H_2.

b: Possible examples are $\{2,3\}$ for hypergraph H_1, $\{2,3,4\}$ for H_2.

11.1.16. No; see Fig. A11.1.1.

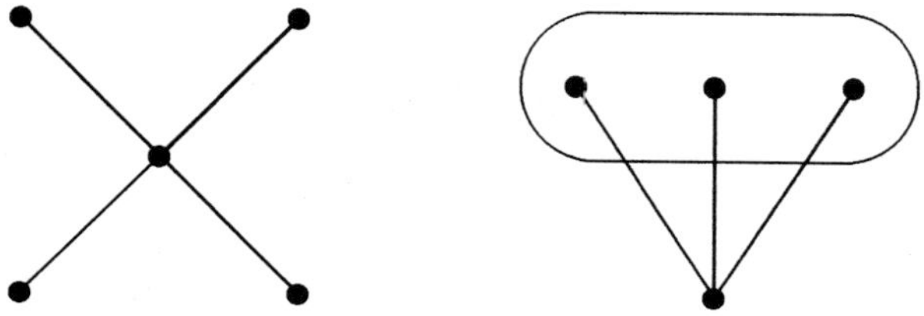

Figure A11.1.1: To Exercise 11.1.16

11.1.17. a: Any graph. b: Any connected graph. c: Any disconnected graph. <u>Hint</u>: For a graph G without isolated vertices consider the line graph $L(G^*)$.

11.1.18. K_n.

11.1.19. The hypergraph H_n defined in Exercise 11.1.18.

11.1.20. Yes.

11.1.21. <u>Hint</u>: Prove that $K(H^*) \cong K(L(H))$.

11.1.22. <u>Hint</u>: For a graph H without isolated vertices, the dual hypergraph H^* is a linear hypergraph. Further, see the hint to the Exercise 11.1.17.

11.1.23. <u>Hint</u>: Use the Exercise 11.1.21.

11.1.25. <u>Hint</u>: It is sufficient to prove the following: every clique of $(H)_2$ is contained in some edge of H if and only if for any triple $e_1, e_2, e_3 \in \mathcal{E}H$ there exists $e_0 \in \mathcal{E}H$ such that $(e_1 \cap e_2) \cup (e_1 \cap e_3) \cup (e_2 \cap e_3) \subseteq e_0$.

The necessity part of the latter statement is evident. The sufficiency may be proved by induction over the number of vertices of a clique C using the equality $C = (C_1 \cap C_2) \cup (C_1 \cap C_3) \cup (C_2 \cap C_3)$, where $C_i = C \backslash v_i$ and v_1, v_2, v_3 are three different vertices of C.

11.1.27. a: <u>Hint</u>: Suppose that $\mathcal{E}(v) \subseteq \mathcal{E}(v')$, X is a maximal independent vertex set such that $v \notin X$. Then $v' \in X$ and $X \backslash \{v'\} \cup \{v\}$ is also an independent set.

b: Clearly, H has a minimum cardinality covering without vertex v. Therefore $\beta_0(H) \ge \beta_0(H - v)$. On the other hand, the inequality $\beta_0(H) \le \beta_0(H - v)$ is evidently valid for any vertex $v \in V$.

c: Hint: Use the Exercises 11.1.27b and 11.1.26c.

d: Solution: For every maximum cardinality matching of a hypergraph H there corresponds some matching of a hypergraph $H-v$. Therefore $\alpha_1(H) \leq \alpha_1(H-v)$. On the other hand, no matching of $H-v$ may contain two edges from $\mathcal{E}(v)$, since v is a hyperpendant vertex of H. This implies that every matching of $H-v$ is a matching of H, i.e., $\alpha_1(H) \geq \alpha_1(H-v)$.

11.1.28. Hint: Let G be a graph with a covering $C = (L_1, \ldots, L_s)$ described in the statement of the exercise. Then v belongs to at most k complete graphs L_i from C. If it belongs to l such graphs then we add $k-l$ copies of K_1 with the vertex set $\{v\}$ to C. Doing the same for all vertices of G, we obtain a covering $C' = (L_1, \ldots, L_t)$. Consider the hypergraph $H = (V, \mathcal{E})$ such that $V = VG$ and $\mathcal{E} = \{VL_1, \ldots, VL_t\}$. Prove that $L(H^*) \cong G$.

Conversely, let $VH = \{v_1, \ldots, v_n\}$ be the vertex set of a hypergraph H. Consider the covering of the line graph $L(H)$ by the graphs $L_1, \ldots, L_n$, where L_i, $i = 1, 2, \ldots, n$, is a subgraph of $L(H)$ induced by the set of edges of H incident to v_i.

11.1.29. Using the previous exercise, prove that the graph shown at Fig. A11.1.2 is the required one for any t.

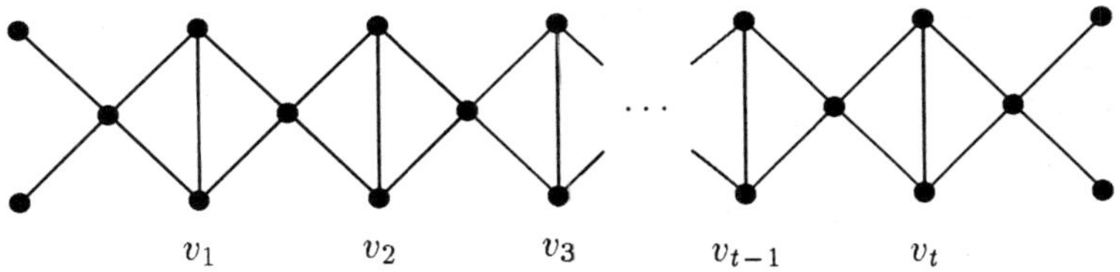

Figure A11.1.2: To Exercise 11.1.29

11.1.31. See the similar statement for graphs in Chapter 4 "Connectivity".

11.1.32. Only the sufficiency part is true. See, e.g., Fig. A11.1.3.

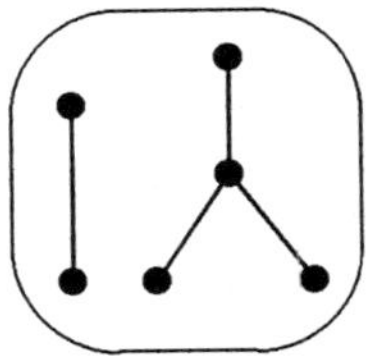

Figure A11.1.3: To Exercise 11.1.32

11.1.33. Only the sufficiency part is true. See, e.g., Fig. A11.1.3.

11.1.34. Yes for H_1, no for H_2.

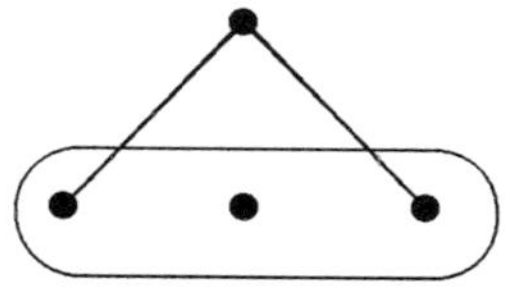

Figure A11.1.4: To Exercise 11.1.38

11.1.35. Solution: a: Let v_0 be a vertex of the hypergraph H of degree $\delta(H)$. The deletion of all edges of $\mathcal{E}(v_0)$ from H clearly increases the number of components, therefore $\lambda(H) \leq \delta(H)$.

b: The following relations are evident:

$$r(H)|\mathcal{E}(H)| \geq \sum_{e \in \mathcal{E}H} |e| = \sum_{v \in VH} |\mathcal{E}(v)| \geq |H|\delta(H).$$

Therefore $\delta(H) \leq |\mathcal{E}H|r(H)/|H|$.

11.1.36. a: $\chi(H_n^k) = \lceil n/(k-1) \rceil$. b: $\chi(H_1) = 2$; $\chi(H_2) = 3$; $\chi(H_3) = 2$.

11.1.37. a, b: It either remains unchanged or decreases by one.

11.1.38. The converse statement is false.

Hint: The proof is similar to the proof of the sufficiency for the König theorem (Exercise 1.2.15). An example of a bicolorable hypergraph with a cycle of odd length is shown at Fig. A11.1.4.

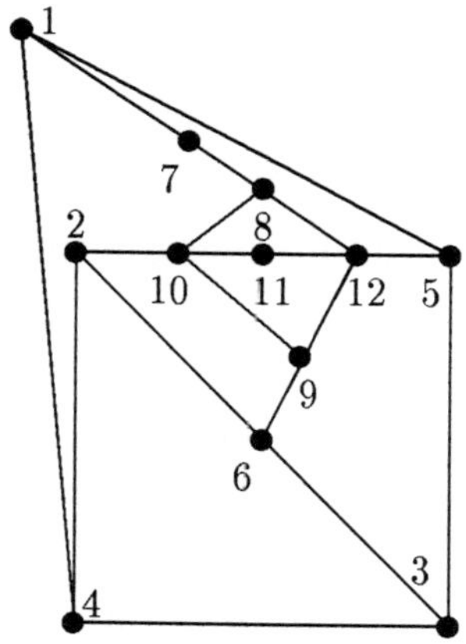

Figure A11.1.5: To Exercise 11.2.2

11.2 Hypergraph Realizations

11.2.1. Yes for H_1, no for H_2.

11.2.2. Yes for H_1, H_2, H_4 no for H_3. A planar realization for the hypergraph H_2 is shown at Fig. A11.1.5.

11.2.3. Hint: a: A required realization of the hypergraph $\tilde{H}$ may be obtained from a planar realization $R = (V, E)$ of H, respectively, by

- deletion of all edges from E that belong to no edge of $\tilde{H}$;

- contraction of all edges from E that belong to the contracted edge of H;

- addition of a new vertex and the edge connecting it with some vertex of the given edge of the hypergraph;

- contraction of some edge from E incident to the deleted vertex.

b: Use the item a.

11.2.4. Solution: A planar realization for the hypergraph shown at Fig. A11.2.1 exists (see Exercise 11.2.2). But the deletion of the vertex v produces the

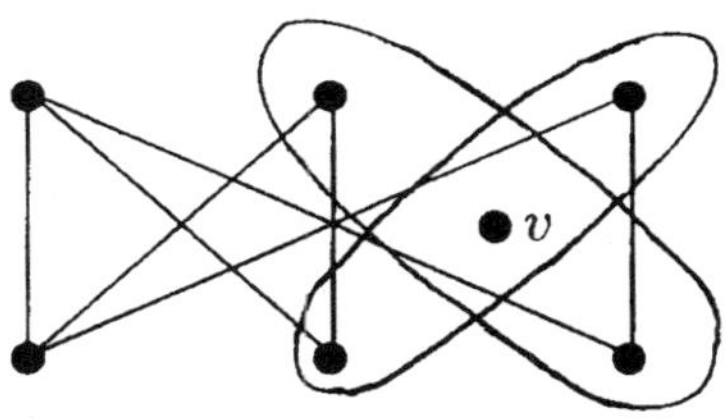

Figure A11.2.1: To Exercise 11.2.4

hypergraph which is the graph $K_{3,3}$.

11.2.5. Solution: Suppose that $H = (V, \mathcal{E})$. Let $E(v)$ denote the set of edges of the graph $K(H) = (V \cup \mathcal{E}, E)$ incident to $v \in \mathcal{E}$. Selecting an edge $e(v)$ from every set $E(v)$, $v \in \mathcal{E}$, and contracting it, we obtain a planar realization of H.

An example when a planar realization of a hypergraph exists but its König representation is nonplanar is H_4 from Fig. 11.2.3.

11.2.6. a: Hint: Use the Theorem 11.2.1. b: A connected hypergraph H with at most three edges is realizable by tree if either $L(H) \ncong C_3$ or $L(H) \cong C_3$ and H has a vertex belonging to all three edges.

11.2.7. Solution: Without loss of generality we may consider only connected hypergraphs, and taking the previous exercise into account, we may consider only four-edge hypergraphs. Let $H = (V, \mathcal{E})$ be a four-edge hypergraph and let $H' = (V', \mathcal{E}')$ be a connected part of H with $|\mathcal{E}'| = 3$. Let us construct a realization R' of H' by tree or cycle. Select a vertex in R' belonging to an edge $e \in \mathcal{E} \backslash \mathcal{E}'$ and connect it with all remaining vertices from e. The resulting graph is a planar realization of H.

11.2.8. Hint: Use the similar statement for graphs.

11.2.9. Yes. Hint: The König representation of a hypergraph H is isomorphic to that of H^*. Now you may use the Exercise 11.2.5.

11.2.10. No. Solution: The hypergraph H_4 from Fig. 11.2.3 has a planar realization, but H_4^* is $K_{3,3}$.

11.2.11. Hint: Let P be a path graph which is a realization of the hypergraph obtained from H by the deletion of two vertices v_1, v_2. Then the required planar realization of H may be obtained by the addition of v_1, v_2 to P and connecting v_i, $i = 1, 2$, with the vertices of P which are adjacent to v_i in H. If v_1, v_2 are adjacent in H then we add the edge v_1v_2.

11.2.13. Solution: The required realization may be obtained by the addition of the two vertices deleted from H to the cycle graph C which is a realization of H' and connecting them by edges with all vertices of C with which they are adjacent in H.

11.2.14. No. Hint: See the hypergraph H_2 from Fig. 11.2.2.

11.2.16. Hint: Necessity: Let $f : VH \to \{1, 2, \ldots, k\}$ be a proper coloring of a k-colorable hypergraph H. Consider the graph R with the same vertex set VH whose edge set is all pairs of differently colored vertices that belong to the same edge of H. Clearly, R is k-colorable and is a realization of H.

Sufficiency: Clearly, every proper coloring of a realization of a hypergraph is a proper coloring of the hypergraph itself.

11.2.17. 2. Hint: Use the Exercise 11.2.16.

11.2.18. Hint: Use the Exercise 11.2.16 and the Four Color Conjecture.

11.2.19. a: No. b: Yes.

Solution: a: Should there exist a strict realization of the hypergraph H, it would contain at least $2 \cdot 4 = 8$ edges, which is impossible for a four-vertex graph.

b: See Fig. A11.2.2.

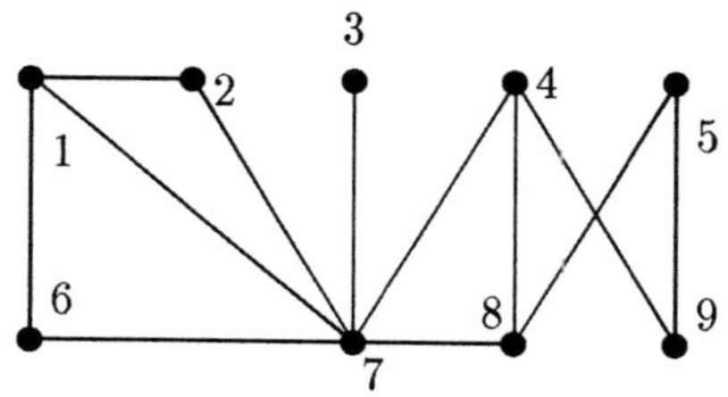

Figure A11.2.2: To Exercise 11.2.19b

11.2.20. See Fig. A11.2.3.

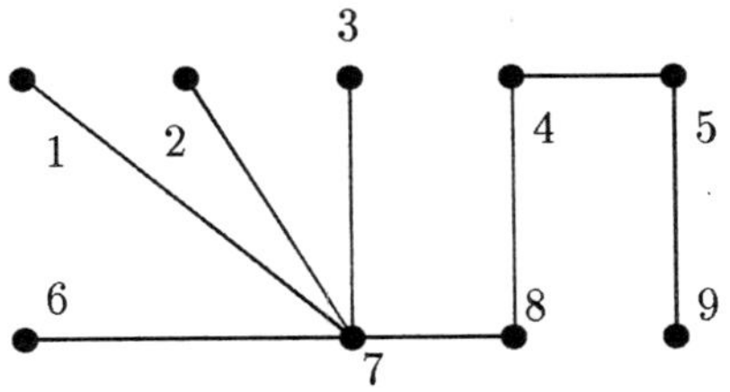

Figure A11.2.3: To Exercise 11.2.20

11.2.22. a: Hint: A realization of a hypergraph H with $t(H)$ edges is obtained from a realization of $H - v$ with $t(H - v)$ edges by the addition of a vertex v and an edge vv', where v' is such that $\mathcal{E}(v) \subseteq \mathcal{E}(v')$.

b: Hint: Use the item a.

11.2.23. Hint: A possible proof is by induction over the cardinality of the set of pairwise adjacent edges in the following way. Let $\{e_1, \ldots, e_k\}$ be a set of pairwise adjacent edges. Consider three sets:

$$V_1 = \bigcap_{i=1}^{k-1} e_i \neq \emptyset, \; V_2 = \bigcap_{i=2}^{k} e_i \neq \emptyset, \; V_3 = e_1 \bigcap e_k \neq \emptyset$$

and prove that $V_1 \cap V_2 \cap V_3 \neq \emptyset$.

11.2.25. Hint: To prove the necessity, use the Exercise 11.2.24 and the Theorem 11.2.1. The sufficiency may be proved by contradiction.

Bibliography

[1] Berge C. *Graphs and Hypergraphs,* North-Holland Publ. Co., 1973.

[2] Christofides N. *Graph Theory: An Algorithmic Approach,* Academic Press, 1975.

[3] Evstigneev V.A., Melnikov L.S. *Problems and Exercises in Graph Theory and Combinatorics,* Novosibirsk University Publ., Novosibirsk, 1981 (in Russian).

[4] Gavrilov G.P., Sapozhenko A.A. *Collected Exercises in Discrete Mathematics,* Nauka, Moscow, 1977 (in Russian).

[5] Harary F. *Graph Theory,* Addison-Wesley Publ.Co., 1969.

[6] Melnikov O., Tyshkevich R., Yemelichev V., and Sarvanov V. *Lectures on Graph Theory,* BI-Wissenschaftsverlag, Mannheim, 1994.

[7] Ore O. *Theory of Graphs,* AMS, Providence, Rhode Island, 1962.

[8] Swamy M.N.S, Thulasiraman K. *Graphs, Networks and Algorithms,* Wiley, 1981.

[9] Tutte W.T. *Graph Theory,* Addison-Wesley Publ.Co., 1984.

[10] Wilson R.J. *Introduction to Graph Theory,* Oliver and Boyd, Edinburgh, 1972.

[11] Zykov A.A. *Fundamentals of Graph Theory,* BCS Associates, Moscow, Idaho, USA, 1990. (Transl. from Russian original.)

*

Index

Notations

$\times$ – sign of Cartesian product (of graphs, sets)

$\cong$ – sign of isomorphism (of graphs, digraphs, etc.)

$\cup$ – sign of union (of graphs, matroids, sets)

$\cap$ – sign of intersection (of graphs, matroids, sets)

$(A, B; E)$ – bipartite graph with parts A, B and edge set E

AG – set of arcs of digraph G

$A(G)$ – adjacency matrix of graph G

$\mathrm{Aut}\ _E(G)$ – edge group of graph G

$\mathrm{Aut}\ (G)$ – automorphism group of graph G

$B(G)$ – block graph of graph G

$c(G)$ – minimal number of cliques covering VG

$cc(G)$ – minimal number of complete graphs whose union is G

$\mathcal{B}(S)$ – set of bases of independence system S

C_n – cycle graph of order n

C_n^2 – number of two-element subsets of an n-element set

$cr(G)$ – crossing number of graph G

$\mathcal{C}(S)$ – set of cycles of independence system S

$\deg G$ – degree of regular graph G

$\deg_G v$ – degree of vertex v in graph G

$\deg v$ – degree of vertex v in the considered graph

$d(G)$ – diameter of graph G

$d_G(u, v)$ – distance between vertices u and v in graph G

$d^-(v)$ – indegree of vertex v

$d^+(v)$ – outdegree of vertex v

EG – edge set of graph G

$\mathcal{E}H$ – edge set of hypergraph H

$\mathcal{E}(v)$ – set of edges incident to vertex v of a hypergraph

$e(v)$ – excentricity of vertex v

$\overline{G}$ – complementary graph for graph G

G_b – substrate of digraph G

$G - e$ – subgraph of G produced by deletion of edge e from G

$g(G)$ – girth of G

$G[H]$ – composition of graphs G and G

$G \wedge H$ – conjunction of graphs G and G

$G \diamond H$ – modular product of graphs G and G

$G + H$ – join of graphs G and G

G^p – p-th degree of graph G

$G - v$ – subgraph of G produced by deletion of vertex v and the edges incident to it from G

$G(X)$ – subgraph of G induced by subset $X \subseteq VG$

IG – set of independent subsets of vertices of graph G plus empty set

$I(G)$ – incidence matrix of graph G

$\mathcal{I}(S)$ – set of independent sets of independence system S

$K(G)$ – (1) Kirchhoff matrix of graph G (2) König representation of graph or hypergraph G

$k(G)$ – number of components of graph G

K_n – complete graph of order n

$K_{p,q}$ – complete bipartite graph with parts of cardinalities p and q

$L(H)$ – line graph of graph or hypergraph H

$M(G)$ – matroid of cycles of graph G

$M^*(G)$ – matroid of cuts of graph G

$m(G)$ – number of edges of graph G

M_{ij} – entry of matrix M at position (i, j)

mK_2 – graph of order $2m$ whose edge set is m pairwise nonadjacent edges

$n(G)$ – number of vertices of graph G

$N_G(v)$ – neighborhood of vertex v in graph G

O_n – empty graph of order n

$P(G)$ – deck of graph G

P_n – path graph of order n

$Q(G)$ – clique graph for graph G

$R(G)$ – reachability matrix of graph G

$r(G)$ – radius of graph G

$S(G)$ – graph of skeletons of graph G

$sk(G)$ – skewness of graph G

$t(G)$ – thickness of graph G

$th(G)$ – threshold number of graph G

uv – the edge of a graph connecting the vertices u and v

VH – vertex set of graph or hypergraph H

W_n – wheel graph of order $n + 1$

$|X|$ – cardinality of set X, order of graph, hypergraph or matroid X

$X \cup a$ – shorthand for $X \cup \{a\}$

$X \cap a$ – shorthand for $X \cap \{a\}$

$\mathbf{Z}_n$ – additive group of integer residuals modulo n

$\mathbf{Z}_2$ – two-element field

$\alpha_0(H)$ – independence number of graph or hypergraph H

$\alpha_1(H)$ – matching number of graph or hypergraph H

$\beta_0(H)$ – vertex covering number of graph or hypergraph H

$\beta_1(H)$ – edge covering number of graph or hypergraph H

$\gamma(G)$ – genus of graph G

Γv – the set of heads of arcs outcoming from vertex v

$\Gamma^{-1}v$ – the set of tails of arcs incoming to vertex v

$\Delta(G)$ – maximal degree of vertices of graph G

$\delta(G)$ – minimal degree of vertices of graph G

$\varphi(G)$ – toughness of graph G

$\kappa(H)$ – vertex connectivity number of graph or hypergraph H

$\lambda(H)$ – edge connectivity number of graph or hypergraph H

$\nu(G)$ – cyclic rank of graph G

$\nu^*(G)$ – cocyclic rank of graph G

$\xi(G)$ – arboricity of graph G

$\rho(M)$ – rank of matroid M

$\chi(H)$ – chromatic number of graph or hypergraph G

$\chi'(H)$ – chromatic index of graph G

Kluwer Texts in the Mathematical Sciences

1. A.A. Harms and D.R. Wyman: *Mathematics and Physics of Neutron Radiography.* 1986 ISBN 90-277-2191-2
2. H.A. Mavromatis: *Exercises in Quantum Mechanics.* A Collection of Illustrative Problems and Their Solutions. 1987 ISBN 90-277-2288-9
3. V.I. Kukulin, V.M. Krasnopol'sky and J. Horácek: *Theory of Resonances.* Principles and Applications. 1989 ISBN 90-277-2364-8
4. M. Anderson and Todd Feil: *Lattice-Ordered Groups.* An Introduction. 1988 ISBN 90-277-2643-4
5. J. Avery: *Hyperspherical Harmonics.* Applications in Quantum Theory. 1989 ISBN 0-7923-0165-X
6. H.A. Mavromatis: *Exercises in Quantum Mechanics.* A Collection of Illustrative Problems and Their Solutions. Second Revised Edition. 1992 ISBN 0-7923-1557-X
7. G. Micula and P. Pavel: *Differential and Integral Equations through Practical Problems and Exercises.* 1992 ISBN 0-7923-1890-0
8. W.S. Anglin: *The Queen of Mathematics.* An Introduction to Number Theory. 1995 ISBN 0-7923-3287-3
9. Y.G. Borisovich, N.M. Bliznyakov, T.N. Fomenko and Y.A. Izrailevich: *Introduction to Differential and Algebraic Topology.* 1995 ISBN 0-7923-3499-X
10. J. Schmeelk, D. Takaçi and A. Takaçi: *Elementary Analysis through Examples and Exercises.* 1995 ISBN 0-7923-3597-X
11. J.S. Golan: *Foundations of Linear Algebra.* 1995 ISBN 0-7923-3614-3
12. S.S. Kutateladze: *Fundamentals of Functional Analysis.* 1996 ISBN 0-7923-3898-7
13. R. Lavendhomme: *Basic Concepts of Synthetic Differential Geometry.* 1996 ISBN 0-7923-3941-X
14. G.P. Gavrilov and A.A. Sapozhenko: *Problems and Exercises in Discrete Mathematics.* 1996 ISBN 0-7923-4036-1
15. R. Singh and N. Singh Mangat: *Elements of Survey Sampling.* 1996 ISBN 0-7923-4045-0
16. C.D. Ahlbrandt and A.C. Peterson: *Discrete Hamiltonian Systems.* Difference Equations, Continued Fractions, and Riccati Equations. 1996 ISBN 0-7923-4277-1
17. J. Engelbrecht: *Nonlinear Wave Dynamics.* Complexity and Simplicity. 1997 ISBN 0-7923-4508-8
18. E. Pap, A. Takači and D. Takači: *Partial Differential Equations through Examples and Exercises.* 1997 ISBN 0-7923-4724-2
19. O. Melnikov, V. Sarvanov, R. Tyshkevich, V. Yemelichev and I. Zverovich: *Exercises in Graph Theory.* 1998 ISBN 0-7923-4906-7

KLUWER ACADEMIC PUBLISHERS – DORDRECHT / BOSTON / LONDON